Wavelets from a Statistical Perspective

Wavelets from a Statistical Perspective

Maarten Jansen

CRC Press is an imprint of the
Taylor & Francis Group, an **informa** business
A CHAPMAN & HALL BOOK

First edition published 2022
by CRC Press
6000 Broken Sound Parkway NW, Suite 300, Boca Raton, FL 33487-2742

and by CRC Press
4 Park Square, Milton Park, Abingdon, Oxon, OX14 4RN

CRC Press is an imprint of Taylor & Francis Group, LLC

ISBN: 978-1-032-20067-5 (hbk)
ISBN: 978-1-032-20820-6 (pbk)
ISBN: 978-1-003-26537-5 (ebk)

DOI: 10.1201/9781003265375

Typeset in NimbusSanL-Regu font
by KnowledgeWorks Global Ltd.

Publisher's note: This book has been prepared from camera-ready copy provided by the authors.

Table of contents

List of recurrent symbols

Symbol	: Description	Page
$\mathbf{0}$	: zero vector	
$\emptyset$	: empty set	
∞	: infinity	
$\lfloor x \rfloor$	: integer part of real number x	6
$\lceil x \rceil$	: smallest integer equal or larger than real number x	6
$\langle \cdot, \cdot \rangle$	: scalar product in unitary space	
$\|\cdot\|$	: norm in a general vector space	
$\|f\|_p$	: L_p-norm	
$\|f\|_{B^{\alpha}_{p,q}}$	: Besov semi-norm of f	
∂m	: neighbouring knots	86
$(\downarrow 2)$	: dyadic subsampling matrix	36
$(\uparrow 2)$	: dyadic upsampling matrix	131
$\boldsymbol{d}_j$	: detail (or wavelet) coefficients at level j	21
$d_{j,k}$	: wavelet coefficient at level j, location k	18
$\mathbf{D}_j$	: diagonal rescaling matrix at level j	39
$\boldsymbol{e}_i$	: ith canonical vector, i.e., ith column of $\mathbf{I}$	31
$F(\omega)$	: Fourier transform of $\boldsymbol{f}$ or $f(x)$	7
$\mathbf{G}_j$	: detail matrix in inverse wavelet transform	54
$\widetilde{\mathbf{G}}_j$	: detail matrix in forward wavelet transform	54
$\boldsymbol{\mathcal{G}}_j$	: detail matrix in nondecimated inverse wavelet transform	219
$\widetilde{\boldsymbol{\mathcal{G}}}_j$	: detail matrix in nondecimated forward wavelet transform	216
$G(\omega)$	: discrete time Fourier transform of $\boldsymbol{g}$	168
$\widetilde{G}(\omega)$	: discrete time Fourier transform of $\widetilde{\boldsymbol{g}}$	168
$\mathbf{H}_j$	: refinement (two-scale) matrix in inverse wavelet transform	54
$\widetilde{\mathbf{H}}_j$	: dual refinement (two-scale) matrix, used in forward wavelet transform	54
$\boldsymbol{\mathcal{H}}_j$	: nondecimated refinement matrix	219
$\widetilde{\boldsymbol{\mathcal{H}}}_j$	: dual nondecimated refinement matrix	216
$H(\omega)$	: discrete time Fourier transform of $\boldsymbol{h}$	168
$\widetilde{H}(\omega)$	: discrete time Fourier transform of $\widetilde{\boldsymbol{h}}$	168
$\mathbf{I}_j$	: (also $\mathbf{I}_{n_j}$) identity matrix of size $n_j \times n_j$	
J	: finest resolution level in a wavelet transform	18

Symbol	Description	Page
J^*	: superresolution	61
$\widetilde{\mathbf{J}}_j$	: upsampling matrix	131
$\widetilde{\mathbf{J}}_j^\top$	: subsampling matrix	36
$\widetilde{\mathbf{J}}_j^{o\top}$	: complementary subsampling matrix	36
L	: coarsest resolution level in a wavelet transform	35
$\mathcal{M}$	: index set of admissible models	285
n_1	: number of nonzero coefficients in a wavelet decomposition of a piecewise polynomial	197
n_j	: number of scaling coefficients at level j	25
n'_j	: number of wavelet coefficients at level j	25
p	: number of primal vanishing moments	51
p	: index of model under consideration, element of $\mathcal{M}$	285
$\widetilde{p}$	: number of dual vanishing moments	62
$\mathbf{P}_j$	: prediction (dual lifting) matrix	40
$\widetilde{\mathbf{P}}_L$	: projection matrix	74
$\widetilde{\mathbf{P}}_{L\perp}$	: orthogonal projection matrix	76
PE	: prediction error	286
$\boldsymbol{s}_j$	: vector of scaling coefficients at resolution level j	21
$s_{j,k}$	: scaling coefficient at level j, location k	18
$\mathbf{S}_j$	: shift or translation matrix	130
S_p, S	: admissible submodel	287
SS_E	: sum of squared residuals	287
$\mathbf{U}_j$	: update (primal lifting) matrix	40
$\mathbf{V}_L$	: repeated refinement matrix	74
$\widetilde{\mathbf{V}}_L$	: repeated coarsening matrix	74
$\widetilde{\mathbf{W}}$	: forward wavelet transform matrix	21
$\mathbf{W}$	: inverse wavelet transform matrix	27
$\widetilde{\mathbf{W}}_j$	: one level forward wavelet transform	54
$\mathbf{W}_j$	: one level inverse wavelet transform	54
$\boldsymbol{\delta}$	: Kronecker delta vector (canonical vector)	
δ_k	: Kronecker delta	
$\Delta_{j,k}$	: interknot distance	35
Δ_j	: maximum interknot distance at resolution level j	35
Δ	: equispaced interknot distance	176
$(\Delta_f x)^2$	: width of Heisenberg's uncertainty window for the function $f(x)$	172
$\Delta_h^{(k)} f(x)$	: k-th difference of a function $f(x)$	204
κ_F	: multiscale variance propagation	81
κ	: numerical condition number	145
$\widehat{\kappa}$	: number of selected wavelet coefficients	276
$\widehat{\kappa}_F$	: number of false positives	276
$\mathbf{\Lambda}$	: diagonal matrix containing eigenvalues	156
Λ_λ	: non-studentised version of Mallows' C_p	288
λ	: threshold or smoothing parameter	11

ν_λ, ν_p	: number of (generalised) degrees of freedom	287
$\Phi(\omega)$	: Fourier transform of $\varphi(x)$	163
$\Phi_j(x)$	: row vector of scaling functions at resolution level j	26
$\varphi_{j,k}(x)$	: scaling function at level j, location k	24
$\widetilde{\Phi}_j(x)$	: row vector of dual scaling functions at resolution level j	61
$\Psi_j(x)$	: row vector of wavelet functions at resolution level j	26
$\psi_{j,k}(x)$	: wavelet function at level j, location k	25
$\widetilde{\Psi}_j(x)$	: row vector of dual wavelet functions at resolution level j	62

Introduction

From a statistical perspective, much of science is about finding the optimal balance between bias and variance. The search for an equilibrium between these somehow contradictory objectives is visualised in what is arguably the most important type of curves in statistics, even more so than the straight line so often sought for in regression. The curve, as depicted in Figure 1, plots the combined effect of bias and variance against the value of a tuning parameter. In this case, the tuning parameter is the size of a model. Indeed, the plot in

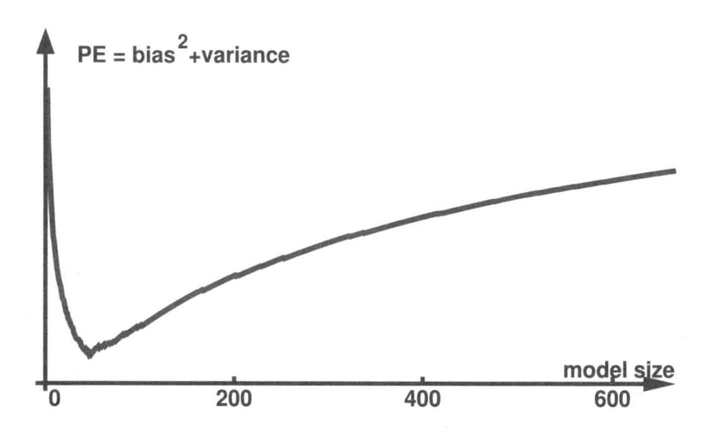

Figure 1
The balance between bias and variance in high-dimensional, sparse model selection. Finding the optimal trade-off between bias and variance is an important theme in a large variety of statistical problems.

Figure 1 arises from the problem of finding the right number of explanatory values from a high-dimensional number of candidate variables. Taking the number too small leaves part of the observations unexplained, which is bias. On the other hand, if the selected model contains too many candidates, the minor ones carry little information and lots of noise, which contributes to the variance.

Fine-tuning the trade-off between conflicting interests is a crucial and often delicate operation, not just in statistical, but also in economical and political decision making, human resource management or, for instance, in balancing between health care needs and economic loss, as we witnessed in the

COVID-19 pandemic crisis. As a matter of fact, it seems fair, though maybe not quite scientifically sound, to state here that the identification of an equilibrium somewhere in the middle arises far more often in daily life than taking a decision along a straight line sloping to the extreme left or right.

The topic of this book

Balancing on the edge between a conservative and permissive approach, false positives and false negatives becomes even more sensitive in today's popular domain of sparse high-dimensional variable selection. By high-dimensional we mean that the number of candidate explanatory variables is large, typically outranging the number of observations. Having fewer observations than unknown candidate parameters to be estimated, it is impossible to find a value for all candidates, unless a subset is selected for estimation while the others are set to zero by default. This approach rests on the assumption of sparsity, stating that most candidates have indeed a zero or near zero, neglectable true value. The degree of sparsity, the size of the selected model that is, plays the role of a tuning parameter. Fine-tuning is a delicate operation, as indeed, in a high-dimensional model, with lots of uncertain parameters, a small deviation from the optimal choice may result already in an unacceptable increase of bias, variance, false positives or missed parameters.

A prototype of a sparse, high-dimensional model, as in Figure 1, is a vector of wavelet coefficients or coefficients from similar decompositions, many of which will be discussed in this text. Depending on the exact wavelet transform algorithm, the number of coefficients equals or exceeds the sample size, but wavelet transforms have been designed to deliver sparsity. With the bulk of the information concentrated in a limited number of coefficients, and the noise spread out more or less evenly, it is possible to remove an important part of the noise by selecting the most significant coefficients only for reconstruction.

The literature on wavelets and statistics, which has boomed in the two decades of the 1990's and 2000's, has mainly focussed on the problem of coefficient selection and estimation. This domain of research, which nowadays seems to have reached convergence (more or less), is discussed in Chapter 12. Somehow less attention has been paid to the subsequent statistical inference, including the often nontrivial construction of confidence bands and statistical testing. This theme has seen older, but also more recent, advanced contributions, while new insights can be expected from the ongoing developments in active fields like post-selection inference. A brief discussion of statistical inference on wavelet smoothing is included in Section 12.6. The last two decades have seen contributions of wavelets in various kinds of statistical modelling, such as random and mixed effect models (Angelini et al., 2003; Claeskens et al., 2011), semiparametric regression models (Chang and Qu, 2004; Wand and Prmerod, 2011; Ruppert et al., 2003; Tsiatis, 2006), generalised linear regression models (Meyer, 2003), single-index models (Park et al., 2005), functional data models (Chang et al., 2014), or a combination of these (Morris and Carroll, 2006), just to list an arbitrary selection. This book has by no means exhaustive or encyclopedic ambitions. Instead, the emphasis in this book lies on yet another issue, which is the design of wavelet transforms for applications in statistics. Although statistical considerations, in particular the bias-variance trade-off, are the leitmotiv in this book, the focus on the design means that the text contains somehow less plain statistical formulas and more general applied mathematics, for use in a wide range of applications in and outside the field of statistics. By this perspective this work offers a complementary discussion to earlier works on wavelets and statistics (Ogden, 1997; Härdle et al., 1998; Vidakovic, 1999; Nason, 2008). Although the emphasis of this book is definitely situated at a basic level of wavelet analysis, this perspective may well be underexposed, thus offering to the reader, hopefully, not only basic understanding and insights but maybe also interesting starting points for the development of new research.

Wavelet transforms provide a tool for decomposing non-sparse data into a sparse representation, the wavelet coefficients. While the problem of fine-tuning between bias and variance is most prominent in the processing of these coefficients, the design of the transform itself involves choosing parameters that are just as well subject to a bias-variance trade-off. The bias is related to the approximation error of a wavelet based projection, defined by a forward wavelet transform, followed by a selection of the coefficients and finally a reconstruction. The approximation error, further explored in Chapter 8, depends on how close the input data vector is to the eigenvectors of the projection matrix. Wavelet projections prefer piecewise smooth data, as explained in Chapters 1 and 8. Whereas the eigenvalue decomposition shows up in the analysis of the bias, Chapter 3 explains how the variance propagation is described by the singular value decomposition of the wavelet projection.

For whom has this book been written?

In an effort to be consistent with its own message, this book balances between two groups of readers. At a general level, the book targets an audience of statisticians, data scientists, and applied mathematicians seeking to learn about wavelet theory and methods from a statistical viewpoint. This general audience also includes students or researchers with a background in signal processing or in artificial intelligence, machine learning, and data mining, interested in the slightly different vocabulary of statistics.

At a more specific level, this book is addressed to statisticians working in nonparametric regression, using kernel or spline based methods. The common key point in these methods is the notion of locality, formalised by kernel bandwidths in local polynomial smoothing, or by compactly supported B-splines. The smoothing operations are localised in time or space, and thus spatially adaptive. Wavelets extend the idea of locality so that operations take place at localised space and at a well specified scale. From the very beginning in Chapter 1, it turns out that the combined space-scale locality leads to sparse data representations and to nonlinear and multiscale processing with applications that outrange those of classical spline or kernel methods. Indeed, as demonstrated in Chapter 4, all splines can be upgraded to spline wavelets, no matter how many knots there are and where these knots are located. In principle, all spline methods can be extended to include a multiscale analysis. Multiscale local polynomial methods are presented in Chapter 11. Both the multiscale local polynomials and multiscale spline approaches illustrate a key point of this book: wavelets can be constructed on nonequidistant data. Although mostly beyond the scope of this text, the construction of wavelets or other multiscale analyses described in this book could be extended towards all sorts of irregular point sets, including graphical data, trees, networks, and so on.

The classical wavelet book hinges on Fourier analysis as a tool in the construction, the analysis, and the proofs of wavelet decompositions. This book will pay attention to Fourier analysis, as a tool for linear data processing or filtering, along with splines and kernels, but also as a way to look at results and methods from a frequency point of view, especially in Chapter 6. Nevertheless, readers not familiar with this tool may skip the Fourier sections, as the actual construction of wavelet methods here does not depend on a Fourier analysis.

The exercises

At several occasions, the discussion is enriched with figures and examples, but also with exercises, the goal of which is to refine the argument, or to make a link with earlier discussions. In line with the construction of the wavelets, these exercises themselves appear at nonequidistant points. Some chapters have more exercises, other have more examples. Most exercises do not involve simulation or implementation. Software can be used, though, to check the exercises numerically, but also to reproduce the figures.

Accompanying software

The plots and images illustrating this book were generated by Matlab^{TM}routines, although this book is not oriented towards one programming language or software. The reason for working with Matlab, rather than R or Python, is mainly historical: the development of the collection of Matlab routines started a quarter of a century ago, and it would not be easy to make a quick switch to another platform, keeping the same level of sophistication. Moreover, unlike for instance R, Matlab is a procedural language, which can be argued to be the right choice for dealing with procedures such as wavelet transforms and wavelet coefficient selection.

The routines can be downloaded for reproduction of the figures from the website `http://homepages.ulb.ac.be/~majansen/software/threshlab.html` which hosts a Matlab package, `ThreshLab`, implementing all the wavelet transforms and all wavelet coefficient processing discussed in this book. No other Matlab toolbox is needed to run this software. As an alternative to Matlab, the software runs also under GNU Octave, which can be downloaded and installed for free.

Acknowledgements

The endeavour of writing this book has given lots of scientifical satisfaction, even in the sometimes long and lonely hours. The text structures and completes more than 25 years of work in the field, the persistence of which has been far from self-evident at several occasions. I have been lucky enough not only to have collaborated with real experts and to have learnt from many peers throughout the years, but also to have enjoyed crucial support at critical moments. Without the people mentioned below, my research and this book would not be here.

It all started in the mid 1990s when Wim Sweldens gave me a very accessible introduction to wavelet theory, clearly pointing to the key notions. The traces of his explanation, and his idea of building your own wavelets at home (Sweldens and Schröder, 1996) can still be found, especially in the first chapter, but also further throughout this book.

Guy Nason, Bernard Silverman (then at Bristol University), and Richard Baraniuk (at Rice University) gave me the opportunity to develop my own direction in the wavelet world. I would like to thank also Véronique Delouille (then at UC Louvain) and Christine De Mol (ULB, Brussels) for encouraging me at crucial moments in my professional life.

I thank Lara Spieker of CRC Press for the swift and constructive support in transforming this manuscript into a published book.

Finally, but of course most importantly, my wife, Gerda Claeskens, and family, Hanne-Sara and Gijsbrecht, master the art of balancing between science and life. It has been wonderful to switch between talking about work and daily issues all those years.

1

Wavelets: nonlinear processing in multiscale sparsity

1.1 Compressing big data

In a scientist's effort to explain a phenomenon through a model, a major challenge lies in filling in the model parameters. In the setting of a simple and known model and without measurement errors, model parameters are easily found by solving small linear systems applied to the observed response. In the real world, where measurement errors are ubiquitous, parameter values can be learnt or estimated from the observations. As data are big these days, there is a good chance that there are more observations to learn from than parameters to be estimated. The linear system for the model parameters is then overdetermined, leading to least squares solutions or other regression techniques.

And yet, not only do we have big data, but the models tend to become high dimensional as well. In many applications, the model itself has to be learnt from the data. This may lead to model or variable selection methods. Another option to start with is to realise that the data may well be less complicated than they appear to be. In that case, a data transformation reorganises the representation of the data so that the available information appears in a more concentrated way. The wavelet transform is a typical example of such an operation, and images are a popular illustration of what a wavelet transform can do. Figure 1.1 shows an grey scale image of 1024×1417 pixels[1]. Each of the 1,451,008 pixels represents an observation, but also a parameter to store on your computer. The pixel representation is not sparse, or economical, as typically adjacent pixels have similar grey values. Figure 1.2 displays another image, which is the wavelet transform of the image in Figure 1.1. Obviously, the wavelet representation is for internal use only. Whenever the image needs to be displayed on a screen or in a book which is not about wavelets, we first apply the inverse wavelet transform to get back Figure 1.1. There are many wavelet transforms to choose from, but the one

[1] Readers familiar with old school wavelets might be surprised to see an image where at least one dimension is not dyadic, a power of two, that is. As a matter of fact, this poses no problem at all for the wavelet transforms discussed in this book. No special treatment whatsoever will be necessary.

DOI: 10.1201/9781003265375-1

Figure 1.1
A grey scale image of 1024×1417 pixels.

used in this case is quite popular for image processing. While its details will be developed later, we mention here for further reference, it is known as the Cohen-Daubechies-Feauveau-2,2 or CDF-2,2 wavelet transform. The CDF-2,2 is used in the JPEG-2000 image coding system, in particular for lossless compression. The transformed image has the same size as the original, but it contains just a small version of the picture in Figure 1.1, along with large blocks dominated by white colours. The pixels in the small version of the original are termed **scaling coefficients**, whereas the others are known as **wavelet** or **detail coefficients**. Figure 1.2 has also thick black lines between the blocks. These lines are not part of the transformed image itself; they are superimposed to mark the boundaries between the blocks. In the large detail blocks, white pixels correspond to zero coefficients. The convention white for zero is a bit unusual, because typically zero stands for black. We use white for zero here since it is visually more attractive and it requires less ink to print. Indeed, there are many coefficient values close to zero. With so many coefficients close to zero, it may be tempting not to store the smallest values at all, in order to reduce the dimension of the data. In Figure 1.3, all detail blocks have been erased. The image has been reconstructed by applying the inverse transform to the small version in the left upper block in combination with zero detail coefficients. The outcome is clearly too blurred for practical use: most details, most texture is lost. In Figure 1.4, we see what happens if we keep just the largest ten percent of the detail coefficients: although some

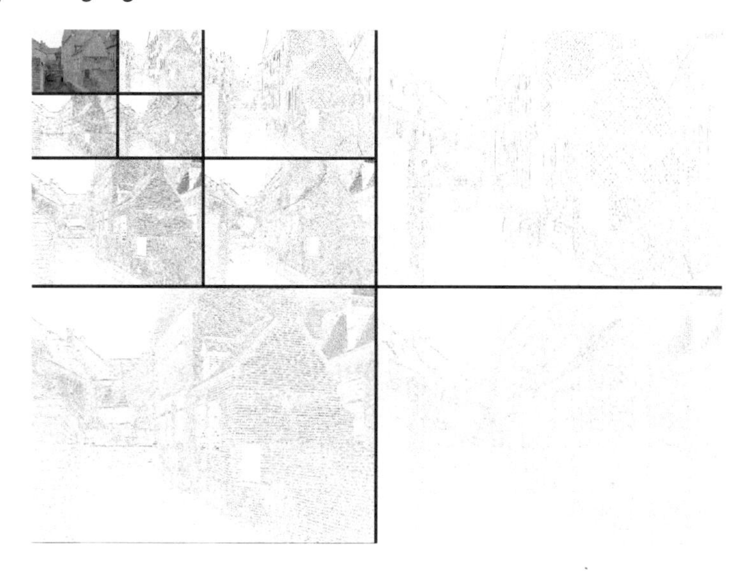

Figure 1.2
A wavelet transform (CDF2,2) of the image in Figure 1.1. White pixels correspond to zero coefficients.

Figure 1.3
A reconstruction of the image from the scaling coefficients only.

Figure 1.4
A reconstruction of the image from the scaling coefficients and the largest ten percent of the detail coefficients.

parts are visually not on the same level as the original image, many details are restored.

1.2 Wavelets among other methods for data processing

A data representation allowing us to compress the data is also an interesting tool in data processing, because compressibility means that the underlying model is not as complicated as it appears to be[2]. The Fourier analysis, presented in Section 1.2.1, is a prominent example. Whereas in many textbooks about wavelets it takes a central role in the construction of wavelet transforms, it is presented here for reasons of comparison. This section can be skipped without the risk of missing the key concepts of wavelet theory.

[2]Later, in Section 3.1.3, it will become clear that compressibility is not enough for statistical processing: denoising is more difficult than compression. Compression is about bias only; it says nothing about variance.

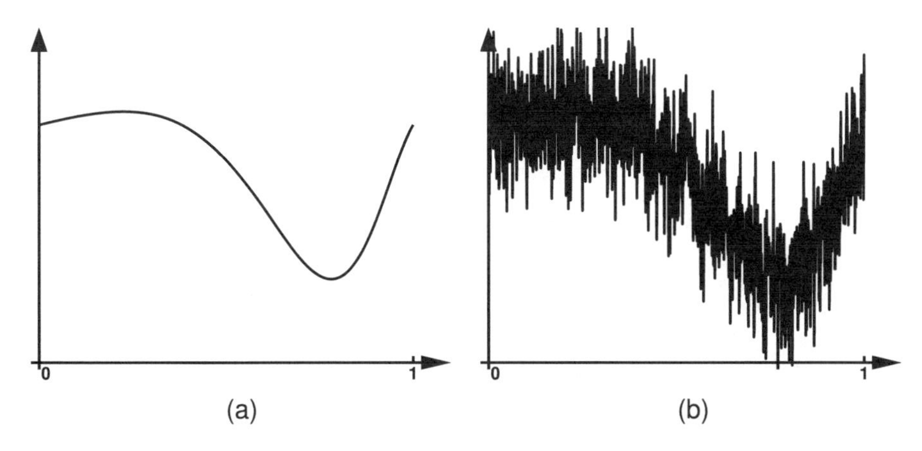

Figure 1.5
Test signal and $n = 1000$ observations with additive normal noise.

1.2.1 The Fourier transform

Consider the signal $f(x)$ for $x \in [0, 1]$, shown in Figure 1.5(a). The signal is observed with errors, ϵ_i as depicted in Figure 1.5(b), for which we assume the additive model

$$Y_i = f(x_i) + \epsilon_i, \text{ with } i = 1, 2, \ldots, n. \tag{1.1}$$

In vector notation, this is $\boldsymbol{Y} = \boldsymbol{f} + \boldsymbol{\varepsilon}$, where all three vectors are n-tuples. We suppose that no parametric model of $f(x)$ is available. For the time being, the errors are assumed to have zero mean, but we impose no further model for the noise. In our search for a good estimator of the true signal, we decompose[3] $f(x)$ as an infinite linear combination of complex sinusoids $\exp[i(2\pi)kx] = \cos(2\pi kx) + i\sin(2\pi kx)$,

$$f(x) = \sum_{k=-\infty}^{\infty} a_k \exp[i(2\pi)kx]\,. \tag{1.2}$$

This *expansion* is termed a **Fourier series**. The complex functions $\psi_k(x) = \exp[i(2\pi)kx]$ are orthogonal on $[0, 1]$ in the sense that whenever $k \neq l$,

$$\int_0^1 \psi_k(x)\overline{\psi_l(x)}dx = 0. \tag{1.3}$$

[3]A decomposition involves issues of convergence. In particular, the equality in (1.2) should be understood as a series converging to a function. This convergence is not for granted, and depends on the function, but also on the way convergence is defined: pointwise, in integrated squared distance, uniformly and so on. For the moment we happily ignore these issues, assuming that convergence is all set.

The integral in (1.3) should be interpreted as an inner product. As we are working with complex functions here, the second factor $\psi_l(x)$ takes a complex conjugate, denoted by $\overline{\psi_l(x)}$. The functions are normalised, as $\psi_k(x)\overline{\psi_k(x)} = 1$, integrating to one. Thanks to the orthogonality, multiplication with $\overline{\psi_l(x)}$ on both sides of (1.2), followed by integration leads to the *analysis*[4],

$$a_l = \int_0^1 f(x)\overline{\psi_l(x)}dx = \int_0^1 f(x)\exp\left[-i(2\pi)lx\right]dx. \tag{1.4}$$

In practice, $f(x)$ is observed, at best without errors, in n points x_t, for $t = 0, 1, \ldots, n-1$. We assume for now that these points are equidistant, i.e., $x_t = t/n$ and we set $f_t = f(x_t)$. The integral in (1.4) can be discretised as follows

$$F_l = \frac{1}{n}\sum_{s=0}^{n-1} f_s \exp\left[-i(2\pi)ls/n\right]. \tag{1.5}$$

This is the **Discrete Fourier Transform**, DFT, for which there exists a famous implementation of $\mathcal{O}(n\log(n))$ computational complexity, the Fast Fourier Transform. Because of periodicity, it holds that $F_l = F_{l+n}$. As a result, n observations are linearly transformed into n unique Fourier coefficients.

Exercise 1.2.1 *For an arbitrary choice of n, write the linear transform matrix $\mathbf{F}$ with complex entries that maps the vector of f_s, with $s = 0, 1, \ldots, n-1$ onto the vector of F_l, with $l = 0, 1, \ldots, n-1$. Check numerically that this transform matrix is invertible.*

Defining[5] $\lfloor x \rfloor$ the smallest integer equal or smaller than x and $\lceil x \rceil$ the smallest integer equal or larger than x, we keep the coefficients F_l with $l \in \{-\lceil n/2 \rceil + 1, -\lceil n/2 \rceil + 2, \ldots, \lfloor n/2 \rfloor\}$, although other subsets of n consecutive coefficients are equally possible. The orthogonality in (1.3) has a discrete equivalent, as for $s \neq t$,

$$\sum_{l=0}^{n-1} \exp\left[i(2\pi)lt/n\right]\exp\left[-i(2\pi)ls/n\right] = \sum_{l=0}^{n-1}\exp\left[i(2\pi)l(t-s)/n\right] = 0.$$

This is because the terms are complex roots of the unity. It is well known and straightforward to verify that they sum up to zero[6]. So, if in n equidistant points x_t we impose that

$$f(x_t) = \sum_{k=-\lceil n/2 \rceil+1}^{\lfloor n/2 \rfloor} F_k \exp\left[i(2\pi)kx_t\right]; \tag{1.6}$$

[4]By analysis or decomposition we mean going from a function to the Fourier, wavelet, or spline coefficients. This is the forward transform. When the outcome of a forward transform is an infinite sum of coefficients with basis functions, this is referred to as an expansion. Analysis contrasts with synthesis, which is the inverse transform, realising the reconstruction of the function from its decomposition.

[5]For an integer n it holds that $\lfloor n/2 \rfloor + \lceil n/2 \rceil = n$.

[6] Indeed, let $z = \exp\left[i(2\pi)(t-s)/n\right]$, then the terms in the sum are z^l, so the sum is found as a truncated geometric series, $\sum_{l=0}^{n-1} z^l = (z^n - 1)/(z-1)$, which is zero because $z^n = 1$.

then the coefficient F_l can be found by 1.5.

Exercise 1.2.2 *Check that the vector of observations $f(x_t)$ can be found from the vector of DFT coefficients F_l by matrix-vector multiplication with the matrix $n\mathbf{F}^\top$, where $\mathbf{F}^\top$ is the transpose of the matrix $\mathbf{F}$, found in Exercise 1.2.1.*

The analysis above started from a function $f(x)$ on the continuous axis of x and is then discretised. We could also work with discrete observations f_i from the beginning and compute the inner product with observations from a sinusoid $\exp(i\omega x)$ with an arbitrary pulsation ω,

$$F(\omega) = \sum_{s=0}^{n-1} f_i \overline{\exp(i\omega x_s)} = \sum_{s=0}^{n-1} f_i \exp\left[-i\omega s/n\right]. \tag{1.7}$$

As an independent variable, the pulsation ω is often replaced by the frequency, which is $\omega/2\pi$. The analysis in (1.8), one possible variant of the **Discrete Time Fourier Transform** (DTFT), can be further discretised at $\omega_l = (2\pi)l$, leading to the DFT in (1.5). Most publications replace x_s by nx_s in (1.7), allowing for n to grow arbitrarily large, leading to

$$F(\omega) = \sum_{s=-\infty}^{\infty} f_s \exp(-i\omega s). \tag{1.8}$$

The orthogonality of the complex oscillations on $[-\pi, \pi]$ (or any other interval of length 2π) leads to the reconstruction

$$f_k = \frac{1}{2\pi} \int_{-\pi}^{\pi} F(\omega) \exp(i\omega k) d\omega. \tag{1.9}$$

It is remarkable that the Fourier transform of a discretised signal (1.8), a function of the pulsation ω, has the same form as a Fourier expansion (1.2), except for the minus sign and the constant 2π in the exponent. This illustrates the fact that in a Fourier analysis, the roles of the input and output functions can easily be switched. This can be seen in the fully continuous Fourier transform. The analysis is[7]

$$F(\omega) = \frac{1}{2\pi} \int_{-\infty}^{\infty} f(x) \exp(-i\omega x) dx, \tag{1.10}$$

while, under mild conditions, the synthesis can be proven to be

$$f(x) = \int_{-\infty}^{\infty} F(\omega) \exp(i\omega x) d\omega. \tag{1.11}$$

The symmetry in (1.10) and (1.11) can be linked to the orthogonality of the building sinusoids in (1.3).

[7]The factor $1/(2\pi)$ in front is a matter of convention and normalisation. In the literature, this can be replaced by 1, or by $1/\sqrt{2\pi}$. Obviously, the choice has an impact on subsequent formulae, starting with the inverse transform.

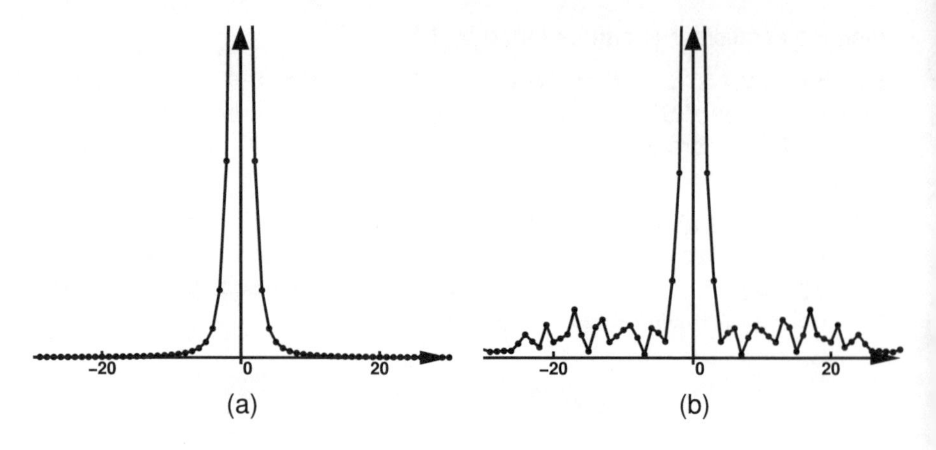

Figure 1.6
(a) Magnitudes of the centre of the Fourier spectrum of the discretised test signal ($n = 1000$). (b) Magnitudes of the centre of the Fourier spectrum of the observations with noise in Figure 1.5(b).

1.2.2 Using Fourier analysis in linear data processing

Figure 1.6 shows parts of the magnitudes of the Fourier coefficients of the observations in Figure 1.5, i.e., the values of F_l defined in (1.5). The vector of coefficients F_l is termed the **Fourier spectrum**. For the sake of clarity, the plots zoom in on the central coefficients, i.e., for $l \in \{-30, -29, \ldots, +30\}$. The sequence of F_l is plotted as a polyline with dots. Figure 1.6(a) shows the coefficients from the observations $F(x_s)$ without noise, whereas 1.6(b) shows the coefficients from the DFT of the observations in Figure 1.5(b). The spectrum in Figure 1.6 shows the typical fast decay of Fourier coefficients beyond a few large pulsations near the centre $l = 0$. As coefficients for l away from zero are dominated by the noise, Figure 1.7(a) investigates what happens if all coefficients with $|l| > 5$ in the spectrum of 1.6(b) are replaced by zero. The synthesis plotted in Figure 1.7(b), using (1.2) or (1.6), yields a good estimation of the signal without noise.

The data processing depicted in Figure 1.7 consists of three steps: forward Fourier transform, processing the spectrum by putting the high frequency coefficients to zero, inverse Fourier transform on the processed spectrum. These three steps constitute a prototype of what is known as a **linear filter** in signal processing. It is linear, because the forward and inverse transforms are linear, and so is the processing step in between, as it satisfies the superposition principle. Indeed, let D_k denote the function that maps a vector (i.c. a spectrum) to the vector where all elements with index $|l| > k$ are replaced by zero. Then obviously, $D_k(\alpha_1 \boldsymbol{x}_1 + \alpha_2 \boldsymbol{x}_2) = \alpha_1 D_k(\boldsymbol{x}_1) + \alpha_2 D_k(\boldsymbol{x}_2)$. The function D_k can be represented by a matrix multiplication, involving a

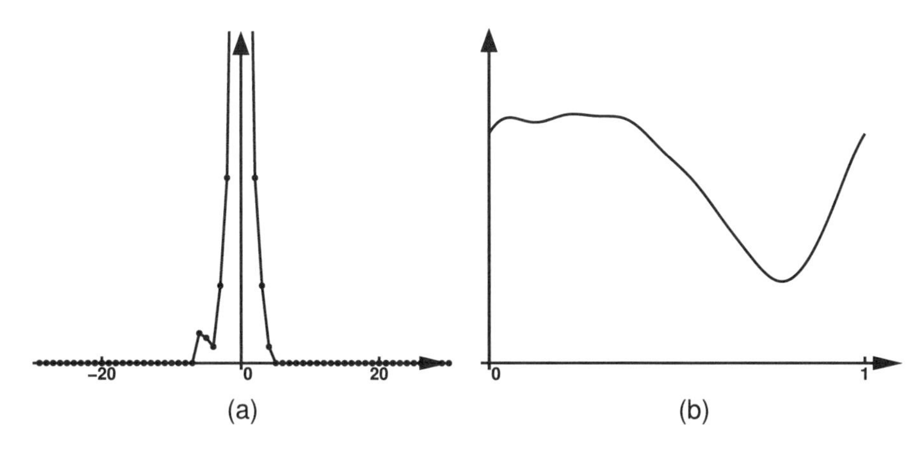

Figure 1.7
(a) Filtered Fourier spectrum from Figure 1.7(b), all coefficients with index $|l| > 5$ replaced by zero. (b) Synthesis from (a), using (1.6).

diagonal matrix $\mathbf{D}_k$. The diagonal elements are either zero or one. The ones are for the coefficients to be preserved, i.e., with indices $|l| \leq k$. The three step operation is then represented by the matrix $n\mathbf{F}^\top\mathbf{D}_k\mathbf{F}$, where $\mathbf{F}$ stands for the Fourier analysis as in (1.5) and $n\mathbf{F}^\top$ is the Fourier synthesis in (1.6).

A function, process, or operation is termed a filter if it is meant to enhance the quality of the data by removing undesired features. While this is of course a broad and a rather vague definition, practice in signal processing tends to think about filters as operations that take away unwanted frequencies or otherwise can be looked at from the frequency perspective. In our case, the high frequencies are filtered out, which explains why $n\mathbf{F}^\top\mathbf{D}_k\mathbf{F}$ is termed a **lowpass filter**.

As the sinusoid basis functions are orthogonal, the Fourier filtering procedure can also be understood as an orthogonal (i.e., least squares) projection onto the basis formed by the $2k+1$ central sinusoids, $\{\exp\left[i(2\pi)lx\right], l = -k, -k+1\ldots, k\}$, where we propose $k = 5$ in this example. The complex basis functions can be linearly transformed into the real basis $\{1, \sin(2\pi lx), \cos(2\pi lx); l = 1, 2, \ldots, k\}$. Among all sinusoids with period $2\pi/l$, these have the lowest frequencies. Orthogonal projection onto this basis generates smooth, slowly fluctuating functions out of observations with heavily fluctuating noise. In statistical terms, the lowpass filter acts as a non-parametric regression.

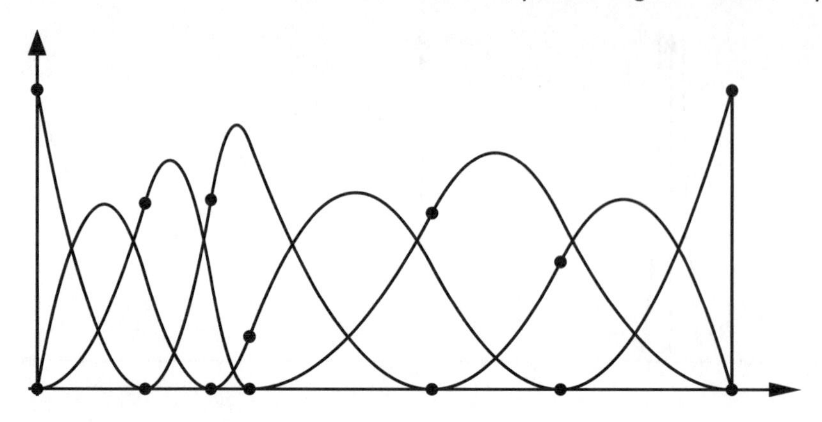

Figure 1.8
Basis of quadratic B-splines. These basis functions are quadratic polynomials on each of the intervals between the knots — indicated by the dots, while at the knots they have a continuous derivative. The basis functions have compact support of three adjacent intervals.

1.2.3 From sines to splines

Another class of popular bases in nonparametric regression are B-spline bases, of which an example is plotted in Figure 1.8. The unknown function $f(x)$ is approximated by a piecewise polynomial function, named a spline function. To this end, the interval $[0,1]$ is partitioned into subintervals, separated by knots. On each subinterval, the approximation takes the form of a polynomial of a given degree, while at the knots, continuity conditions hold. A detailed definition follows in Chapter 4. The vector space of splines is generated by B-splines, basis functions that have compact support, denoted here by $\varphi_k(x)$. The approximation then takes the form

$$\widehat{f}(x) = \sum_{k=1}^{n_b} \widehat{\beta}_k \varphi_k(x). \tag{1.12}$$

The number of basis functions n_b in the decomposition (1.12) depends on the number of knots and the degree of the polynomials. Details will follow in Chapter 3. The coefficients $\widehat{\beta}_k$ may follow, for instance, from an orthogonal projection onto the basis, i.e., from the least squares regression

$$\widehat{\boldsymbol{\beta}} = \arg\min_{\boldsymbol{\beta}} \sum_{i=1}^{n} \left[Y_i - \sum_{k=1}^{n_b} \beta_k \varphi_k(x_i) \right]^2. \tag{1.13}$$

The resulting curve $\widehat{f}(x)$ is termed a **regression spline**.

In contrast to the sinusoid basis of a Fourier analysis, B-splines are **local** basis functions, having compact support. Compactly supported basis functions are flexible tools for curve fitting when the curves have intermittent degrees of smoothness or curvature. On the other hand, sinusoids are said to be **local in frequency**, meaning that the basis functions are pure waves. Wavelets will combine the spatial locality of splines with the local frequency of sinusoids. Locality in frequency and space (or time) will be the key to sparse representations and nonlinear processing of data that contain jumps. The nonlinearity lies at the very heart of the success of wavelet based methods.

1.2.4 Regression splines, smoothing splines, P-splines

The spline regression approach in (1.13) depends on the number and the positions of the knots. Unless the knots are fixed in advance, for instance to be equidistant, positioning the knots leaves the user with a complicated parameter optimisation problem. As an alternative, knots can be chosen to coincide with each of the covariate locations x_i, $i = 1, 2, \ldots, n$. Now, the degrees of freedom in the least squares problem (1.13) outnumber the parameter vector. Any interpolating spline has optimal, zero that is, sum of least squares. For data with noise, as in Figure 1.5(b), interpolation is not an option. For data without noise, the least squares problem (1.13) is underdetermined and hence singular. Therefore, the least squares problem is regularised by adding a term that controls the smoothness of the outcome

$$\widehat{\boldsymbol{\beta}} = \arg\min_{\boldsymbol{\beta}} \sum_{i=1}^{n} \left[Y_i - \sum_{k=1}^{n_b} \beta_k \varphi_k(x_i) \right]^2 + \lambda \int_0^1 \left[\sum_{k=1}^{n_b} \beta_k \varphi_k''(x) \right]^2 dx. \tag{1.14}$$

The function $\widehat{f}(x) = \sum_{k=1}^{n_b} \widehat{\beta}_k \varphi_k(x)$ with coefficients from (1.14) is known as a **smoothing spline**. The regularisation term in (1.14) quantifies the global curvature of the output, as a measure of the smoothness of the fitting curve. The tuning parameter λ controls the balance between smoothness and closeness to the observations. Large values of λ generate smooth estimators $\widehat{f}(x)$, small values of λ make estimators close to the observations. Estimators close to the observations tend to have large variances, because they depend heavily on the errors. Smooth estimators have less variance but they are more biased. The smoothness-closeness balance is also a bias-variance balance.

The smoothing spline approach, adopting a regularisation, can be combined with flexible knot placing of the regression spline setting. In the context of a B-spline basis, the combination is known as **P-splines** (Eilers and Marx, 1996).

1.2.5 Kernels and local polynomial smoothing

Instead of working in a basis of local functions, a nonparametric regression can also be based on a local fitting approach, centred around each point of observation. The simplest member of this class of methods is probably the Nadaraya-Watson estimator

$$\widehat{f}(x) = \frac{\sum_{i=1}^{n} Y_i K\left(\frac{x_i - x}{h}\right)}{\sum_{i=1}^{n} K\left(\frac{x_i - x}{h}\right)}. \tag{1.15}$$

In this expression, $K(u)$ is a kernel function, which is a positive function, mostly taken to be symmetric. The value of h, the **bandwidth**, is a parameter to be fine-tuned. Although the construction (1.15) takes the form of a linear combination of basic building blocks, the functions $K\left(\frac{x-x_i}{h}\right)$ are by no means basis functions. This can be understood by considering (1.15) as the local constant case, i.e., $\widetilde{p} = 1$, of the **local polynomial estimator**[8] $\widehat{f}(x) = \widehat{\beta}_0(x)$, where

$$\widehat{\boldsymbol{\beta}}(x) = \arg\min_{\boldsymbol{\beta}} \sum_{i=1}^{n} \left[Y_i - \sum_{k=0}^{\widetilde{p}-1} \beta_k (x_i - x)^k \right]^2 K\left(\frac{x_i - x}{h}\right). \tag{1.16}$$

1.3 An illustrative example of a wavelet decomposition

Section 1.2 has presented a couple of curve fitting methods, based on non-parametric regression. Each of the methods operates linearly and in a local way, either by using a kernel function or by using a basis of functions that are local by having a compact support or a basis of functions that are local in frequency. The test function in Figure 1.5 is, however, smooth on the whole axis. For data with jumps or edges, such as images, a linear approach would fail and locality in either support or frequency is not enough. Instead we will use nonlinear operations in a basis that is local in both support and in frequency. This basis of localised waves is termed a wavelet basis.

1.3.1 A test function with jumps and cusps

In order to understand the power of a wavelet analysis, we consider a simple yet typical example. Figure 1.10 depicts $n = 2000$ equidistant observations of

[8] At this point, it may seem a bit awkward to let $\widetilde{p}$ stand for a local polynomial of degree $\widetilde{p} - 1$. In particular, $\widetilde{p} = 1$ refers to a local constant regression. The reason for this convention lies in the interpretation that can be given to $\widetilde{p}$ in a wavelet context, namely the number of dual vanishing moments. This $\widetilde{p}$ is the number of power functions exactly reproduced by a polynomial regression: x^q, for $q = 0, 1, \ldots, \widetilde{p} - 1$. The same convention is adopted in Definition 4.2.1 of splines.

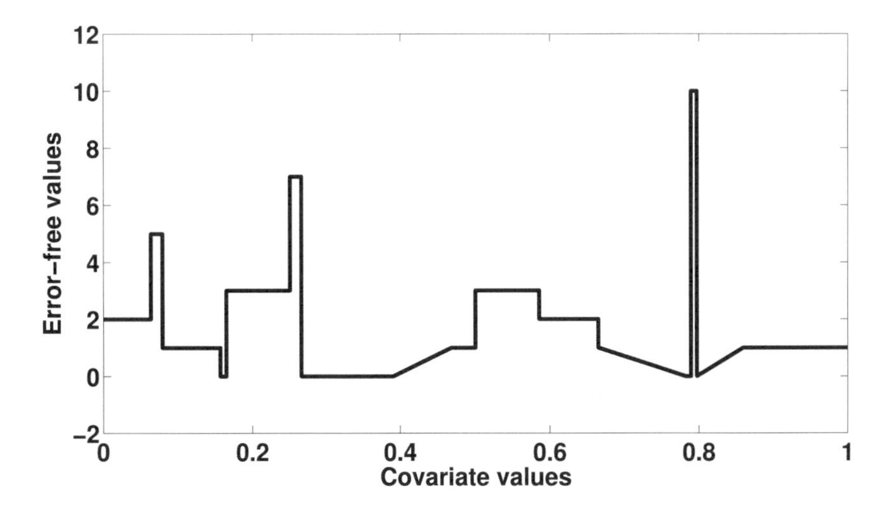

Figure 1.9
Skyline test function.

the function `skyline`, plotted in Figure 1.9, and defined as

$$f(x) = \begin{cases} 2 & \text{if} \quad x \leq 1/16 \\ 5 & \text{if} \quad 1/16 < x \leq 5/64 \\ 1 & \text{if} \quad 5/64 < x \leq 5/32 \\ 0 & \text{if} \quad 5/32 < x \leq 21/128 \\ 3 & \text{if} \quad 21/128 < x \leq 1/4 \\ 7 & \text{if} \quad 1/4 < x \leq 17/64 \\ 0 & \text{if} \quad 17/64 < x \leq 25/64 \\ \frac{64}{5}x - 5 & \text{if} \quad 25/64 < x \leq 15/32 \\ 1 & \text{if} \quad 15/32 < x \leq 1/2 \\ 3 & \text{if} \quad 1/2 < x \leq 75/128 \\ 2 & \text{if} \quad 75/128 < x \leq 85/128 \\ \frac{20}{3} - \frac{128}{15}x & \text{if} \quad 85/128 < x \leq 25/32 \\ 0 & \text{if} \quad 25/32 < x \leq 1615/2048 \\ 10 & \text{if} \quad 1615/2048 < x \leq 51/64 \\ 16x - \frac{51}{4} & \text{if} \quad 51/64 < x \leq 55/64 \\ 1 & \text{if} \quad 55/64 < x \leq 1 \end{cases} \tag{1.17}$$

The observations are subject to additive, homoscedastic, independent normal errors, that is according to the model (1.1), where $x_i = i/n$, $\varepsilon_i \sim N(0, \sigma^2)$, and all ε_i, with $i \in \{1, 2, \dots, n\}$ are mutually independent.

The objective is to estimate the n-vector $\boldsymbol{f}$ with components $f_i = f(x_i)$. The function $f(x)$ is said to be **piecewise smooth**, with an unknown number of points with a singularity. The singularities can be jumps (discontinuities),

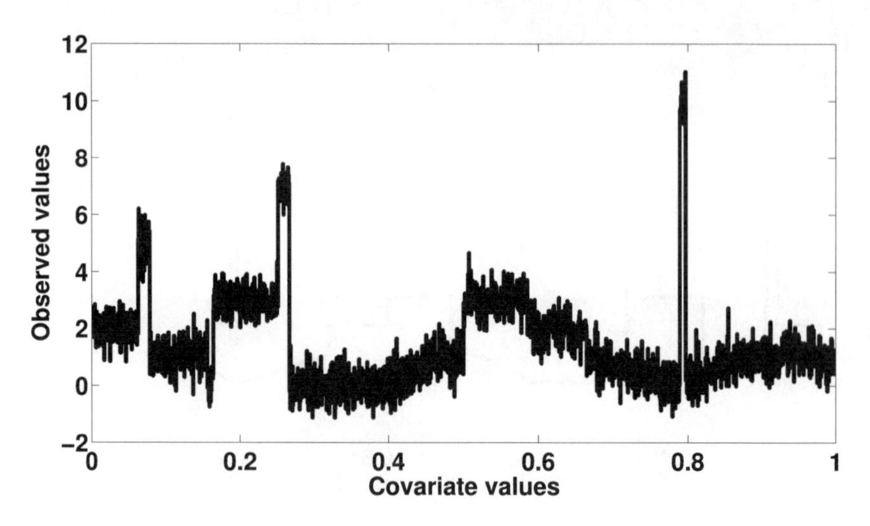

Figure 1.10
Noisy observations of the skyline test function. Sample size is 2000.

cusps, kinks (jumps in the first derivative), or others. The precise locations of the singularities are unknown.

1.3.2 Linear smoothing

The presence of singularities poses problems to any **linear** estimator, such as least squares regression, smoothing splines, kernel or local polynomial regression, or any lowpass filtering. Indeed, all these methods can be written as a matrix, that is, $\widehat{\boldsymbol{f}} = \mathbf{A}_{\boldsymbol{x}}\boldsymbol{Y}$. Here, $\boldsymbol{Y}$ denotes the column vector of observations Y_i, with $i = 1, 2, \ldots, n$ and the index $\boldsymbol{x}$ refers to the column vector of observational points x_i, again with $i = 1, 2, \ldots, n$. The estimation matrix $\mathbf{A}_{\boldsymbol{x}}$ depends on $\boldsymbol{x}$ but not on $\boldsymbol{Y}$. Therefore it cannot possibly adapt to the presence of singularities.

Example 1.3.1 *For the linear Fourier filtering in Section 1.2.2, we have* $\mathbf{A}_{\boldsymbol{x}} = n\mathbf{F}^\top\mathbf{D}_k\mathbf{F}$.

Focussing on one component of the estimator $\widehat{f_\ell} = \sum_{i=1}^n A_{\ell,i}Y_i$, the bias $E(\widehat{f_\ell}) - f_\ell$ is close to zero if

$$f_\ell \approx E(\widehat{f_\ell}) = \sum_{i=1}^{n} A_{\ell,i}E(Y_i) = \sum_{i=1}^{n} A_{\ell,i}f_i.$$

Now let the number of observations $n \to \infty$, while we keep x_ℓ fixed. Then a zero bias is easy to achieve if f is continuous at x_ℓ and the adjacent x_i

are tending towards x_ℓ. It suffices, for instance, to choose $A_{\ell,i} = 0$ for $|i - \ell|$ above a certain value and to impose $\sum_{i=1}^{n} A_{\ell,i} = 1$. This construction fails whenever x_ℓ tends to a jump. The bias is now of order $\mathcal{O}(1)$.

1.3.3 A Fourier analysis of data with discontinuities

As an illustration of linear smoothing on data with jumps, this section elaborates on filtering in a Fourier series. This discussion can be skipped without losing the main thread of the chapter. In Figure 1.11 the Fourier transform method of Section 1.2.2 is applied to the data in Figure 1.10. As the test signal has sudden jumps, there is an important contribution at high frequencies in the Fourier spectrum. Cutting off high frequencies will therefore always leave artificial oscillations, while the discontinuities are blurred.

Elaboration on the bad performance of Fourier analysis on data with jumps actually reveals three problems. The first is the oscillations in the approximations, the second is the slow pointwise and least square convergence, the third is the impossibility of uniform convergence. For a closer look at these problems we consider the sawtooth function $f(x) = 1/2 - x$ defined on $[0, 1]$, for which the coefficients in the expansion (1.2) in the basis of complex sinusoids is found by direct application of (1.4). For $l = 0$ we find the average value $a_0 = 0$, while for $l \neq 0$, this becomes

$$a_k = \int_0^1 \left[\frac{1}{2} - x\right] \exp\left[-i(2\pi)kx\right] dx = \frac{\exp\left[-i(2\pi)k\right]}{-i(2\pi)k} = \frac{-1}{-i(2\pi)k},$$

leading to the expansion

$$f(x) = \sum_{k=-\infty,k\neq 0}^{\infty} \frac{1}{i(2\pi)k} \exp\left[i(2\pi)kx\right] = \sum_{k=1}^{\infty} \frac{1}{\pi k} \sin\left[(2\pi)kx\right].$$

Exercise 1.3.2 *Prove the second equality in the expression above by checking that* $a_{-k} = -a_k$.

The truncated series with 10 and 20 terms are depicted in Figure 1.12. The plot reveals that the function $f(x)$ actually has discontinuities at the end points of the interval. This is because the basis functions are infinitely far running sinusoids. Hence, the expansion into that basis implicitly involves a periodical extension of the function $f(x)$ to form a sawtooth wave, thereby introducing the jumps.

Exercise 1.3.3 *Consider another example of a function with discontinuities: the characteristic or indicator function* $f(x) = \chi_{[1/4,3/4)}(x)$*, i.e., the piecewise constant function with value one on the interval* $[1/4, 3/4)$ *and zero outside that interval. Periodic extension does not create discontinuities in the end points, the result being now a square wave. Find the coefficients* a_k *in the Fourier series (1.2) using Expression (1.4). In particular, prove that* $a_{2k} = 0$

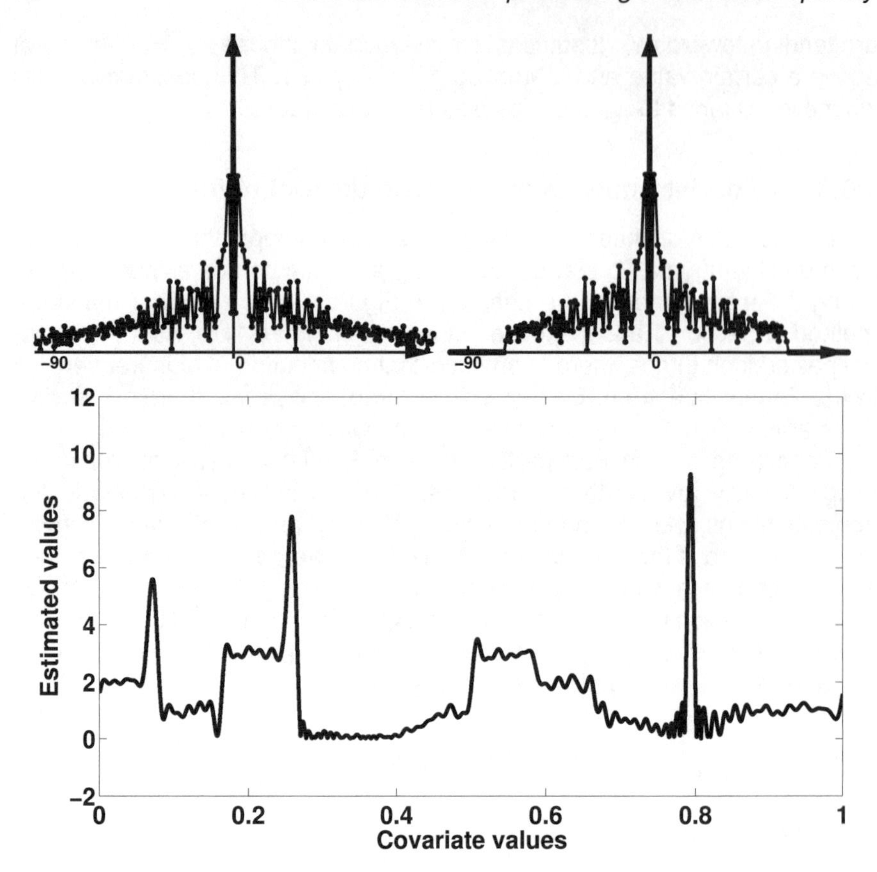

Figure 1.11
Fourier based linear filtering of the observations in Figure 1.10. Top left: zoom on the DFT of the observations without the noise. Top right: selected Fourier coefficients from the DFT of the observations with noise: out of 2000 coefficients, 141 are selected. Bottom: reconstruction from selected Fourier coefficients.

for $k \neq 0$, while $a_{2k-1} = (-1)^k/[(2k-1)\pi]$. Using the fact that $a_k = a_{-k}$, the Fourier series can be written as a cosine series, i.e.,

$$f(x) = a_0 + \sum_{k=1}^{\infty} 2a_k \cos\left[(2\pi)kx\right].$$

The approximation of a sawtooth wave by the smooth sinusoidal waves introduces the earlier observed oscillating behaviour. Comparison of the two

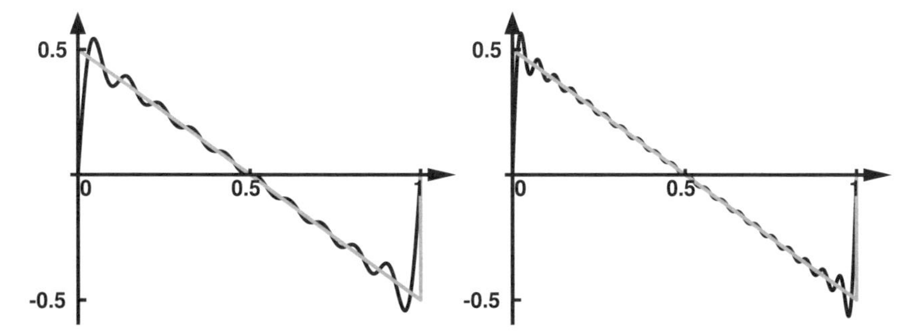

Figure 1.12
Truncated Fourier series approximation of a piecewise continuous function. The approximation is wiggly, convergence is slow.

panels in Figure 1.12 reveals that the magnitude of the oscillation next to the discontinuities in 0 and 1 does not converge. This follows from a development of the error function $R_n(x) = f_n(x) - f(x)$, where

$$f_n(x) = \sum_{k=1}^{n} \frac{1}{\pi k} \sin\left[(2\pi)kx\right].$$

In order to find the local maxima of the error function, its derivative is computed,

$$R_n'(x) = 1 + 2\sum_{k=1}^{n} \cos\left[(2\pi)kx\right] = \frac{\sin\left[(n+1/2)(2\pi)x\right]}{\sin\left[(2\pi)x/2\right]}.$$

The last equality is Dirichlet's identity, which can easily be proven by an induction argument. The limit of the derivative in $x = 0$ equals $2n+1$, so the error is an increasing function in $x = 0$. Furthermore, the first maximum beyond $x = 0$ is reached in $x_n^* = 1/(2n+1) \Rightarrow R_n'(x_n^*) = 0$. The error in that point is given by

$$\begin{aligned} R_n\left(\frac{1}{2n+1}\right) &= R_n(0) + \int_0^{1/(2n+1)} R_n'(u)du \\ &= -\frac{1}{2} + \int_0^1 \frac{\sin(\pi v)}{\sin\left(\pi v/(2n+1)\right)} \frac{dv}{(2n+1)}. \end{aligned}$$

When the number of terms increases, the denominator within the integral tends to

$$\lim_{n\to\infty} \sin\left(\pi v/(2n+1)\right)(2n+1) = \lim_{t\to 0} \sin(\pi v t)/t = \pi v,$$

leading to

$$\lim_{n\to\infty} R_n\left(\frac{1}{2n+1}\right) = -\frac{1}{2} + \int_0^1 \frac{\sin(\pi v)}{\pi v} dv = 0.0895.$$

The magnitude of the maximum oscillation does not converge to zero. This is known as the **Gibbs phenomenon**.

The plot also illustrates the **slow convergence** of the L_2 error norm of the n-term approximation

$$\|R_n\|_2^2 = \int_0^1 \left[f(x) - f_n(x)\right]^2 dx,$$

where in general

$$f_n = \sum_{k=-n}^{n} a_k \exp\left[i(2\pi)kx\right].$$

The slow convergence for functions with one or more discontinuities is due to the **slow asymptotic decay** of the Fourier coefficients, $a_k = \mathcal{O}\left(1/|k|\right)$. Thanks to the orthogonality of the basis, we can write

$$\|R_n\|_2^2 = \sum_{|k|>n} |a_k|^2 = \mathcal{O}\left(2\sum_{k=n+1}^{\infty} \frac{1}{k^2}\right) = \mathcal{O}\left(n^{-1}\right). \tag{1.18}$$

Moreover, the continuous approximation cannot possibly converge uniformly. Indeed, in the immediate neighbourhood of the jump, the order of the approximation error of any continuous function must be at least half of the jump height.

1.3.4 A multilevel approach

A major problem in smoothing data with discontinuities is that we do not know the number and the locations of the singularities. Therefore we start off by a most prudent smoothing operation. The operation involves nothing but the pairwise averages of successive observations. We first rename the observations as $s_{J,k} = Y_{k+1}$ for $k = 0, 1, \ldots, n-1$. The meaning and the goal of the index J will become clear below, together with the reason for the renaming. Next, we compute the means as in

$$s_{J-1,k} = (s_{J,2k} + s_{J,2k+1})/2,$$

for $k = 0, 1, \ldots, \lfloor n/2 \rfloor$. If n is odd, we just keep the last element, $s_{J-1,\lceil n/2\rceil} = s_{J,n}$ We have defined an operation that maps the n-tuple s_J onto the $\lceil n/2\rceil$-tuple s_{J-1}. This operation is termed the *moving average filter*. Going from n to $\lceil n/2\rceil$ elements, there must be some loss of information, i.e., loss of details in the process. In order to capture the details, we also compute the differences between the original values and the averages, in what is called the *moving difference filter.* It is sufficient to compute the offsets coming from the original values with an odd index, $s_{J,2k+1}$. This is because the offsets of the even indexed values $s_{J,2k}$ are just the opposites. So we compute

$$d_{J-1,k} = s_{J,2k+1} - s_{J-1,k} = -(s_{J,2k} - s_{J-1,k}).$$

If n is odd, the operation does not involve the last element $s_{J,n}$; hence $\boldsymbol{d}_{J-1}$ contains $\lfloor n/2 \rfloor$ elements. Together, the moving average filter and the moving difference filter form a **filterbank**, which basically means that the combination of these two filters induces no loss of information. Indeed, from the pair $s_{J-1,k}$ and $d_{J-1,k}$, the original pair[9] $s_{J,2k}$ and $s_{J,2k+1}$ can be restored by $s_{J,2k} = s_{J-1,k} - d_{J-1,k}$ and $s_{J,2k+1} = s_{J-1,k} + d_{J-1,k}$.

For most values of k, the detail offset $d_{J-1,k}$ will be dominated by the errors. A large magnitude of $d_{J-1,k}$, however, is probably not due to noise, but to the presence of a big jump in $f_{2k+1} - f_{2k} = E(s_{J,2k+1}) - E(s_{J,2k})$. This way, **the large detail offsets point towards the locations of the singularities**.

The moving average and difference filters operate at very small scale, between immediate neighbours, that is. It may therefore not discriminate well between jumps in the vector $\boldsymbol{f}$ and noisy fluctuations in $\boldsymbol{\varepsilon}$. What could distinguish a jump in $\boldsymbol{f}$ from the noisy fluctuations, though, is that jumps in $\boldsymbol{f}$ typically have a wider range, meaning that they are separated from each other by more than just two neighbours. As a result, it may be interesting to apply a less local smoothing next to the moving average. Such a broader averaging would further attenuate the noise, i.e., reduce its variance, making the jumps in $\boldsymbol{f}$ to appear more prominently in the offsets. A broader smoothing is realised by averaging the averages again,

$$s_{J-2,k} = (s_{J-1,2k} + s_{J-1,2k+1})/2,$$

or for general $j = J - 1, \ldots, L$,

$$s_{j,k} = (s_{j+1,2k} + s_{j+1,2k+1})/2. \quad (1.19)$$

The index j can be understood as the **resolution level**. It is straightforward to verify that

$$s_{j,k} = \frac{1}{2^{J-j}} \sum_{i=1}^{2^{J-j}} Y_{2^{J-j}k+i}.$$

The resolution level j corresponds to the **scale** 2^{J-j}. In a Haar transform, the scale is the number of observations involved in the average $s_{j,k}$. The scale is a power of two; therefore it is called *dyadic.* As we proceed across several, successive dyadic scales, the whole analysis is known as a **multiresolution**

[9] In general filterbanks, the reconstruction does not need to proceed by pairs. We would be more than happy to reconstruct the n-*vector* $\boldsymbol{s}_J$ from the $\lceil n/2 \rceil$-vector $\boldsymbol{s}_{J-1}$ and the $\lfloor n/2 \rfloor$-vector $\boldsymbol{d}_{J-1}$.

analysis, a notion which will be formalised in Section 5.5. With each average $s_{j,k}$ corresponds a difference

$$d_{j,k} = s_{j+1,2k+1} - s_{j,k} = -(s_{j+1,2k} - s_{j,k}). \tag{1.20}$$

The difference between one observation and the mean value is straightforwardly written as half the difference between the two successive observations

$$d_{j,k} = \frac{1}{2}(s_{j+1,2k+1} - s_{j+1,2k}). \tag{1.21}$$

Expressions (1.19) and (1.21) constitute a matrix block in one step of a multilevel decomposition

$$\begin{bmatrix} s_{j,k} \\ d_{j,k} \end{bmatrix} = \begin{bmatrix} \frac{1}{2} & \frac{1}{2} \\ -\frac{1}{2} & \frac{1}{2} \end{bmatrix} \begin{bmatrix} s_{j+1,2k} \\ s_{j+1,2k+1} \end{bmatrix}.$$

A small difference indicates that the average is taken over slightly fluctuating values, while a large difference suggests the presence of a jump. In that case, it would be better not to use $s_{j,k}$ as a local estimator of the noise-free values, but rather stay with the values $s_{j+1,2k}$ and $s_{j+1,2k+1}$, or even with values at *finer scales* $j+2$ or higher.

In most publications, zero resolution $j = 0$ is fixed to be the level where n observations are reduced to a single scaling coefficient $s_{0,0}$. With this definition, the resolution of the observations is $J = \lceil \log_2(n) \rceil$. Obviously, $L = 0$ is the lowest or coarsest possible resolution level. In many applications, the Haar transform is not computed up to this coarse resolution. This is because at coarse resolutions, most coefficients are not zero. There is often no sparsity at coarse scales. Therefore, it is sufficient to compute the transform up to scale, say, $L = 3$ or $L = 5$, or up to another level, very much depending on the application. The indexing of resolution levels and scales is pretty much dependent on the publication. We have fixed here a system for this book, with resolution levels ranging between $0 \leq L$ for the coarsest scales and $J > L$ for the finest scales. We have named 2^{J-j} the scale of resolution level j, but this could have been 2^{-j} as well; there is no standardisation. Readers may find publications where fine resolutions are given by small numbers j, and coarse resolutions by large numbers j, the opposite of the convention followed in this book.

1.3.5 The Haar transform

Given the observations Y_i, $i = 1, \ldots, n$, we have computed $s_{j,k}$ and $d_{j,k}$ at successive levels $j = J-1, J-2, \ldots, L$ and at locations $k = 0, 1, \ldots, 2^j - 1$. We do not need to store all the intermediate scaling coefficients $s_{j+1,k}$ for reconstruction. Indeed, as we have seen, the values $s_{j,k}$ and $d_{j,k}$ at level j are enough to recursively reconstruct the values $s_{j+1,k}$ at the finer scale,

$$\begin{aligned} s_{j+1,2k} &= s_{j,k} - d_{j,k} \\ s_{j+1,2k+1} &= s_{j,k} + d_{j,k}. \end{aligned} \tag{1.22}$$

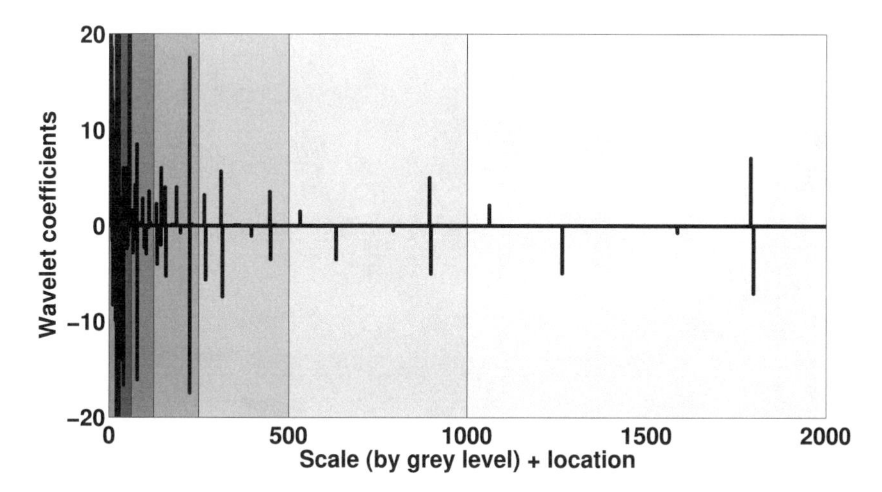

Figure 1.13
Haar transform of the skyline test function. The different shading corresponds to the successive scales of the coefficients.

Therefore, the complete vector of observations can be reconstructed from the transformed vector $\boldsymbol{w}_L$ which contains the coarsest scale coefficients $\boldsymbol{s}_L = (s_{L,0}, s_{L,1}, \ldots, s_{L,2^L-1})^\top$, together with the detail coefficients at all scales $\boldsymbol{d}_j = (d_{j,0}, d_{j,1}, \ldots, d_{j,2^j-1})^\top$, that is

$$\boldsymbol{w}_L = [\boldsymbol{s}_L \, \boldsymbol{d}_L \, \boldsymbol{d}_{L+1} \, \ldots \boldsymbol{d}_{J-1}]. \tag{1.23}$$

The linear transform that maps $\boldsymbol{Y}$ onto $\boldsymbol{w}_L$ is known as the Haar transform. Its matrix[10], is denoted as $\widetilde{\mathbf{W}}$ so

$$\boldsymbol{w}_L = \widetilde{\mathbf{W}} \cdot \boldsymbol{Y}. \tag{1.24}$$

Figures 1.13 and 1.14 depict the Haar transformed vectors of $\boldsymbol{f}$ and $\boldsymbol{Y}$ respectively. It is clear that the Haar transform of $\boldsymbol{f}$, $\boldsymbol{v}_L = \widetilde{\mathbf{W}} \cdot \boldsymbol{f}$, contains many zeros and near-zeros. The transformed vector $\boldsymbol{v}_L$, is said to be **sparse**. A few large coefficients point towards the locations and scales of the main features of the vector $\boldsymbol{f}$. The vector $\boldsymbol{v}_L$ can be **compressed** for storage in a computer. This is the principle behind the JPEG-2000 image compression standard, where the pixels of an image play the role of the signal $\boldsymbol{f}$. The large coefficients in $\boldsymbol{v}_L$ are easy to estimate from the noisy version in $\boldsymbol{w}_L$, as the noise

[10] See Section 1.3.8 for a concrete example of the Haar transform matrix.

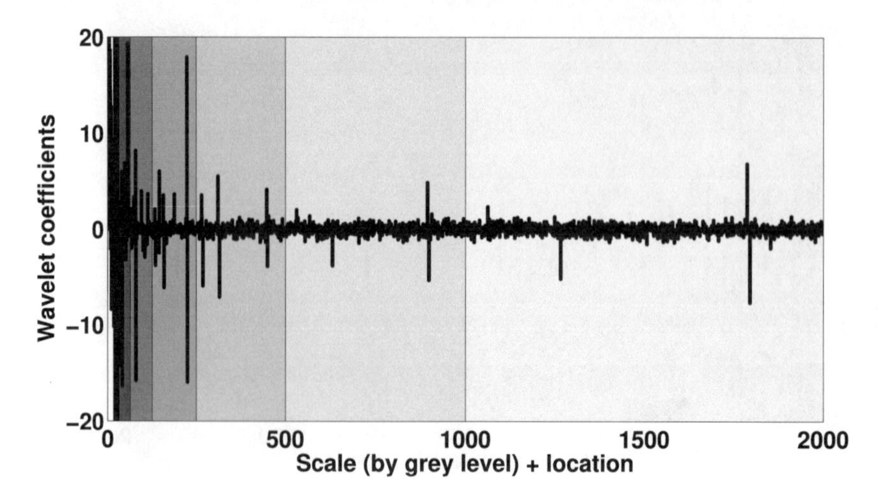

Figure 1.14
Noisy Haar transform of the skyline test function.

has spread out equally over all coefficients[11]. In order to estimate $\boldsymbol{v}_L$ from $\boldsymbol{w}_L$, we could select the largest observations in $\boldsymbol{w}_L$, that is, all coefficients with an absolute value above a well chosen threshold λ. This procedure is illustrated in Figures 1.15 and 1.16. The former figure shows the thresholded coefficients, whereas the latter depicts the output vector from applying the inverse transform on the thresholded coefficients. We observe that the reconstruction preserves the peaks and jumps of the input pretty well, although the overall picture may seem a bit blocky. The good reconstruction of jumps is a typical achievement of a nonlinear approach within a sparse data representation. Thresholding, selecting the *largest* coefficients, is indeed a nonlinear method, in contrast to selecting the *first* coefficients according to prefixed ordering, or all coefficients up to a prefixed scale. Comparing the result in Figure 1.16 with the linear Fourier approach in 1.11, the conclusion may be a mixed one: the wavelet approach leads definitely to sharper reconstruction of the jumps with less oscillations in between. On the other hand, the reconstruction is quite blocky. In order to understand and reduce the blockiness, we add an interpretation of the decomposition in the next section.

[11] In particular, the noise on the wavelet coefficients is homoscedastic, meaning that the variance is the same in all coefficients. For this homoscedasticity to hold true, we need that the noise at the input is homoscedastic, but also uncorrelated. Moreover, the transform has to be orthogonal. See also page 269. The Haar transform is orthogonal if properly normalised, see Exercise 1.3.5. Alternatively, if the noise is heteroscedastic, and if the structure of the covariance matrix is known, then the coefficients can be standardised with the square roots of the diagonal elements of that covariance matrix.

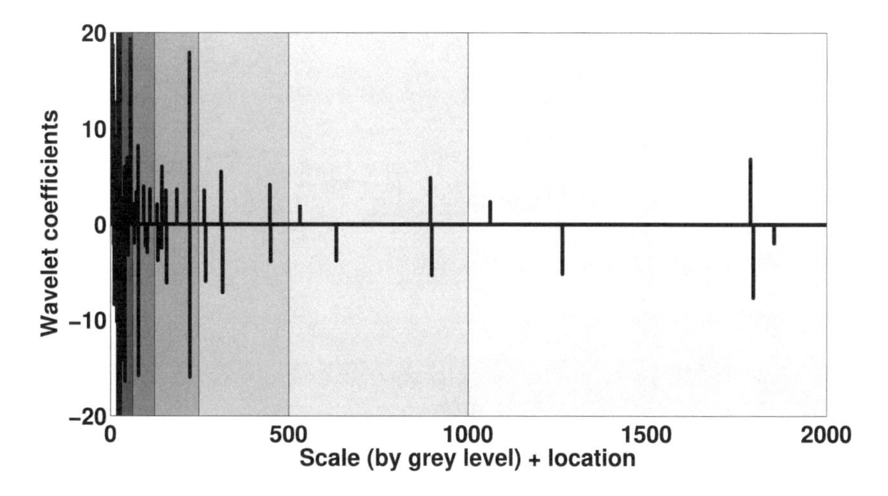

Figure 1.15
Thresholded noisy Haar coefficients.

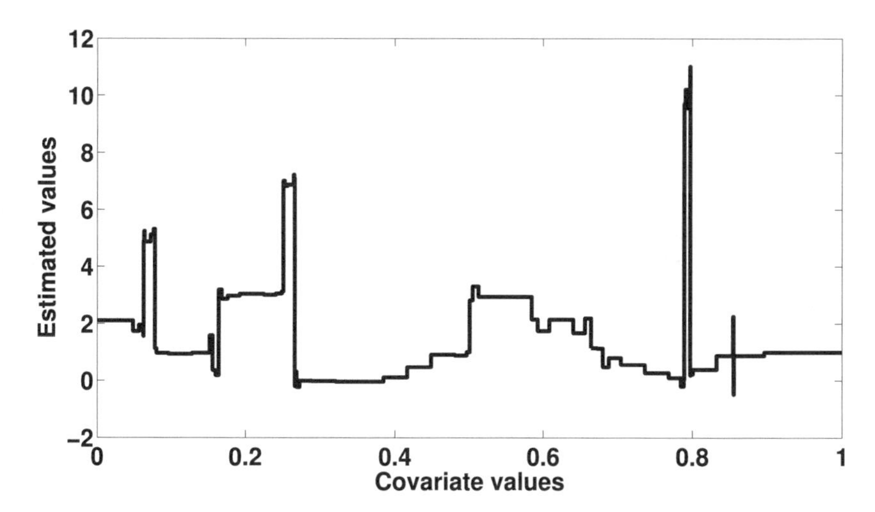

Figure 1.16
Reconstruction from thresholded Haar coefficients.

1.3.6 The Haar basis

Suppose we are given a very short vector of eight observations $\boldsymbol{Y}$. Figure 1.17, upper left, displays these observations as a piecewise constant function, a staircase that is, where the height of stair i corresponds to the value of Y_i. This is of course a blocky representation, but we are at the finest scale:

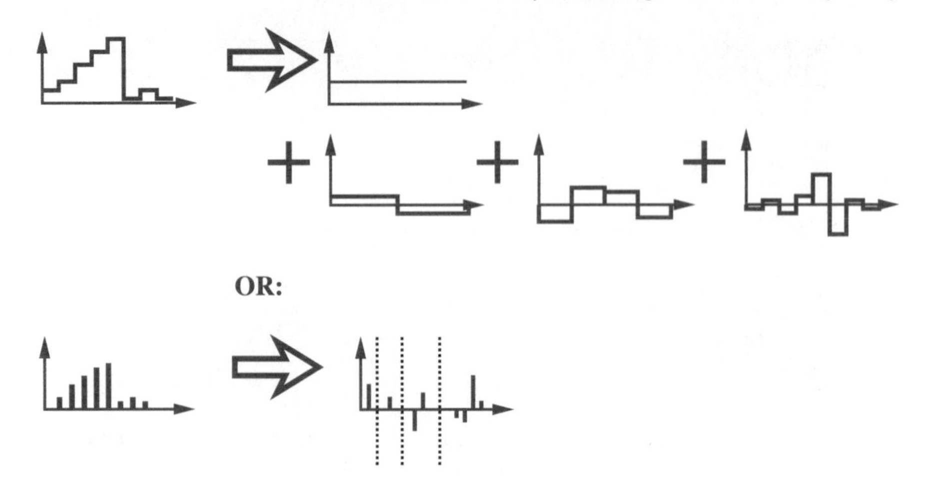

Figure 1.17
The Haar transform, seen as a decomposition of a piecewise constant function. On the bottom line, the corresponding vector transformation. The vertical dashed lines indicate the scales.

when the number of observations is large, the visual perception is dominated by the smoothness of the underlying function $f(x)$. The Haar transform, however, preserves the staircase form when proceeding to coarser scales. Indeed, as illustrated in Figure 1.17, it is straightforward to represent the averages as a staircase with broader stairs. At the same time, difference coefficients can be represented by a double stair with zero mean: the first half indicates how much the first observation of a pair is above the average, while the second half is exactly the opposite number, indicating how much the second observation is below the average.

Let $s_J(x)$ denote the blocky function that represents the observations $Y_i = s_{J,i-1}$; then this function can be written as

$$s_J(x) = \sum_{i=0}^{n-1} s_{J,i}\varphi_{J,i}(x), \tag{1.25}$$

where the basis functions $\varphi_{J,i}(x)$ are characteristic functions or indicator functions on the subintervals $[i/n, (i+1)/n)$, i.e., $\varphi_{J,i}(x) = \chi_{[i/n,(i+1)/n)}(x)$. The functions $\varphi_{J,i}(x)$ appear in Figure reffig:haarbasis on the left hand side. These basis functions are known as the Haar **scaling functions**. Likewise, the coefficients $s_{J,i}$ are named **scaling coefficients**. The scaling functions are all equally rescaled translations of a single **father function** $\varphi_{J,i}(x) = \varphi(nx - i)$. The Haar father function equals $\varphi(x) = \chi_{[0,1)}(x)$. The Haar decomposition rewrites this function in terms of coarse scale indicator

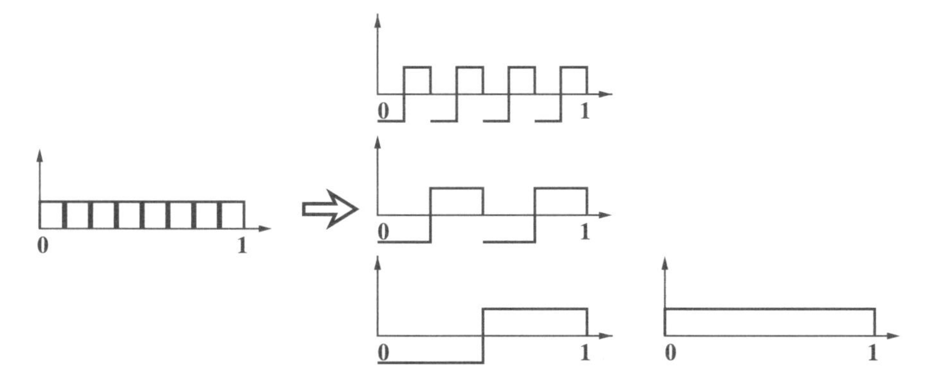

Figure 1.18
The Haar transform maps coefficients in a scaling basis onto coefficients in a wavelet basis.

functions plus refining differences, i.e.,

$$s_J(x) = \sum_{k=0}^{n_L-1} s_{L,k}\varphi_{L,k}(x) + \sum_{j=L}^{J-1}\sum_{k=0}^{n'_j-1} d_{j,k}\psi_{j,k}(x). \qquad (1.26)$$

In this expression, the integers n_j and n'_j are the number of scaling functions and wavelet functions at scale j. The coarse scaling functions are again rescaled indicator functions, $\varphi_{L,k}(x) = \varphi(2^L x - k)$, whereas the functions $\psi_{j,k}(x)$ are all dilations (rescalings) and translations of a single **mother function** $\psi(x)$, defined as

$$\psi(x) = \chi_{[1/2,1[}(x) - \chi_{[0,1/2)}(x).$$

This mother function has integral zero, i.e.,

$$\int_{-\infty}^{\infty} \psi(x)\,dx = 0. \qquad (1.27)$$

The zero integral is a common property for a wide class of generalisations of the Haar decomposition. A function with zero integral has positive and negative parts that exactly compensate for each other. Such a function shows at least some degree of fluctuation. The Haar mother function looks like a single, blocky wave. The zero integral thus explains the name of the basis functions, which are known as **wavelets**. The corresponding coefficients $d_{j,k}$ are detail coefficients or wavelet coefficients. The diminutive -lets refers to the fact that the basis functions are localised; the waves do not roll forever. This is in contrast to sine waves. The function $\sin(\omega x)$ has infinite support and it has a single, constant frequency. Its integral over the real axis is not defined. This is related to the fact that its squared integral diverges. For wavelets, we impose that

$$\int_{-\infty}^{\infty} \psi^2(x)\, dx < \infty. \tag{1.28}$$

Most observations are about phenomena that are square integrable. Indeed, the integral of a squared function has often a physical interpretation related to the "total energy" in the system. From this perspective, it may be a bit unnatural to adopt a decomposition, such as the Fourier transform, whose basic building block has infinite energy. As all wavelet basis functions, the Haar basis is **local in two dimensions**. Indeed, the basis functions are situated at a specific time or space, and at the same time, they are local in scale. Here, scale could be seen as a rough version of inverse frequency, where fine or small scales correspond to high frequencies.

1.3.7 Wavelet basis functions

The Haar basis transform rewrites a fine scale decomposition (1.25) into a multiscale decomposition (1.26). The same expressions hold for any other wavelet transform. Whereas the Haar transform represents discrete data as a staircase, every wavelet transform has its own interpretation of the fine scale coefficients, given by the basis functions.

Let $\Phi_j(x)$ and $\Psi_j(x)$ be the row vectors of all scaling and wavelet basis functions at level j, i.e., $\Phi_j(x) = \begin{bmatrix} \varphi_{j,0}(x) & \varphi_{j,1}(x) & \ldots & \varphi_{j,n_j}(x) \end{bmatrix}$, and likewise for $\Psi_j(x)$. Then one step of the forward wavelet transform corresponds to going from the representation on the left to one on the right in

$$\Phi_{j+1}(x)\boldsymbol{s}_{j+1} = \Phi_j(x)\boldsymbol{s}_j + \Psi_j(x)\boldsymbol{d}_j. \tag{1.29}$$

The full wavelet transform runs over several levels j; this corresponds to the following basis transformation

$$\Phi_J(x)\boldsymbol{s}_J = \Phi_L(x)\boldsymbol{s}_L + \sum_{j=L}^{J-1} \Psi_j(x)\boldsymbol{d}_j. \tag{1.30}$$

This is a compact vector based notation for the expression that follows from the equivalence of the two representations in (1.25) and (1.26). The forward wavelet transform finds the representation on the right hand side, starting from the one on the left hand side. The inverse wavelet transform proceeds from the right to the left.

One of the topics in wavelet theory is to find out which representations

$\Phi(x)$ and $\Psi(x)$ are possible and which is the corresponding wavelet decomposition. In order to do so, the Section 1.3.8 develops a general framework, again starting from the example of the Haar transform.

1.3.8 The Haar transform matrix

Let $\Psi(x) = [\Phi_L(x)\,\Psi_L(x)\,\Psi_{L+1}(x)\,\ldots\,\Psi_{J-1}]$ denote the row vector of basis functions after the Haar decomposition. Then the basis transform (1.30)

$$\Phi_J(x)\boldsymbol{s}_J = s_J(x) = \Psi(x)\boldsymbol{w}_L, \tag{1.31}$$

with $\boldsymbol{w}_L$ as in (1.23). Let $\mathbf{W}$ denote the inverse wavelet transform matrix, i.e., $\mathbf{W} = \widetilde{\mathbf{W}}^{-1}$; then $\boldsymbol{s}_J = \mathbf{W}\boldsymbol{w}_L$, and so $\Phi_J(x)\mathbf{W}\boldsymbol{w}_L = \Psi(x)\boldsymbol{w}_L$, from which we find

$$\Phi_J(x)\mathbf{W} = \Psi(x). \tag{1.32}$$

The *columns* of $\mathbf{W}$ are coefficients in the decomposition of a wavelet function in terms of fine scaling functions. Using Haar wavelets, these fine scaling functions are fine scale characteristic functions, which makes the retrieval of $\mathbf{W}$ particularly straightforward.

$$\mathbf{W} = \begin{bmatrix} 1 & -1 & -1 & 0 & -1 & 0 & 0 & 0 \\ 1 & -1 & -1 & 0 & 1 & 0 & 0 & 0 \\ 1 & -1 & 1 & 0 & 0 & -1 & 0 & 0 \\ 1 & -1 & 1 & 0 & 0 & 1 & 0 & 0 \\ 1 & 1 & 0 & -1 & 0 & 0 & -1 & 0 \\ 1 & 1 & 0 & -1 & 0 & 0 & 1 & 0 \\ 1 & 1 & 0 & 1 & 0 & 0 & 0 & -1 \\ 1 & 1 & 0 & 1 & 0 & 0 & 0 & 1 \end{bmatrix}.$$

The columns of $\mathbf{W}$ can be seen as a discrete version of the blocky waves in $\Psi(x)$. This is no coincidence, but an instance of a general process known as **subdivision**, further explored in Section 2.2.5.

The inverse of $\mathbf{W}$, $\widetilde{\mathbf{W}}$, can be found to be

$$\widetilde{\mathbf{W}} = \begin{bmatrix} \frac{1}{8} & \frac{1}{8} & \frac{1}{8} & \frac{1}{8} & \frac{1}{8} & \frac{1}{8} & \frac{1}{8} & \frac{1}{8} \\ -\frac{1}{8} & -\frac{1}{8} & -\frac{1}{8} & -\frac{1}{8} & \frac{1}{8} & \frac{1}{8} & \frac{1}{8} & \frac{1}{8} \\ -\frac{1}{4} & -\frac{1}{4} & \frac{1}{4} & \frac{1}{4} & 0 & 0 & 0 & 0 \\ 0 & 0 & 0 & 0 & -\frac{1}{4} & -\frac{1}{4} & \frac{1}{4} & \frac{1}{4} \\ -\frac{1}{2} & \frac{1}{2} & 0 & 0 & 0 & 0 & 0 & 0 \\ 0 & 0 & -\frac{1}{2} & \frac{1}{2} & 0 & 0 & 0 & 0 \\ 0 & 0 & 0 & 0 & -\frac{1}{2} & \frac{1}{2} & 0 & 0 \\ 0 & 0 & 0 & 0 & 0 & 0 & -\frac{1}{2} & \frac{1}{2} \end{bmatrix}.$$

Exercise 1.3.4 *Decompose the forward Haar transform matrix above as* $\widetilde{\mathbf{W}} = \widetilde{\mathbf{W}}_0\widetilde{\mathbf{W}}_1\widetilde{\mathbf{W}}_2$*, where* $\widetilde{\mathbf{W}}_j$ *performs one level of the multiscale transform, mapping* $\boldsymbol{w}_{j+1}$*, as defined in (1.23), onto* $\boldsymbol{w}_j$*. Write the matrices* $\widetilde{\mathbf{W}}_j$*, for* $j = 0, 1, 2$.

The inverse transform matrix can be written as

$$\widetilde{\mathbf{W}} = \mathrm{diag}\left(\left[\begin{array}{cccccccc} \frac{1}{8} & \frac{1}{8} & \frac{1}{4} & \frac{1}{4} & \frac{1}{2} & \frac{1}{2} & \frac{1}{2} & \frac{1}{2} \end{array}\right]\right)\mathbf{W}^{\top},$$

where $\mathrm{diag}(\boldsymbol{x})$ is the diagonal matrix with the elements of vector $\boldsymbol{x}$ on the diagonal. Otherwise stated, $\mathbf{W}\mathbf{W}^{\top}$, $\mathbf{W}^{\top}\mathbf{W}$, $\widetilde{\mathbf{W}}\widetilde{\mathbf{W}}^{\top}$, and $\widetilde{\mathbf{W}}^{\top}\widetilde{\mathbf{W}}$ are diagonal vectors. Up to a normalisation, the Haar transform is *orthogonal.*

Exercise 1.3.5 *An orthogonal Haar transform is found by normalising the columns of* $\mathbf{W}$ *and* $\widetilde{\mathbf{W}}$. *This is*

$$\mathbf{W}_{\mathrm{norm}} = \mathbf{W}\mathrm{diag}\left(\left[\begin{array}{cccccccc} \frac{1}{2\sqrt{2}} & \frac{1}{2\sqrt{2}} & \frac{1}{2} & \frac{1}{2} & \frac{1}{\sqrt{2}} & \frac{1}{\sqrt{2}} & \frac{1}{\sqrt{2}} & \frac{1}{\sqrt{2}} \end{array}\right]\right),$$

and $\widetilde{\mathbf{W}}_{\mathrm{norm}} = \mathbf{W}^{\top}_{\mathrm{norm}}$. *Find the orthogonal matrix mapping the decompostion at level* $j+1$ *into the decomposition at level* j.

The orthogonality of a Haar decomposition can also be read from the Haar wavelet basis functions. It is obvious that non-overlapping blocky waves are orthogonal to each other. When two Haar basis functions overlap, this overlap always occurs within the subinterval where the coarsest of the two is a constant, so that the zero integral of the finest function makes that

$$\int_{-\infty}^{\infty} \psi_{j,k}(x)\psi_{j',k'}(x)\, dx = 0. \tag{1.33}$$

1.4 The key properties of a wavelet method

The example in Section 1.3 has illustrated that wavelet basis functions are **local** (or localised) in time or space and, at the same time, in scale. Intuitively speaking, scale is more or less the inverse of frequency. Wavelet bases thus combine features of Fourier analysis, which is local in frequency, with ideas from kernel or spline based nonparametric methods, which operate on a local basis in time or space. The double locality lies at the basis of two fundamental and unique characteristics of a wavelet decomposition: **sparsity** and **multiresolution**.

Sparsity enables **nonlinear** processing on **piecewise** smooth data, with jumps or other singularities, that is. A prototype of nonlinear processing is coefficient **thresholding**, which selects the *largest* coefficients. Both Fourier based methods and kernel or spline based nonparametric methods are mostly linear. The double locality thus opens a wide range of options beyond the combined approaches of Fourier and kernels or splines.

The **multiscale** or **multiresolution** nature of the wavelet decomposition is well suited for the analysis of piecewise smooth data, as the unknown locations of singular points and ranges between them makes those data have a natural multiresolution aspect. Multiscale processing, such as *level-dependent* or *across-scale* processing, are specific strategies that can be applied in addition to sparsity based methods.

Wavelets have **zero integrals**. This property is related to the fact that they are designed to form bases for square integrable functions, see (1.28). A more fundamental argument for the zero integral has to do with the numerical condition of the wavelet transform, in particular in combination with nonlinear processing. This will be explored in Section 3.1.1.

Although wavelets are linked to nonlinear processing, the wavelet transform *itself* is a linear transform. Its transformation matrix is denoted by $\widetilde{\mathbf{W}}$ throughout this book. The inverse is denoted as $\mathbf{W} = \widetilde{\mathbf{W}}^{-1}$. In theory, a matrix operation on a vector s of length n requires $\mathcal{O}(n^2)$ floating point operations, but the Haar transform procedure as presented in Section 1.3 is completed in a linear number of computations. This is the **fast wavelet transform**. The fast wavelet transform is slightly faster than the fast Fourier transform, which is of $\mathcal{O}(n \log(n))$ complexity.

With each proper wavelet transformation matrix $\widetilde{\mathbf{W}}$, we can associate a scaling basis $\Phi_J(x)$ with elements $\varphi_{J,i}(x)$ and a wavelet basis $\Psi(x)$ with elements $\varphi_{L,k}(x)$ and $\psi_{j,k}(x)$. This has been illustrated for the Haar transform in Section 1.3. Chapter 3 will explain how we can find the associated bases $\Phi_J(x)$ and $\Psi(x)$ for other $\widetilde{\mathbf{W}}$ than the Haar transform. Finding the basis functions is crucial for reasons explained in Section 1.6.3.

1.5 Intrinsic multiscale problems

From Section 1.4 it is clear that time series with singularities at unknown, variable locations are prototypes of data that are well described by a sparse multiscale analysis. It is not surprising that finding the exact locations of the singularities in a **change point** problem is a typical task in a wavelet analysis, especially when there is an unknown number of multiple change points. In principle, the algorithm searching for these points has to check all possible locations at all possible scales in order to identify the points and the length of the intervals between them.

In other problems, the multiscale aspect may be a bit hidden, or there may not be an immediate connection to sparsity. A typical example is the problem of **density estimation**, which can be treated as a multiscale problem, even if the unknown density is smooth. This is because, by definition, the density of the available observations to estimate from is generally not a constant. As a result, the optimal bandwidth h in a kernel density estimation, which depends on the local density of the observations, varies along the axis of abscissae. While working with local bandwidths is one solution to this problem, another may be to consider a bandwidth as a scale parameter and use a multiscale framework to solve the problem.

For problems involving smoothly varying responses, a latent multiscale component may justify the use of wavelets. In other problems, such as the one illustrated in Figure 1.5, methods like spline or kernel smoothing or Fourier analysis may well be the better alternative.

1.6 Design of a wavelet transform

The Haar transform, discussed in this chapter, is the oldest and simplest wavelet transform. More advanced wavelet decompositions have been developed, with more parameters to choose from and with interesting characteristics in theory and for applications. The design or choice of a wavelet analysis involves a couple of recurrent issues, the most important ones of which are listed below.

1.6.1 Perfect reconstruction (PR)

The Haar transform computes averages and differences, and from these two, the input can easily be reconstructed. Instead of averages, other forms of local smoothing can be used for the computation of scaling coefficients. The question is how the detail offsets or wavelet coefficients should be defined so that the set of detail and scaling coefficients together allows to reconstruct the fine scaling input. This question is basically about the inversion of a linear transform. It is a purely algebraic matter, and as such, it is fairly easy to control. One particular wavelet design method, known as the **lifting scheme**, introduced in Chapter 2.1, offers perfect reconstruction automatically.

1.6.2 Fast decomposition and reconstruction

In the ideal case, both the forward and inverse wavelet transform matrices $\widetilde{\mathbf{W}}$ and $\mathbf{W}$ are structured in a way that the wavelet decomposition and reconstruction are computed in a linear computational complexity. This is the **fast wavelet transform**, mentioned already in Section 1.4, and further developed throughout the following chapters. In the literature, the fast transform is also referred to as Mallat's repeated **filterbank** algorithm (Mallat, 1989; Meyer,

1992; Vetterli and Herley, 1992; Strang and Nguyen, 1996). As illustrated with the Haar transform in Section 1.3.4, the fast wavelet transform relies on the multiscale nature of the wavelet transform, proceeding scale by scale. For the sake of a fast decomposition and reconstruction, each step from one scale to the following should go by sparse matrices. Again the lifting scheme provides a method that guarantees fast transforms in both directions.

1.6.3 The characteristics of the basis functions

In the approximation or estimation of a function, it is a rule of thumb that the building blocks should roughly have the same characteristics as the function. For instance, if the function has a natural piecewise constant behaviour, the Haar decomposition is an excellent tool for analysis and reconstruction.

For other wavelet transforms, it is important to know how the corresponding basis functions look; in particular, it is crucial to have an idea about the **smoothness** of the basis functions. It determines the smoothness of any reconstruction within that basis. In general, it is far from trivial to find the smoothness class of the basis functions starting from the operations used in the transform. For instance, a transform based on polynomial interpolation or smoothing almost never leads to polynomials or piecewise polynomials as basis functions.

A first step in the analysis of the smoothness of the basis functions is to find the inverse transform matrix $\mathbf{W}$, as in Section 1.3.8. The ith column of the matrix $\mathbf{W}$ can be found by taking $\boldsymbol{w}_L = \boldsymbol{e}_i$, where $\boldsymbol{e}_i$ is the ith canonical vector, i.e., the ith column of the $n \times n$ identity matrix. Otherwise stated, $\mathbf{W}$ is found by applying the reconstruction algorithm on $\boldsymbol{w}_L = \mathbf{I}$, where $\mathbf{I}$ is the identity matrix. As mentioned in Section 1.3.8, the columns of $\mathbf{W}$ look like discrete versions of the basis functions. Indeed, these columns contain the coefficients of the decomposition

$$\psi_{j,k}(x) = \sum_{\ell=1}^{n} W_{\ell,m}\varphi_{J,\ell}(x). \tag{1.34}$$

In this expression, we assumed that the wavelet function at scale j, location k, occupies the mth position in the vector of functions $\Psi(x)$. The set of all n decompositions (1.34) is summarised into the matrix version in (1.32). All expressions (1.34) decompose functions in terms of fine scale basis functions. If the fine scale is sufficiently fine, then the details of the basis function at that scale are small compared with the scale of the functions $\psi_{j,k}(x)$. Loosely speaking, through the eyes of $\psi_{j,k}(x)$ the details of $\varphi_{J,\ell}(x)$ are hardly visible. All fine scale basis functions look almost like an indicator function on a very short subinterval.

Using the inverse transform in (1.22), the **refinement** can continue even up to an arbitrarily fine **superresolution**, beyond the level J at which the data have been observed, that is. One can add zero details, i.e., $\boldsymbol{d}_j = \mathbf{0}$

for $j = J+1, J+2, \ldots$ at increasingly fine scales. This infinite refinement is known as **subdivision**, further discussed in Section 2.2.5. The study of subdivision is a central problem in wavelet theory.

In the example of the Haar transform in Section 1.3, the observations are equidistant. The refinement operation is then exactly the same at all levels j, all places k. Only in this case, there is a father function from which all other scaling functions are found by dyadic dilation and translation. The smoothness of the father function follows from subdivision, but since all refinement steps are exactly the same, all the information about the father function follows from one step only. This step decomposes $\varphi(x)$ into basis functions at the next, finer level, more precisely

$$\varphi(x) = \sum_{k \in \mathbb{Z}} h_k \varphi(2x - k). \tag{1.35}$$

This is the central equation of equidistant wavelet theory, sometimes referred to as *first generation wavelets*. It is known as the **two-scale equation**, or **refinement equation**. It is the basis version of the more general subdivision or refinement algorithm.

The smoothness problem can also be posed in the opposite direction: given a desired degree of smoothness, how can we design an appropriate wavelet transform?

1.6.4 The sparsity of the decomposition

Although the Haar decomposition in Figure 1.13 provides a sparse representation of the skyline test function, it still contains small nonzero detail coefficients. These nonzero detail coefficients are not visible on the plot. They are due to the fact that the Haar basis cannot represent the slopes in the skyline test function. If the function $f(x)$ is anything other than a horizontal line, then the Haar decomposition consists of a coarse scale approximation of the line by a piecewise constant function $\Phi_L(x)\boldsymbol{s}_L$, and small detail coefficients at all finer scales to correct for the approximation error at each scale. Since the nonzero coefficients are small, they are mostly missed in a procedure estimating the true function from observations with noise. This explains the blocky reconstruction in Figure 1.16. The blocky reconstruction is visually unpleasant but also inaccurate. In order to improve the accuracy, it would be interesting to go for scaling functions that exactly represent lines and possibly also higher order polynomials. We thus impose that there exists a vector $\boldsymbol{s}_L$ so that

$$x^q = \Phi_L(x)\boldsymbol{s}_L,$$

for $q = 0, 1, \ldots, \widetilde{p} - 1$ and for arbitrary L, except for very small values where boundary effects may be inevitable. The parameter $\widetilde{p}$ is termed the number of **dual vanishing moments**. The dual vanishing moments have a direct impact on the sparsity of the decomposition, thereby playing a central role in approximation and estimation results. On one hand, a large number of vanishing moments reduces the small detail coefficients. The price to pay is that each vanishing moment consumes a degree of freedom in the design. This makes the basis functions have a wider support, and so they are less local. When the data contain a singularity, then there are more basis functions that interfere with it, and so there are more large nonzero coefficients, which means less sparsity. Choosing the number of dual vanishing moments is thus a matter of balancing between a smooth reconstruction of the smooth intervals of a function and a sharp reconstruction of a singularity.

In general, the design of a wavelet decomposition with a desired number of dual vanishing moments is not a difficult task. The condition can be imposed on each step of the multiscale forward transform, for instance, in a lifting scheme.

1.6.5 Numerical condition

The general scheme of a wavelet based data processing routine goes as follows

1. Apply a wavelet transform to the observations $\boldsymbol{w}_L = \widetilde{\mathbf{W}} \cdot \boldsymbol{Y}$, as in (1.24).
2. Perform some action, e.g., an estimation Est_λ, on the wavelet coefficients $\widehat{\boldsymbol{w}}_\lambda = \mathrm{Est}_\lambda(\boldsymbol{w}_L)$.
3. Apply the inverse transform to construct the approximation or estimation $\widehat{\boldsymbol{f}}_\lambda = \mathbf{W}\widehat{\boldsymbol{w}}_\lambda$.

The processing of the wavelet coefficients is often fine-tuned by a tuning parameter λ, a typical example being a threshold in a thresholding scheme. Assessment of the quality of the fine-tuning is based on the effect on the final outcome, measured by some norm of the difference between output and input $\widehat{\boldsymbol{f}}_\lambda - \boldsymbol{Y}$. In order to avoid constant forward and backward transforming of the data in the fine-tuning process, it is interesting that the effect on the outcome can be computed, or at least estimated or bounded from the effect on the wavelet coefficients, i.e., from a norm of the difference $\widehat{\boldsymbol{w}}_\lambda - \boldsymbol{w}_L$. It is then possible to find a good tuning parameter value from the wavelet coefficients only. In this context, **orthogonal transforms** are ideal, as the Euclidean norm of the effect remains unchanged by forward or inverse transforms. Orthogonal wavelet transforms are, however, rare, as orthogonality puts serious constraints on the degrees of freedom. As an example, the only wavelet transform that is orthogonal and at the same time has symmetric scaling functions is the Haar transform. Therefore many applications use non-orthogonal wavelet transforms, as close as possible to an orthogonal transform. The deviation from the orthogonal ideal can be expressed by **Riesz bounds**, which are the minimal and maximal norm of a reconstruction from a normalised vector of wavelet coefficients. Riesz-bounds are closely related to the condition number of a linear transform, used in numerical analysis, and defined in

Section 5.4.6. A necessary condition for Riesz bounds to exist is that the scaling basis has at least one **primal vanishing moment**, introduced in Section 2.1.6. Primal vanishing moments are easy to implement in a wavelet transform, for instance, using the lifting scheme.

Although Riesz bounds and numerical condition describe the propagation of fluctuations throughout a linear operation, these notions are insufficient for an adequate description of the variance inflation in applications involving random observations. Specific descriptions of the variance propagation as well as design of wavelet transforms for control of the variance are developed throughout the chapters in this text, especially in Section 3.1.5.

1.7 Conclusion

This introductory chapter has illustrated the key points of a wavelet analysis by its most simple example, the Haar transform. It has also raised the aspects that are important in the design of more advanced wavelet analyses. The following chapters will formally define these notions and develop wavelet transforms according to these criteria.

2

Wavelet building blocks

Generalising the ideas from Chapter 1, the current chapter presents the general framework for the design of a wavelet transform. A wavelet transform consists of a sequence of steps that operate between two resolution levels. In each step, the elementary building blocks are provided by the lifting scheme, discussed in Section 2.1. At a slightly more abstract level, the operation within each step can be described as a filterbank, discussed in Section 2.2. Although the filterbank operations are purely algebraic, the sequence of resolution levels may continue up to arbitrarily fine, i.e., infinitesimal scales. The repeated refinement is best described by associating to each filterbank a set of functions in continuous variables, as suggested in Section 1.3.6 and further generalised in the current chapter. The association of basis functions to a filterbank proceeds to two central types of equations in wavelet theory, the two-scale and wavelet equations.

2.1 From the Haar transform to the lifting scheme

2.1.1 Nonequidistant grids

In contrast to Section 1.3, this section considers observations in **nonequidistant knots** x_i, $i = 1, \dots, n$. Without loss of generality, we assume that all $x_i \in [0, 1]$. The extension to other bounded intervals, but even to infinite intervals, is straightforward if the appropriate bounds in integrals are used. We associate these points with the finest scale J in $x_{J,k} = x_{k+1}$ for $k = 0, \dots, n_J - 1$, where $n_J = n$. The index shift is necessary for bringing the notations in line with common practice in the wavelet literature. Throughout this chapter, we construct coarse scale grids $\{x_{j,k} | k = 0, \dots, n_j - 1\}$, where $j = L, L+1, \dots, J-1$ denotes the scale, L being the lowest or coarsest scale. The locations $x_{j,k}$ and the number of knots n_j at scale j are design parameters of the multiscale transform. Not all choices of these design parameters work well in practice. For this reason, a **multilevel grid** is defined by putting some basic restrictions on the choice of $x_{j,k}$ and n_j. First, denote $\Delta_{j,k} = x_{j,k+1} - x_{j,k}$, and $\Delta_j = \sup_{k=0,\dots,n_j-2} \Delta_{j,k}$, then the grid at scale j is **regular** or **equidistant** if $\Delta_{j,k}$ does not depend on k, i.e., $\Delta_{j,k} = \Delta_j$.

DOI: 10.1201/9781003265375-2

Otherwise, the grid is nonequidistant or irregular, but we limit the irregularity by the following definition.

Definition 2.1.1 *(multilevel grid) The sequence of grids constitutes a* ***multilevel grid*** *if the following conditions are met:*

1. *The number of grid points is strictly increasing:* $n_L < n_{L+1} < \ldots < n_{J-1} < n_J = n$.
2. *There exist constants* $R \in \mathbb{R}$ *and* $\beta > 0$ *so that the maximum gap at scale* j *is bounded as follows,*

$$\Delta_j \leq R n_j^{-\beta}. \tag{2.1}$$

The latter condition is slightly stricter than found in, for instance, (Daubechies et al., 2001). If we denote the "average" gap at level j by $\overline{\Delta}_j = 1/n_j$, then the condition can be rewritten as $\Delta_j \leq R\overline{\Delta}_j^{\beta}$, meaning that the maximum gap at level j should be uniformly bounded by a power function of the average gap. As a consequence, a refinement of the grid should not leave coarse scale gaps among much finer scale gaps.

A multilevel grid is termed **nested** if $x_{j+1,k} \in \{x_{j,k} | k = 0, \ldots, n_j - 1\}$. Going from fine to coarse scales amounts to leaving out some of the knots. This process is known as **decimation**, **subsampling** or **downsampling**. With $\boldsymbol{x}_j$ the vector of knots at level j, the subsampling matrix $\widetilde{\mathbf{J}}_j^\top$ is the $n_j \times n_{j+1}$ submatrix of the $n_{j+1} \times n_{j+1}$ identity matrix so that

$$\boldsymbol{x}_j = \widetilde{\mathbf{J}}_j^\top \boldsymbol{x}_{j+1}. \tag{2.2}$$

The subsampling operation defines a bipartite **split** of the fine scale grid $\boldsymbol{x}_{j+1}$ into **even** and **odd** indexed subvectors denoted as $\boldsymbol{x}_{j+1,e}$ and $\boldsymbol{x}_{j+1,o}$. The even subvector contains the coarse scale knots, $\boldsymbol{x}_j = \boldsymbol{x}_{j+1,e}$. The terms even and odd refer to the common case of **binary subsampling**, denoted and defined by

$$(\downarrow 2)\boldsymbol{x}_{j+1} = \boldsymbol{x}_j \Leftrightarrow x_{j,k} = x_{j+1,2k}. \tag{2.3}$$

The terms even and odd will be used throughout this text for the sets of knots that are preserved and taken out in a subsampling operation $\widetilde{\mathbf{J}}_j^\top$, even if the subsampling is not binary, $\widetilde{\mathbf{J}}_j^\top \neq (\downarrow 2)$. In that case, the partitioning does not correspond to the mathematical notions of even and odd. If the subsampling is binary at each level, then the multilevel grid is termed **two-nested**. Unless otherwise specified, the remainder of this chapter considers two-nested multilevel grids.

Exercise 2.1.2 *Consider the grid* $\boldsymbol{x}_{j+1}$ *at level* $j+1$ *with* $n_{j+1} = 9$*; then write the matrix* $\widetilde{\mathbf{J}}_j^\top$ *of the binary subsampling operation, so that* $\boldsymbol{x}_j = \widetilde{\mathbf{J}}_j^\top \boldsymbol{x}_{j+1}$*. Also write the matrix* $\widetilde{\mathbf{J}}_j^{o\top}$ *representing the complementary subsampling, mapping* $\boldsymbol{x}_{j+1}$ *onto the subvector of its components not in* $\boldsymbol{x}_j$*. This vector will be denoted by* $\boldsymbol{x}_{j+1,o}$ *in further discussions.*

2.1.2 The unbalanced Haar transform

Just like for the Haar transform in Chapter 1, the observations $\boldsymbol{Y}$ are again represented by a piecewise constant function, denoted $s_J(x)$, but this time the staircase has irregular steps. As in (1.25), the staircase is first decomposed into a basis of scaling functions that are all characteristic functions on intervals $I_{J,i}$, that form a partition of $[0,1]$, i.e.,

$$\varphi_{J,i}(x) = \chi_{I_{J,i}}(x), \tag{2.4}$$

where $\bigcup_{i=0}^{n-1} I_{J,i} = [0,1]$ and $I_{J,i} \cap I_{J,i'} = \emptyset$ for $i \neq i'$. The relation between the partition $I_{J,i}$ and the grid points $x_{J,i}$ may depend on the application. The intervals can be fixed to have $x_{J,i}$ as end points, i.e., $I_{J,i} = [x_{J,i}, x_{J,i+1}]$, or $I_{J,i} = [x_{J,i-1}, x_{J,i}]$, with special precautions for $i = 0$ or $i = n-1$. Alternatively, the interval boundaries may be fixed in the middle, i.e., $I_{J,i} = [(x_{J,i-1} + x_{J,i})/2, (x_{J,i} + x_{J,i+1})/2]$. In any case, the irregularity of the grid is reflected in the partition. Let

$$\Delta_{J,i} = \int_0^1 \varphi_{J,i}(x)dx$$

be the length of the interval. An approximation of $s_J(x)$ at scale $J-1$ can be obtained by putting $j = J-1$ in the following forward transform step

$$\begin{aligned}
I_{j,k} &= I_{j+1,2k} \cup I_{j+1,2k+1}, \\
\Delta_{j,k} &= \Delta_{j+1,2k} + \Delta_{j+1,2k+1}, && (2.5)\\
s_{j,k} &= \frac{\Delta_{j+1,2k}s_{j+1,2k} + \Delta_{j+1,2k+1}s_{j+1,2k+1}}{\Delta_{j,k}}, && (2.6)\\
d_{j,k} &= s_{j+1,2k+1} - s_{j,k}. && (2.7)
\end{aligned}$$

In (2.6), the coarse scaling coefficients are defined as *weighted* averages. The weights are taken so that the integral of the coarse scale staircase equals the integral of the fine scale staircase,

$$\sum_{k=0}^{n_{j+1}-1} s_{j+1,k}\Delta_{j+1,k} = \sum_{k=0}^{n_j-1} s_{j,k}\Delta_{j,k}. \tag{2.8}$$

The weights in the calculation of the coarse scale coefficients thus depend on the grid locations at the current scale. In a general wavelet transform, the calculation of the details as in (2.7) may depend explicitly on that grid as well. For the definition of the grid at level j, several options can be considered. The simplest and most often used definition is a binary downsampling of the fine scale grid, defined in Section 2.1.1. An alternative option is to define the coarse scale grid in a similar way as the coarse scaling coefficients,

$$x_{j,k} = \frac{\Delta_{j+1,2k}x_{j+1,2k} + \Delta_{j+1,2k+1}x_{j+1,2k+1}}{\Delta_{j,k}}. \tag{2.9}$$

The **inverse transform** is straightforward, undoing in the reverse direction the computations of detail and coarse scaling coefficients,

$$s_{j+1,2k+1} = s_{j,k} + d_{j,k} \quad (2.10)$$

$$s_{j+1,2k} = \frac{\Delta_{j,k}s_{j,k} - \Delta_{j+1,2k+1}s_{j+1,2k+1}}{\Delta_{j+1,2k}}. \quad (2.11)$$

This requires knowledge of all the $\Delta_{j,k}$, which can be recomputed separately, using the same recursion (2.5) as in the forward scheme. Indeed, the integrals $\Delta_{j,k}$ depend on the grid only, not on the observations. The recursive calculation of $\Delta_{j,k}$ is presented here for use in future extensions of the Haar transform. When using Haar, $\Delta_{j,k}$ can be found straightforwardly as the length of $I_{j,k}$.

Since the finest scaling coefficients are associated with characteristic basis functions in the representation of a staircase, the unbalanced Haar transform can be framed as a basis transform, proceeding from one side to the other in $\Phi_J(x)\boldsymbol{s}_J = \Phi_L(x)\boldsymbol{s}_L + \sum_{j=L}^{J-1}\Psi_j(x)\boldsymbol{d}_j$. In this expression, $\Phi_j(x)$ and $\Psi_j(x)$ are row vectors of basis functions, as in introduced in (1.30). The basis function $\varphi_{L,k}(x)$ can be written as a linear combination of $\Phi_J(x)$, i.e., $\varphi_{L,k}(x) = \Phi_J(x)\boldsymbol{s}_J$. In order to find the coefficients $\boldsymbol{s}_J$, the vector $\boldsymbol{s}_L$ in the coarse scale approximation $\Phi_L(x)\boldsymbol{s}_L$ is taken to be $\boldsymbol{s}_L = \boldsymbol{e}_{L,k}$, with $\boldsymbol{e}_{L,k}$ being a canonical vector (which is a Kronecker delta, or the kth column of the identity matrix) at level L, while all detail coefficients are zero, $d_{j,k} = 0$ for all j and k. Running the inverse transform then yields the coefficients $\boldsymbol{s}_J$ of that function, $\varphi_{L,k}(x)$, at the finest resolution. In the case of the unbalanced Haar transform, application of (2.10) and (2.11) leads to $s_{j,l} = 1$ if $l = 2^{j-L}k, 2^{j-L}k+1, \ldots, 2^{j-L}(k+1)-1$ and zero otherwise. As a result, the scaling functions are characteristic functions at all scales,

$$\varphi_{L,k}(x) = \sum_{l=2^{J-L}k}^{2^{J-L}(k+1)-1} \chi_{I_{J,l}}(x) = \chi_{I_{L,k}}(x).$$

Given that the scaling functions are characteristic functions, $\varphi_{j+1,2k+i}(x) = \chi_{I_{j+1,2k+i}}(x)$, it is sufficient to apply one step of the inverse transform with $d_{j,k} = 1$ and $s_{j,k} = 0$ to arrive at $s_{j+1,2k+1} = 1$ and $s_{j+1,2k} = -\Delta_{j+1,2k+1}/\Delta_{j+1,2k}$. This means that the function, represented with a single nonzero $d_{j,k} = 1$ at level j, can also be written in terms of scaling functions at the finer level $j+1$ using the coefficients $-\Delta_{j+1,2k+1}/\Delta_{j+1,2k}$ and 1,

$$1 \cdot \psi_{j,k}(x) = \varphi_{j+1,2k+1}(x) - \frac{\Delta_{j+1,2k+1}}{\Delta_{j+1,2k}}\varphi_{j+1,2k}(x).$$

The basis function $\psi_{j,k}(x)$, with integral zero, consists of two blocks with unequal heights. It is known as an **unbalanced Haar wavelet**, illustrated in Figure 2.1.

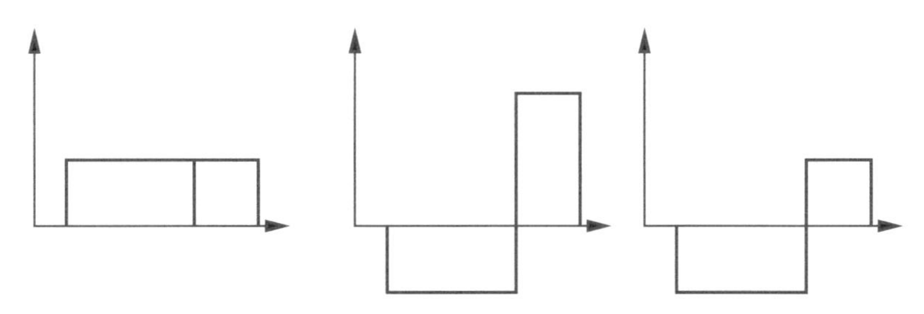

Figure 2.1
Unbalanced versus balanced Haar basis functions. Left panel: two adjacent scaling functions on irregularly spaced knots at level $j+1$. The scaling functions are the same for both the balanced and the unbalanced transforms. Middle: balanced wavelet function at level j. This function has a nonzero integral. Right: unbalanced wavelet function at level j.

The balanced and unbalanced Haar wavelet transforms operate from the **same scaling basis** of characteristic functions. The wavelet basis functions are different. In the unbalanced case, the function values of the wavelet basis functions depend on the interknot distances $\Delta_{j,k}$.

The wavelet functions in a balanced transform have values ± 1. On irregular knots, these functions have a **nonzero integral**. In some applications, it makes sense to apply the balanced transform, even if the knots are irregularly spaced. As a matter of fact, the choice between balanced and unbalanced transform is an example of a **bias-variance** trade-off. Indeed, suppose that the fine scale coefficients in s_{j+1} carry uncorrelated, homoscedastic noise; then the **variance** of s_j in (2.6) is minimised if $\Delta_{j,2k} = \Delta_{j,2k+1}$, i.e., if we replace the unbalanced transform by its balanced counterpart, ignoring the actual positions of the knots. The price to pay is of course an increase in **approximation error** or **bias** at level j, as

$$\int_{I_{j,k}} \left| s_{j,k}\varphi_{j,k}(x) - \left(s_{j+1,2k}\varphi_{j+1,2k}(x) + s_{j+1,2k+1}\varphi_{j+1,2k+2}(x) \right) \right| dx$$

is minimised if $\Delta_{i,l} = \int_{I_{i,l}} \varphi_{i,l}(x)dx$, for $(i,l) \in \{(j,k),(j+1,2k),(j+1,2k+1)\}$. The minimisation of the bias is linked to the vanishing integral of the wavelet basis function.

2.1.3 The Haar transform as a lifting scheme

The algorithm of the forward Haar transform presented in Section 2.1.2 fits into a general scheme, known as the **lifting scheme**, which gradually adds all the characteristics of the Haar transform. A general lifting scheme starts with a bipartite **split** of the scaling coefficients at scale $j+1$ into *even* and *odd* indexed subsets. The two subvectors are denoted as $\boldsymbol{s}_{j+1,e}$ and $\boldsymbol{s}_{j+1,o}$, where the even and odd subsets are the same as for the subsampling of the knots in Section 2.1.1.

Next, the even vector is rescaled using a diagonal matrix, leading to $\boldsymbol{s}_{j+1,e}^{[1]} = \mathbf{D}_j \boldsymbol{s}_{j+1,e}$. In the concrete example of the Haar transform, we have $s_{j+1,2k}^{[1]} = (\Delta_{j+1,2k}/\Delta_{j,k}) s_{j+1,2k}$. This rescaling is a preparatory step for an **update** of the even values by a linear transform of the odd values. Let $\mathbf{U}_j$ denote the matrix of the update; then, in general, the operation takes the form

$$\boldsymbol{s}_{j+1,e}^{[2]} = \boldsymbol{s}_{j+1,e}^{[1]} + \mathbf{U}_j \boldsymbol{s}_{j+1,o}^{[1]}. \tag{2.12}$$

In our case, the matrix $\mathbf{U}_j$ is diagonal and

$$s_{j+1,2k}^{[2]} = s_{j+1,2k}^{[1]} + (\Delta_{j+1,2k+1}/\Delta_{j,k}) s_{j+1,2k+1}.$$

The objective of the update is to filter or smooth the even scaling coefficients. In the case of Haar, one update step is enough to arrive at the coarse scaling coefficients, so

$$\boldsymbol{s}_j = \boldsymbol{s}_{j+1,e}^{[2]}.$$

The computation of the detail or difference in (2.7) is an example of a **prediction** step,

$$\boldsymbol{s}_{j+1,o}^{[3]} = \boldsymbol{s}_{j+1,o}^{[2]} - \mathbf{P}_j \boldsymbol{s}_{j+1,e}^{[2]}, \tag{2.13}$$

The idea is to predict the odd branch by a linear combination of coefficients on the even branch, and subtract the prediction. The offsets between the coefficients on the odd branch and the predictions from the even branch are mostly small. The output from the last prediction step can thus be delivered as wavelet or detail coefficients. In our case, after a rescaling in step $[0]$, an update in step $[1]$, one prediction in step $[2]$ is sufficient to find the detail coefficients, i.e.,

$$\boldsymbol{d}_j = \boldsymbol{s}_{j+1,o}^{[3]},$$

where

$$s_{j+1,2k+1}^{[3]} = s_{j+1,2k+1}^{[2]} - s_{j+1,2k}^{[2]}. \tag{2.14}$$

Expressions (2.12) and (2.13) should be completed by stating that one half of the coefficients does not change. In particular, the odd scaling coefficient after the update step $s_{j+1,2k+1}^{[2]}$ is the same as before that step, $s_{j+1,2k+1}^{[1]}$, and the same holds for the rescaling step, so we get

$$s_{j+1,2k+1}^{[2]} = s_{j+1,2k+1}^{[1]} = s_{j+1,2k+1}^{[0]} = s_{j+1,2k+1}.$$

In a similar way, we have that $s_{j+1,2k}^{[3]} = s_{j+1,2k}^{[2]}$.

Exercise 2.1.3 *Consider the equidistant grid $\boldsymbol{x}_{j+1}$ at level $j+1$ with $n_{j+1}=9$, and $x_{j+1,k}=k$. Write the matrices $\mathbf{D}_j$, $\mathbf{U}_j$, and $\mathbf{P}_j$ at level j of the Haar transform, paying attention to the dimensions of these matrices.*

- *Use the subsampling matrices as in Exercise 2.1.2 to define* $\boldsymbol{s}_{j+1,e}=\widetilde{\mathbf{J}}_j^\top \boldsymbol{s}_{j+1}$ *and* $\boldsymbol{s}_{j+1,o}=\widetilde{\mathbf{J}}_j^{o\top} \boldsymbol{s}_{j+1}$.
- *Rescale the evens, keep the odds:* $\boldsymbol{s}_{j+1,e}^{[1]}=\mathbf{D}_j\boldsymbol{s}_{j+1,e}$ *and* $\boldsymbol{s}_{j+1,o}^{[1]}=\boldsymbol{s}_{j+1,o}$.
- *Update the evens, keep the odds:* $\boldsymbol{s}_{j+1,e}^{[2]}=\boldsymbol{s}_{j+1,e}^{[1]}+\mathbf{U}_j\boldsymbol{s}_{j+1,o}^{[1]}$ *and* $\boldsymbol{s}_{j+1,o}^{[2]}=\boldsymbol{s}_{j+1,o}^{[1]}$.
- *Predict the odds, preserve the evens:* $\boldsymbol{d}_j=\boldsymbol{s}_{j+1,o}^{[2]}-\mathbf{P}_j\boldsymbol{s}_{j+1,e}^{[2]}$ *and* $\boldsymbol{s}_j=\boldsymbol{s}_{j+1,e}^{[2]}$.

Exercise 2.1.4 *The lifting scheme of Exercise 2.1.3 leads to $\boldsymbol{s}_j=\mathbf{D}_j\boldsymbol{s}_{j+1,e}+\mathbf{U}_j\boldsymbol{s}_{j+1,o}$. Rewriting this as $\boldsymbol{s}_j=\mathbf{D}_j(\boldsymbol{s}_{j+1,e}+\mathbf{D}_j^{-1}\mathbf{U}_j\boldsymbol{s}_{j+1,o})$ points towards a lifting scheme where the rescaling takes place after the update. How does the update look in this scheme on a general grid? The rescaling can be shifted further to the end of the scheme, after the prediction. How does the prediction look in that scheme?*

2.1.4 A prediction-first scheme for the same unbalanced Haar transform

Exactly the same unbalanced Haar transform can be realised with an alternative lifting scheme. The first step in the alternative is a prediction of each odd coefficient by its even neighbour on the left.

$$s_{j+1,2k+1}^{[1]}=s_{j+1,2k+1}-s_{j+1,2k}.$$

Next is an update of the even coefficients by adding the rescaled offset.

$$s_{j+1,2k}^{[2]}=s_{j+1,2k}+\frac{\Delta_{j+1,2k+1}}{\Delta_{j,k}}s_{j+1,2k+1}^{[1]}.$$

The updated scaling coefficient already equals the targeted coarse scale value, so we put $s_{j,k}=s_{j+1,2k}^{[2]}$. In order to arrive at the same detail as before, the values on the odd branch need to be rescaled. Indeed,

$$d_{j,k}=s_{j+1,2k+1}-\frac{\Delta_{j+1,2k}s_{j+1,2k}+\Delta_{j+1,2k+1}s_{j+1,2k+1}}{\Delta_{j,k}}=\frac{\Delta_{j+1,2k}}{\Delta_{j,k}}s_{j+1,2k+1}^{[2]},$$

where $s_{j+1,2k+1}^{[2]}=s_{j+1,2k+1}^{[1]}$, because the second step was an update, only affecting the even branch. The rescaling in the last step can be omitted without further impact on the overall transform. This leaves us with two alternatives for the unbalanced Haar transform: an update-first and a prediction-first lifting scheme. Both can serve as starting points for further extensions.

Exercise 2.1.5 *Consider the equidistant grid x_{j+1} at level $j+1$ with $n_{j+1} = 9$, and $x_{j+1,k} = k$. Write the matrices $\mathbf{D}_j$, $\mathbf{U}_j$, and $\mathbf{P}_j$ at level j of the Haar transform, paying attention to the dimensions of these matrices.*

- *Use the subsampling matrices as in Exercise 2.1.2 to define* $\boldsymbol{s}_{j+1,e} = \widetilde{\mathbf{J}}_j^\top \boldsymbol{s}_{j+1}$ *and* $\boldsymbol{s}_{j+1,o} = \widetilde{\mathbf{J}}_j^{o\top} \boldsymbol{s}_{j+1}$.
- *Predict the odds, preserve the evens:* $\boldsymbol{s}_{j+1,o}^{[1]} = \boldsymbol{s}_{j+1,o} - \mathbf{P}_j \boldsymbol{s}_{j+1,e}$ *and* $\boldsymbol{s}_{j+1,e}^{[1]} = \boldsymbol{s}_{j+1,e}$.
- *Update the evens, keep the odds:* $\boldsymbol{s}_{j+1,e}^{[2]} = \boldsymbol{s}_{j+1,e}^{[1]} + \mathbf{U}_j \boldsymbol{s}_{j+1,o}^{[1]}$ *and* $\boldsymbol{s}_{j+1,o}^{[2]} = \boldsymbol{s}_{j+1,o}^{[1]}$.
- *Rescale the odds, keep the evens:* $\boldsymbol{d}_j = \mathbf{D}_j \boldsymbol{s}_{j+1,o}^{[2]}$ *and* $\boldsymbol{s}_j = \boldsymbol{s}_{j+1,e}^{[2]}$.

2.1.5 Triangular hat basis functions

The unbalanced Haar transform takes the positions of the knots into account in the calculation of the coarse scaling coefficients. The positions of the knots are not involved in the calculation of the detail offsets. In terms of basis functions, the detail or wavelet functions are adjusted to the grid of knots in order to have zero integrals. In both lifting schemes of Sections 2.1.2 and 2.1.4, the prediction does not depend on the position of the knots. We now introduce a scheme where the prediction takes the knots into account. Unlike the update in the Haar transform, the goal of knot dependent prediction does not lie in a **bias-variance** trade-off. The prediction will define the scaling basis, and hence it will fix the **smoothness** of both the scaling and wavelet functions, and of all reconstructions or approximations in bases of these functions. At the same time, the prediction determines the degree of **sparsity** of the decomposition.

In a first attempt to construct wavelets that are less blocky than the Haar basis, we propose a prediction operation based on a linear interpolation at the adjacent even points. As for now, the differences between the odd branch and the prediction from the even branch are left unchained throughout the remainder of the transform. As a result, these offsets will be the wavelet or detail coefficients at this level j.

$$d_{j,k} = s_{j+1,2k+1} - (P_{j;k,k} s_{j+1,2k} + P_{j;k,k+1} s_{j+1,2k+2}), \tag{2.15}$$

where

$$\begin{aligned} P_{j;k,k} &= \frac{x_{j+1,2k+2} - x_{j+1,2k+1}}{x_{j+1,2k+2} - x_{j+1,2k}}, \\ P_{j;k,k+1} &= \frac{x_{j+1,2k+1} - x_{j+1,2k}}{x_{j+1,2k+2} - x_{j+1,2k}}. \end{aligned}$$

For inputs that are piecewise linear, all detail coefficients are zero, except for those that are affected by a jump or another singularity. This is in contrast to a Haar transform, where *all* details would be at least small but definitely nonzero, except for constant functions.

The transform can be upgraded using an update, for which we propose, quite arbitrarily at this point, a bidiagonal matrix $\mathbf{U}_j$. As for now, we include no more prediction or update steps. Therefore, in contrast to (2.12), no superscript is needed in the update matrix, and the output will be the scaling coefficients at this scale.

$$s_{j,k} = s_{j+1,2k} + U_{j;k,k} d_{j,k} + U_{j;k,k-1} d_{j,k-1}. \tag{2.16}$$

An appropriate choice for the entries of $\mathbf{U}_j$ will follow from a closer look at the basis functions in Section 2.1.6. The basis functions in their turn follow from **subdivision**, for which we need the inverse transform.

The inverse transform is

$$s_{j+1,2k} = s_{j,k} - U_{j;k,k} d_{j,k} - U_{j;k,k-1} d_{j,k-1}, \tag{2.17}$$

$$s_{j+1,2k+1} = P_{j;k,k} s_{j+1,2k} + P_{j;k,k+1} s_{j+1,2k+2} + d_{j,k}. \tag{2.18}$$

In order to retrieve the scaling basis function $\varphi_{L,\ell}(x)$, we use (1.30) as in Sections 1.3.8 and 2.1.2. Starting off from the canonical vector $\boldsymbol{s}_L = \boldsymbol{e}_{L,\ell}$ on the coarse grid of knots $\boldsymbol{x}_L$, **refinement** proceeds from scale to scale by inserting the knots $\boldsymbol{x}_{j+1,o}$ and adding the coefficients $d_{j,k} = 0$ in the inverse transform (2.18). The refinement of the scaling coefficients then reduces to

$$s_{j+1,2k} = s_{j,k}, \tag{2.19}$$

$$\begin{aligned} s_{j+1,2k+1} &= P_{j;k,k} s_{j+1,2k} + P_{j;k,k+1} s_{j+1,2k+2} \\ &= P_{j;k,k} s_{j,k} + P_{j;k,k+1} s_{j,k+1}. \end{aligned} \tag{2.20}$$

The fine scale evens are just the coarse scale coefficients, while the odds follow from linear interpolation. The process is illustrated in Figure 2.2. From the figure, it is clear that whatever the locations of further refinement points, new fine scaling coefficients will always appear on the **triangular hat function** defined by the initial canonical vector $\boldsymbol{e}_{L,\ell}$ on the initial grid $\boldsymbol{x}_L$. Taking characteristic functions as finest scaling basis, $\varphi_{J,k} = \chi_{I_{J,k}}(x)$ makes $\varphi_{K,k}(x)$ to be a staircase shaped triangular hat as in Figure 2.3.

A more natural choice would be to take triangular hat functions from the beginning, at the finest scale, that is. The triangular hat family is then preserved as the scaling basis at all levels. An example of the basis at a coarse scale is depicted in Figure 2.4. Triangular hats at finest scale and linear interpolating prediction are a perfect match, just as characteristic functions work optimally in combination with the constant extrapolation of a Haar transform. Whereas the basis of characteristic functions represents the discrete vector s_L as a staircase, the triangular hat basis generates all possible **polylines**, i.e., continuously connected line segments with kinks in the knots of

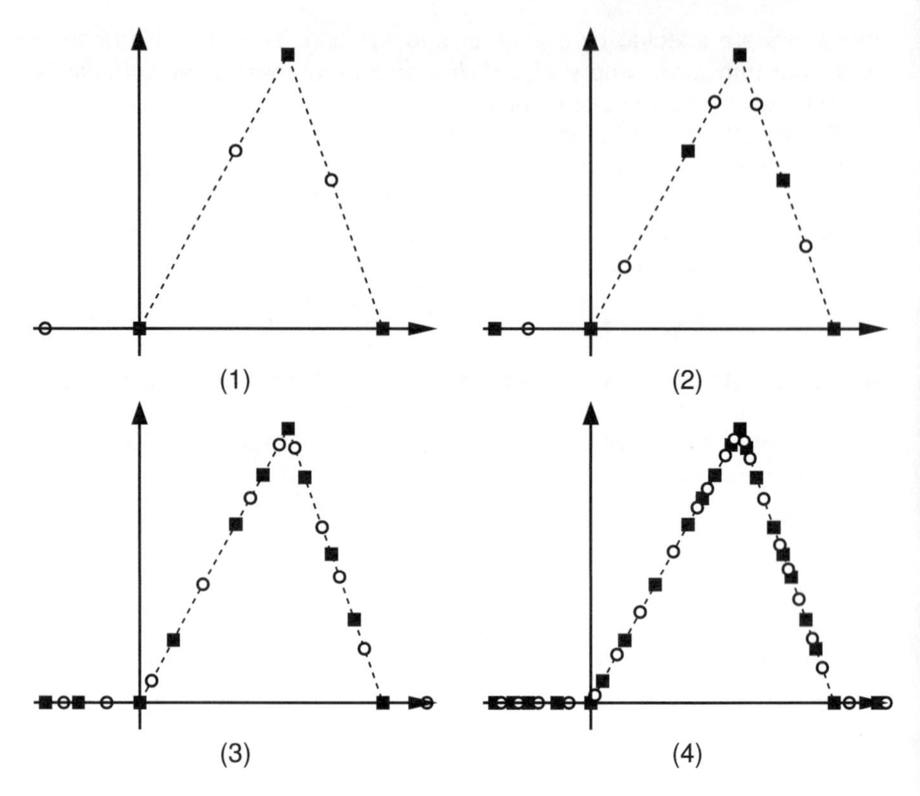

Figure 2.2
Refinement using linear interpolation.

x_L. Moreover, the representation

$$f_L(x) = \sum_{k=0}^{n_L-1} s_{L,k}\varphi_{L,k}(x),$$

with $\varphi_{L,k}(x)$ the triangular basis functions on x_L, interpolates the vector of coefficients s_L, $f_L(x_{L,k}) = s_{L,k}$, that is.

Using polylines instead of staircases removes the blocky artifacts in a reconstruction from processed wavelet coefficients. This is illustrated in Figure 2.5, which repeats the same experiment as in Section 1.3, using exactly the same input. For completeness, we mention that the scheme used in the experiment of Figure 2.5 includes a bidiagonal update matrix $\mathbf{U}_j$. This update will be developed in Section 2.1.6. On the downside of using polylines instead of staircases are the less sharp edges and jumps. This is because, unlike the Haar functions, the hat basis functions in Figure 2.4 have no vertical transitions for sharp representations of discontinuities.

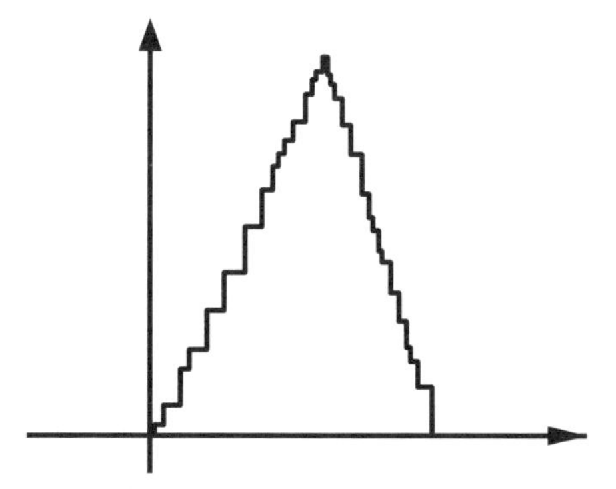

Figure 2.3
Coarse scaling basis function for linear interpolating prediction starting off from characteristic functions at the finest scale. Characteristic functions are not the natural choice of scaling functions here.

2.1.6 Wavelets in a triangular hat basis

Section 2.1.5 has proposed a wavelet transform with linear polynomial interpolation as prediction $\mathbf{P}_j$, followed by an update $\mathbf{U}_j$, as yet unspecified. Whereas the update does not appear in the refinement of the coarse scaling basis, (2.19) and (2.20), the wavelet basis depends on the choice of the update. The basis function $\psi_{j,k}(x)$ follows from an inverse transform with $d_{j,k} = 1$ while all other coefficients, scaling, and details, are zero. It is not necessary to follow the refinement process all the way up to the finest scale. Indeed, after going through the inverse update step in (2.17), we find $s_{j+1,2k} = -U_{j;k,k}$. We also have, slightly less straightforwardly, that $s_{j+1,2k+2} = -U_{j;k+1,k}$. Since $d_{j,k} = 1$ is the only nonzero detail, all subsequent refinement steps involve inverse predictions (2.18) only, just like in Section 2.1.5. We would arrive at exactly the same output if there were no update at all and if we initialise the refinement of (2.18) with the following three values:

$$
\begin{aligned}
s_{j,k} &= -U_{j;k,k} \\
s_{j,k+1} &= -U_{j;k+1,k} \\
d_{j,k} &= 1.
\end{aligned}
$$

Note that we initialise at scale j. For instance, $s_{j,k} = -U_{j;k,k}$ leads immediately to $s_{j+1,2k} = -U_{j;k,k}$, which is exactly the effect of the update on this coefficient. The outcome of the refinement is the superposition of the effect of the three inputs. We know that $s_{j,k} = -U_{j;k,k}$ leads to $-U_{j;k,k}\varphi_{j,k}(x)$, and $s_{j,k+1} = -U_{j;k+1,k}$ leads to $-U_{j;k+1,k}\varphi_{j,k+1}(x)$. The only new element here is the effect of $d_{j,k} = 1$. In absence of other nonzeros, this value proceeds

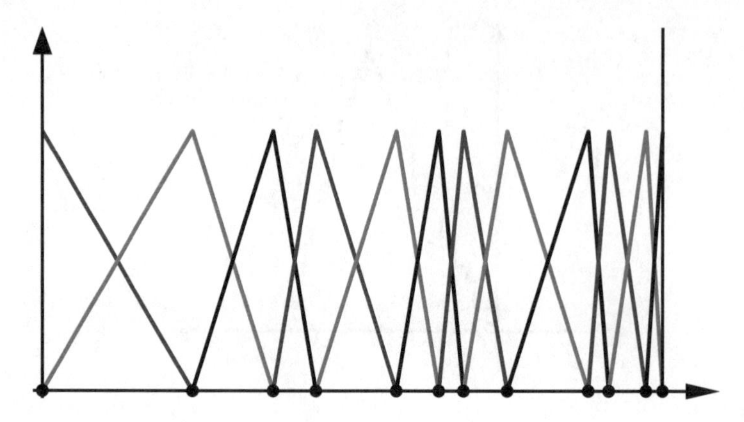

Figure 2.4
Scaling basis functions $\varphi_{L,k}(x)$ with $k = 0, 1, 2, \ldots$ for refinement with linear interpolating prediction.

unchanged to $s_{j+1,2k+1}$. From there, the normal refinement applies, as if it were initialised at scale $j+1$ with $s_{j+1,2k+1} = 1$.

All together, the function $\psi_{j,k}(x)$ whose decomposition requires a single nonzero detail coefficient $d_{j,k}$ can alternatively be represented in terms of scaling functions at scales $j+1$ and j

$$\psi_{j,k}(x) = \varphi_{j+1,2k+1}(x) - U_{j;k,k}\varphi_{j,k}(x) - U_{j;k+1,k}\varphi_{j,k+1}(x). \tag{2.21}$$

This expression allows us to control the properties of $\psi_{j,k}(x)$ by choosing appropriate values for the update coefficients $U_{j,l,k}$. For reasons already mentioned in Section 1.4 and further explored in Section 3.1.1, we have almost no choice but to impose that

$$\int_0^1 \psi_{j,k}(x)dx = 0,$$

leading to one equation for the pair of unknowns $U_{j;k,k}$ and $U_{j;k+1,k}$. A second equation may follow from the condition that

$$\int_0^1 x\psi_{j,k}(x)dx = 0.$$

This is the **second vanishing moment**, the zero integral being the first vanishing moment. This condition is far less mandatory, but it is a classical one. Vanishing moments of $\Psi_j(x)$ are called **primal** vanishing moments, in contrast to **dual** vanishing moments, that will be introduced in Section 2.3. The set of equations in the unknowns $U_{j;k,k}$ and $U_{j;k+1,k}$ contain the values of

$$M_{j,k}^{[q]} = \int_0^1 x^q \varphi_{j,k}(x)dx. \tag{2.22}$$

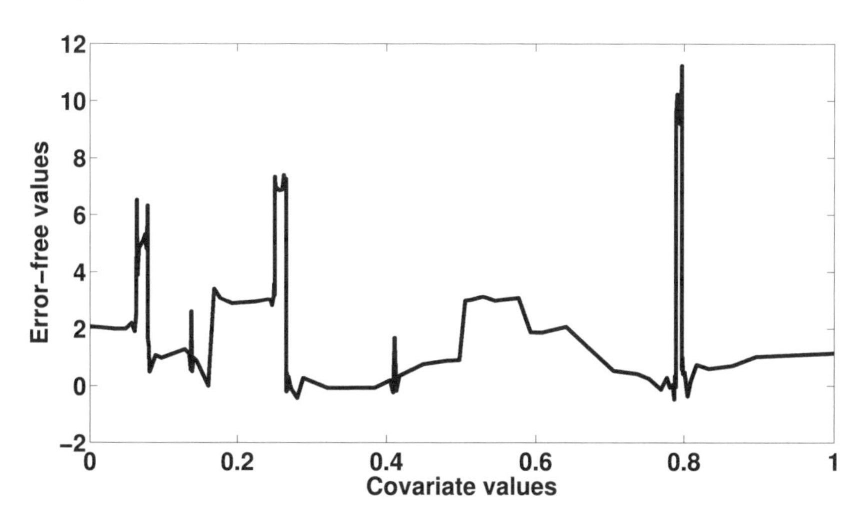

Figure 2.5
Reconstruction from thresholded coefficients obtained by linear interpolating prediction and a bidiagonal update matrix. To be compared with the reconstruction from a Haar transform in Figure 1.16.

More precisely, integration of (2.21) leads to

$$0 = M^{[q]}_{j+1,2k+1} - U_{j;k,k} M^{[q]}_{j,k} - U_{j;k+1,k} M^{[q]}_{j,k+1},$$

with $q \in \{0, 1\}$. For the triangular hat basis, it is possible to write explicit formulas for these **scaling moments** as a function of the grid locations $x_{j,k}$. In general wavelet transforms, the lifting scheme provides recursive formulas, as we will see in Section 2.1.7.

Exercise 2.1.6 *Let $\varphi_{j,k}(x)$ be a scaling function in the triangular hat basis. Then show that the moments are given by*

$$\begin{aligned} M^{[0]}_{j,k} &= (x_{j,k+1} - x_{j,k-1})/2, \\ M^{[1]}_{j,k} &= (x_{j,k+1} - x_{j,k-1})(x_{j,k-1} + x_{j,k} + x_{j,k+1})/6. \end{aligned}$$

Exercise 2.1.7 *Find the prediction and update coefficients of a triangular hat wavelet transform with two primal vanishing moments on an equidistant grid of knots.*

The wavelet transform just introduced, using the triangular hat refinement and adding two primal vanishing moments, is a simple yet important and popular representative of two larger families of wavelet transforms. The first family is the Deslauriers-Dubuc interpolating class, discussed in Section 3.2.

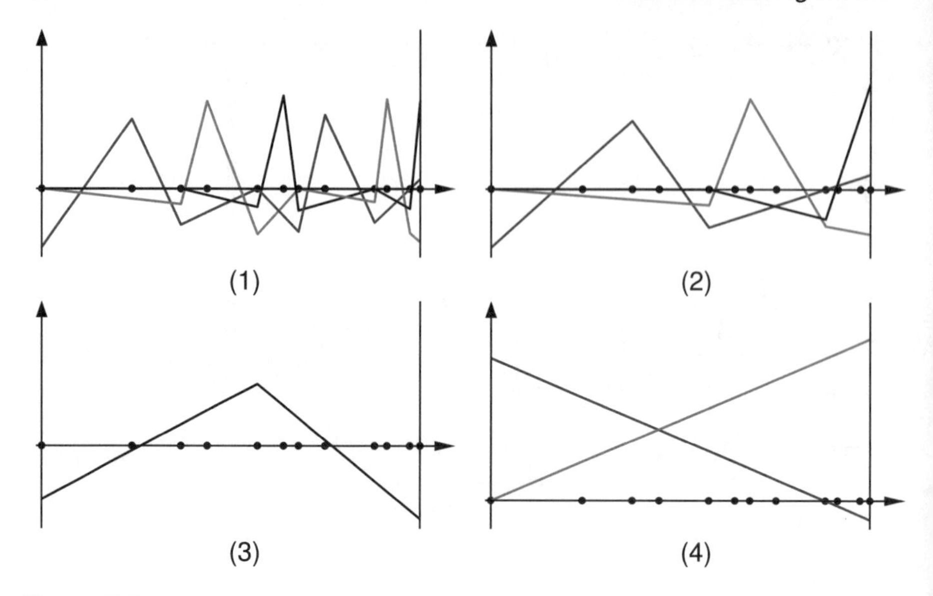

Figure 2.6
(1),(2),(3): Wavelet basis functions at scales $J-1$, $J-2$, and $J-3$, following from linear interpolating prediction in $\mathbf{P}_j$ and a bidiagonal update $\mathbf{U}_j$, where all nonzeros of $\mathbf{U}_j$ are devoted to two primal vanishing moments. Figure (4) contains the course scaling functions $\varphi_{L,k}(x)$ for $k = 1, 2$. These basis functions are coarse scale versions of the triangular hat functions in Figure 2.4, cut off at the boundaries.

The second family is that of B-spline wavelets, discussed in Section 4.2. On equidistant grids, such as images, the triangular hat with two primal vanishing moments is nothing less than the famous CDF-2,2 wavelet, used in the illustration of Section 1.1, and adopted as the JPEG-2000 lossless compression standard. In some literature, it is referred to as the Le Gall-Tabatabai wavelet, after the authors who proposed the refinement in the context of image processing (La Gall and Tabatabai, 1988).

2.1.7 A general two-step lifting scheme

This section discusses general lifting schemes with one update step and one prediction step.

The matrices of the lifting operations are denoted as $\mathbf{U}_j$ and $\mathbf{P}_j$. These matrices are typically bandlimited, thus ensuring the linear computational complexity of the transform and the spatial locality of the information carried by the coefficients. Indeed, if the lifting involves a full matrix, then each coefficient at the output depends on all input coefficients. Therefore, the design of lifting steps typically starts from fixing the degrees of freedom, i.e., the number of nonzero elements in the columns of the lifting matrices.

2.1.7.1 Prediction-first scheme

We first consider a scheme whose forward transform starts with a prediction, followed by an update

$$\begin{aligned} \boldsymbol{d}_j &= \boldsymbol{s}_{j+1,o} - \mathbf{P}_j \boldsymbol{s}_{j+1,e}, &(2.23)\\ \boldsymbol{s}_j &= \boldsymbol{s}_{j+1,e} + \mathbf{U}_j \boldsymbol{d}_j. &(2.24)\end{aligned}$$

The inverse transform first undoes the update, followed by the prediction

$$\begin{aligned} \boldsymbol{s}_{j+1,e} &= \boldsymbol{s}_j - \mathbf{U}_j \boldsymbol{d}_j, &(2.25)\\ \boldsymbol{s}_{j+1,o} &= \boldsymbol{d}_j + \mathbf{P}_j \boldsymbol{s}_{j+1,e}. &(2.26)\end{aligned}$$

The basis functions that can be associated with this transform are found by substituting the expressions (2.23) and (2.24) into (1.29), involving the row vectors of basis functions $\Phi_j(x)$ and $\Psi_j(x)$, introduced in Section 1.3.7. The outcome of the substitution reads

$$\Phi_{j+1}(x)\boldsymbol{s}_{j+1} = \Phi_j(x)(\boldsymbol{s}_{j+1,e} + \mathbf{U}_j \boldsymbol{d}_j) + \Psi_j(x)(\boldsymbol{s}_{j+1,o} - \mathbf{P}_j \boldsymbol{s}_{j+1,e}).$$

After rearranging terms and splitting the vector $\Phi_{j+1}(x)$ into an even and odd part, this becomes,

$$\begin{aligned} \Phi_{j+1,e}(x)\boldsymbol{s}_{j+1,e} + \Phi_{j+1,o}(x)\boldsymbol{s}_{j+1,o} &= \Phi_j(x)(\mathbf{I} - \mathbf{U}_j\mathbf{P}_j)\boldsymbol{s}_{j+1,e} + \Phi_j(x)\mathbf{U}_j\boldsymbol{s}_{j+1,o}\\ &+ \Psi_j(x)\boldsymbol{s}_{j+1,o} - \Psi_j(x)\mathbf{P}_j\boldsymbol{s}_{j+1,e}.\end{aligned}$$

This must hold for any $\boldsymbol{s}_{j+1,e}$, and for any $\boldsymbol{s}_{j+1,o}$. That means that terms in both vectors must sum up to zero, i.e.,

$$\begin{aligned} \Phi_{j+1,e}(x) &= \Phi_j(x)(\mathbf{I} - \mathbf{U}_j\mathbf{P}_j) - \Psi_j(x)\mathbf{P}_j,\\ \Phi_{j+1,o}(x) &= \Phi_j(x)\mathbf{U}_j + \Psi_j(x).\end{aligned}$$

From the latter equation, we obtain

$$\Psi_j(x) = \Phi_{j+1,o}(x) - \Phi_j(x)\mathbf{U}_j, \tag{2.27}$$

which is the general form of (2.21). Substitution into the former equation leads to

$$\begin{aligned} \Phi_{j+1,e}(x) &= \Phi_j(x) - \Phi_j(x)\mathbf{U}_j\mathbf{P}_j - \Phi_{j+1,o}(x)\mathbf{P}_j + \Phi_j(x)\mathbf{U}_j\mathbf{P}_j\\ &= \Phi_j(x) - \Phi_{j+1,o}(x)\mathbf{P}_j.\end{aligned}$$

So the **two-scale** or **refinement** equation becomes

$$\Phi_j(x) = \Phi_{j+1,e}(x) + \Phi_{j+1,o}(x)\mathbf{P}_j. \tag{2.28}$$

This equation is another instance of the central expression in wavelet analysis, first introduced in (1.35). In this scheme, the properties of $\Phi_j(x)$, in particular its smoothness, depend on the prediction $\mathbf{P}_j$ only.

Substitution into the previously obtained $\Psi_j(x) = \Phi_{j+1,o}(x) - \Phi_j(x)\mathbf{U}_j$ leads to the expression for the wavelet basis function. This expression is known as the **wavelet equation**,

$$\Psi_j(x) = \Phi_{j+1,o}(x)(\mathbf{I} - \mathbf{P}_j\mathbf{U}_j) - \Phi_{j+1,e}(x)\mathbf{U}_j. \tag{2.29}$$

This wavelet equation generalises the expression (2.21) in the triangular hat case.

Given the basis functions, we can define their **moments**. The integrals in the expressions below operate componentwise on the elements of the column vectors of functions $\Phi_j^\top(x)x^q$ and $\Psi_j^\top(x)x^q$,

$$\boldsymbol{M}_j^{(q)} = \int_0^1 \Phi_j^\top(x)x^q dx, \tag{2.30}$$

$$\boldsymbol{O}_j^{(q)} = \int_0^1 \Psi_j^\top(x)x^q dx. \tag{2.31}$$

Integrating (2.28) leads to

$$\boldsymbol{M}_j^{(q)} = \boldsymbol{M}_{j+1,e}^{(q)} + \mathbf{P}_j^\top \cdot \boldsymbol{M}_{j+1,o}^{(q)}. \tag{2.32}$$

Integrating (2.29) leads to

$$\boldsymbol{O}_j^{(q)} = -\mathbf{U}_j^\top \cdot \boldsymbol{M}_{j+1,e}^{(q)} + \left(\mathbf{I} - \mathbf{U}_j^\top\mathbf{P}_j^\top\right) \cdot \boldsymbol{M}_{j+1,o}^{(q)}. \tag{2.33}$$

Using the former expression to eliminate $\boldsymbol{M}_{j+1,e}^{(q)}$ from the latter, we can write

$$\boldsymbol{O}_j^{(q)} = \boldsymbol{M}_{j+1,o}^{(q)} - \mathbf{U}_j^\top \cdot \boldsymbol{M}_j^{(q)}. \tag{2.34}$$

This can be seen as the second step in a lifting scheme whose first step is (2.32). The first step transforms even fine scaling coefficients into coarse scaling coefficients, so this step is an update. The second step defines the details, so this step is prediction. We thus have a **dual** update-first lifting scheme with prediction operator $\mathbf{U}_j^\top$ and update operator $\mathbf{P}_j^\top$, while the original lifting schema was a prediction-first one.

For reasons explored in Section 3.1.1, the design of the update lifting step follows often from imposing that $\boldsymbol{O}_j^{(q)} = \mathbf{0}$. This is what we did in Section 2.1.6 with the triangular hat basis. The resulting expression $\boldsymbol{M}_{j+1,o}^{(q)} = \mathbf{U}_j^\top \cdot \boldsymbol{M}_j^{(q)}$ can be seen as a set of equations in the unknown elements of $\mathbf{U}_j$. The moments $\boldsymbol{M}_j^{(q)}$ are supposed to be known, which implies that the design of $\mathbf{P}_j$ precedes that of $\mathbf{U}_j$.

2.1.7.2 Update-first (in the forward transform)

An update-first wavelet transform has the general form

$$\boldsymbol{s}_j = \boldsymbol{s}_{j+1,e} + \mathbf{U}_j\boldsymbol{s}_{j+1,o}, \tag{2.35}$$

$$\boldsymbol{d}_j = \boldsymbol{s}_{j+1,o} - \mathbf{P}_j\boldsymbol{s}_j. \tag{2.36}$$

The inverse transform is given by

$$\boldsymbol{s}_{j+1,o} = \boldsymbol{d}_j + \mathbf{P}_j \boldsymbol{s}_j, \tag{2.37}$$
$$\boldsymbol{s}_{j+1,e} = \boldsymbol{s}_j - \mathbf{U}_j \boldsymbol{s}_{j+1,o}. \tag{2.38}$$

The two-scale equation follows in a way similar as for (2.28).

$$\Phi_j(x) = \Phi_{j+1,e}(x)\big(\mathbf{I} - \mathbf{U}_j\mathbf{P}_j\big) + \Phi_{j+1,o}(x)\mathbf{P}_j. \tag{2.39}$$

The wavelet equation, to be compared with (2.29), is

$$\Psi_j(x) = \Phi_{j+1,o}(x) - \Phi_{j+1,e}(x)\mathbf{U}_j. \tag{2.40}$$

The dual transform for the moments is now

$$\boldsymbol{O}_j^{(q)} = \boldsymbol{M}_{j+1,o}^{(q)} - \mathbf{U}_j^\top \boldsymbol{M}_{j+1,e}^{(q)},$$
$$\boldsymbol{M}_j^{(q)} = \boldsymbol{M}_{j+1,e}^{(q)} + \mathbf{P}_j^\top \boldsymbol{O}_j^{(q)}.$$

Imposing p primal vanishing moments leads to the set of equations for $q = 0, 1, \ldots, p-1$

$$\boldsymbol{M}_{j+1,o}^{(q)} = \mathbf{U}_j^\top \boldsymbol{M}_{j+1,e}^{(q)}, \tag{2.41}$$

where the nonzero elements of $\mathbf{U}_j$ are the unknowns.

From (2.29) and (2.40), we conclude that the update matrix appears in the expression of the wavelet basis $\Psi_j(x)$. This is remarkable, because the update step does not change the values of the corresponding wavelet coefficients $\boldsymbol{d}_j$. This observation illustrates the fact that each wavelet coefficient has a **value** and an **interpretation**, given by the corresponding basis function. Lifting steps can thus be designed with regard to the values or to the basis functions. The effect of an update on the interpretation of the detail coefficients is made concrete for the Haar transform in Section 3.1.1.

The duality between a coefficient's *value* and its *interpretation* is a key understanding in wavelet theory. The notion of duality is further developed in Section 2.3.

2.1.8 Lifting in diagrams

For some readers, it may help to visualise the operations of a lifting scheme in a flow chart. Others, who prefer to think in terms of matrices, rather than diagrams, can easily skip this section.

Figure 2.7 depicts the flow in one scale of a forward Haar decomposition, implemented using a lifting scheme with the update on the fine scale side. This is the decomposition developed in Section 2.1.2, see (2.12) and (2.13). The update matrix $\mathbf{U}_j$ in the figure is diagonal, with $U_{j;k,k} = \Delta_{j+1,2k+1}/\Delta_{j,k}$. The rescaling matrix $\mathbf{D}_j$ is always diagonal. Indeed, the inverse scheme contains $\mathbf{D}_j^{-1}$, which could be a full matrix if $\mathbf{D}_j$ were not diagonal, thereby

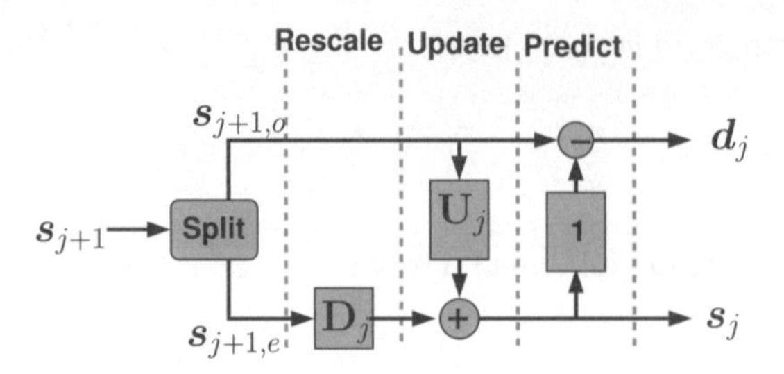

Figure 2.7
Lifting scheme for the Haar transform, update on the fine scale side; see (2.12) and (2.13).

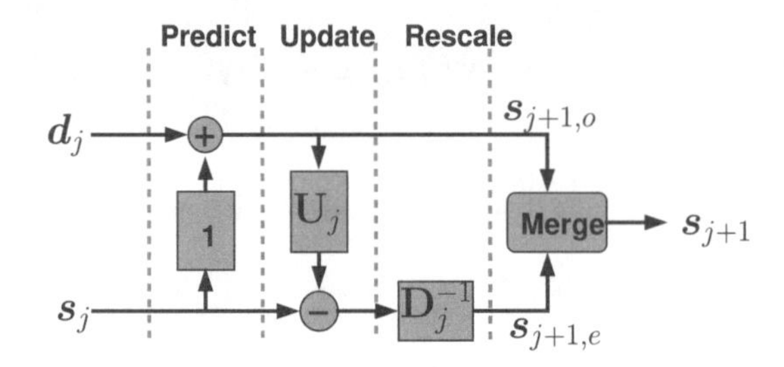

Figure 2.8
Inverse lifting scheme for the reconstruction from a Haar transform, update on the fine scale side.

making a fast reconstruction impossible. In Figure 2.7, the matrix $\mathbf{D}_j$ has elements $D_{j;k,k} = \Delta_{j+1,2k}/\Delta_{j,k}$. Figure 2.8 presents the inverse for this version.

The alternative version of the Haar transform, developed in Section 2.1.4, is depicted in Figures (2.8) and (2.10). The update and rescale matrices happen to have the same entries as in Figure 2.8, but they appear on other locations in the scheme.

Other versions for the same transform, or rescaled versions of it, are possible. Especially the value and place of rescaling steps in the scheme may vary in the literature.

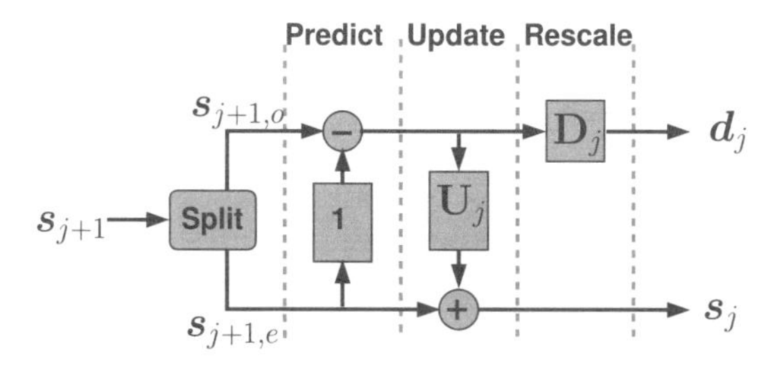

Figure 2.9
Lifting scheme for the Haar transform, update on the coarse scale side; see Section 2.1.4.

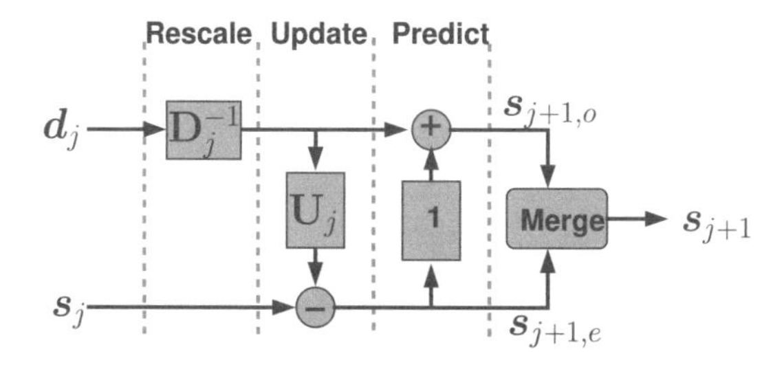

Figure 2.10
Inverse lifting scheme for the reconstruction from a Haar transform, update on the coarse scale side.

2.2 Filterbanks

This section takes the construction of a wavelet transform to a slightly higher level than the lifting steps. It considers the mapping from one scale to another as the combination of two linear operations: one linear operation maps fine scaling coefficients onto coarse scaling coefficients; the other linear operations transform fine scaling coefficients into detail offsets. The combination of the two operations is named a **filterbank**. When filterbanks are constructed from a lifting scheme, then the inverse operation follows straightforwardly from the inverse lifting scheme. At a more abstract level, there is more challenge in the inverse of a filterbank. In particular, it is far from trivial to design a filterbank that both has desired properties and an inverse with interesting properties. The problem of finding forward and inverse filterbanks is known as the **perfect reconstruction** condition.

2.2.1 The filterbank notation of a wavelet transform

Through the lifting construction, we have set up a linear transform and its inverse mapping the vectors $\boldsymbol{s}_{j+1}$ onto $\boldsymbol{s}_j$ and $\boldsymbol{d}_j$ and vice versa. A forward transform at level j, such as the prediction-first two-step lifting scheme in (2.23) and (2.24), or the update-first alternative in (2.35) and (2.36), are specific cases in a more general framework provided by the matrix formulae

$$\boldsymbol{s}_j = \widetilde{\mathbf{H}}_j^\top \boldsymbol{s}_{j+1} \tag{2.42}$$

$$\boldsymbol{d}_j = \widetilde{\mathbf{G}}_j^\top \boldsymbol{s}_{j+1}. \tag{2.43}$$

Clearly, the matrices $\widetilde{\mathbf{H}}_j$ and $\widetilde{\mathbf{G}}_j$ are rectangular. Together, they form the square matrix $\widetilde{\mathbf{W}}_j$:

$$\widetilde{\mathbf{W}}_j = \begin{bmatrix} \widetilde{\mathbf{H}}_j^\top \\ \widetilde{\mathbf{G}}_j^\top \end{bmatrix}. \tag{2.44}$$

This matrix $\widetilde{\mathbf{W}}_j$ in its turn is part of the full wavelet decomposition, introduced in (1.24),

$$\widetilde{\mathbf{W}} = \prod_{j=L}^{J-1} \begin{bmatrix} \widetilde{\mathbf{W}}_j & \mathbf{0}_{jj'} \\ \mathbf{0}_{jj'}^\top & \mathbf{I}_{j'} \end{bmatrix}. \tag{2.45}$$

Here, $\mathbf{I}_{j'}$ is the identity matrix of size $n - n_{j+1}$, where n_{j+1} is the number of scaling coefficients at scale $j+1$. Moreover $\mathbf{0}_{jj'}$ is an $n_{j+1} \times (n - n_{j+1})$ zero matrix.

Exercise 2.2.1 *Consider the decomposition in Exercise 1.3.4, page 27. What are the matrices $\widetilde{\mathbf{H}}_j$ and $\widetilde{\mathbf{G}}_j$ for $j = 0, 1, 2$?*

The inverse transform is represented by the matrices $\mathbf{H}_j$ and $\mathbf{G}_j$ in the reconstruction formula

$$\boldsymbol{s}_{j+1} = \mathbf{H}_j \boldsymbol{s}_j + \mathbf{G}_j \boldsymbol{d}_j. \tag{2.46}$$

The inverse wavelet transform is then written as

$$\mathbf{W} = \prod_{j=J-1}^{L-1} \begin{bmatrix} \mathbf{W}_j & \mathbf{0}_{jj'} \\ \mathbf{0}_{jj'}^\top & \mathbf{I}_{j'} \end{bmatrix}, \tag{2.47}$$

where

$$\mathbf{W}_j = \begin{bmatrix} \mathbf{H}_j & \mathbf{G}_j \end{bmatrix}. \tag{2.48}$$

Note that the order of the factors in the product of (2.47) is from high indices on the left to low indices on the right.

The implementation of a wavelet transform $\widetilde{\mathbf{W}}$ as in (2.45) and its inverse W as in (2.47) is known as a **filterbank**. The matrices (filters) $\widetilde{\mathbf{H}}_j$ and $\widetilde{\mathbf{G}}_j$ together form a bank which allows a fast implementation and a sparse structure with a sparse inverse. Sparse matrices with sparse inverses are nontrivial. As an example, the inverse of a bandlimited matrix is in general a full matrix [1]. A matrix-vector product with such a matrix is fast, whereas the *inverse problem*, i.e., the reconstruction of the original vector from the product, is computationally intense, and often numerically ill conditioned.

2.2.2 From lifting to filterbanks and back

The lifting scheme consisting of one prediction and one update step from Section 2.1.7, has a forward transformation given by (2.23) and (2.24), which can be developed into

$$\begin{aligned}\begin{bmatrix} \boldsymbol{s}_j \\ \boldsymbol{d}_j \end{bmatrix} &= \begin{bmatrix} \mathbf{I} & \mathbf{U}_j \\ \mathbf{0} & \mathbf{I} \end{bmatrix} \begin{bmatrix} \mathbf{I} & \mathbf{0} \\ -\mathbf{P}_j & \mathbf{I} \end{bmatrix} \begin{bmatrix} \boldsymbol{s}_{j+1,e} \\ \boldsymbol{s}_{j+1,o} \end{bmatrix} \\ &= \begin{bmatrix} \mathbf{I} - \mathbf{U}_j\mathbf{P}_j & \mathbf{U}_j \\ -\mathbf{P}_j & \mathbf{I} \end{bmatrix} \begin{bmatrix} \boldsymbol{s}_{j+1,e} \\ \boldsymbol{s}_{j+1,o} \end{bmatrix} \\ &= \begin{bmatrix} (\mathbf{I} - \mathbf{U}_j\mathbf{P}_j)\boldsymbol{s}_{j+1,e} + \mathbf{U}_j\boldsymbol{s}_{j+1,o} \\ \boldsymbol{s}_{j+1,o} - \mathbf{P}_j\boldsymbol{s}_{j+1,e} \end{bmatrix}.\end{aligned}$$

Identification with

$$\begin{bmatrix} \boldsymbol{s}_j \\ \boldsymbol{d}_j \end{bmatrix} = \begin{bmatrix} \widetilde{\mathbf{H}}_{j,e}^\top & \widetilde{\mathbf{H}}_{j,o}^\top \\ \widetilde{\mathbf{G}}_{j,e}^\top & \widetilde{\mathbf{G}}_{j,o}^\top \end{bmatrix} \begin{bmatrix} \boldsymbol{s}_{j+1,e} \\ \boldsymbol{s}_{j+1,o} \end{bmatrix}$$

leads to $\widetilde{\mathbf{H}}_{j,e}^\top = \mathbf{I} - \mathbf{U}_j\mathbf{P}_j$; $\widetilde{\mathbf{H}}_{j,o}^\top = \mathbf{U}_j$; $\widetilde{\mathbf{G}}_{j,e}^\top = -\mathbf{P}_j$; and $\widetilde{\mathbf{G}}_{j,o}^\top = \mathbf{I}$. In these expressions, the matrix $\widetilde{\mathbf{H}}_{j,e}$ stands for the submatrix of $\widetilde{\mathbf{H}}_j$ containing the even rows of $\widetilde{\mathbf{H}}_j$, and $\widetilde{\mathbf{H}}_{j,o}$ the submatrix with the odd rows.

Exercise 2.2.2 *Using the Expressions (2.15) and (2.16), find the matrices $\widetilde{\mathbf{H}}_j$ and $\widetilde{\mathbf{G}}_j$ for a transform with one bidiagonal prediction and one bidiagonal update. Check with the expressions above.*

Denote by $\widetilde{\mathbf{J}}_j^{o\top}$ the complementary subsampling matrix, defined by $\boldsymbol{s}_{j+1,o} = \widetilde{\mathbf{J}}_j^{o\top}\boldsymbol{s}_{j+1}$, while $\widetilde{\mathbf{J}}_j^\top$ is the subsampling matrix, introduced in (2.2). Then we have

$$\boldsymbol{s}_j = \widetilde{\mathbf{H}}_j^\top \boldsymbol{s}_{j+1} = \left[(\mathbf{I} - \mathbf{U}_j\mathbf{P}_j)\widetilde{\mathbf{J}}_j^\top + \mathbf{U}_j\widetilde{\mathbf{J}}_j^{o\top}\right]\boldsymbol{s}_{j+1} \tag{2.49}$$

$$\boldsymbol{d}_j = \widetilde{\mathbf{G}}_j^\top \boldsymbol{s}_{j+1} = \left[\widetilde{\mathbf{J}}_j^{o\top} - \mathbf{P}_j\widetilde{\mathbf{J}}_j^\top\right]\boldsymbol{s}_{j+1}. \tag{2.50}$$

[1] A (near) square matrix $\mathbf{A}$ is bandlimited if all entries $A_{i,j}$ are zero if $|i-j| > r$ for some integer r. Matrix-vector or matrix-matrix products with bandlimited matrices are computationally faster by one order of magnitude than with full matrices.

The subsampling matrix $\widetilde{\mathbf{J}}_j^\top$ and its complement $\widetilde{\mathbf{J}}_j^{o\top}$ satisfy the identities

$$\widetilde{\mathbf{J}}_j^{o\top}\widetilde{\mathbf{J}}_j = \mathbf{0} \tag{2.51}$$

$$\widetilde{\mathbf{J}}_j^{o\top}\widetilde{\mathbf{J}}_j^o = \mathbf{I}_{n_j'} \tag{2.52}$$

$$\widetilde{\mathbf{J}}_j^\top\widetilde{\mathbf{J}}_j = \mathbf{I}_{n_j} \tag{2.53}$$

$$\widetilde{\mathbf{J}}_j\widetilde{\mathbf{J}}_j^\top + \widetilde{\mathbf{J}}_j^o\widetilde{\mathbf{J}}_j^{o\top} = \mathbf{I}_{n_{j+1}} \tag{2.54}$$

By right multiplication with $\widetilde{\mathbf{J}}_j$ in (2.50) and with $\widetilde{\mathbf{J}}_j^o$ in (2.49) we obtain expressions for the lifting matrices $\mathbf{P}_j$ and $\mathbf{U}_j$ for a given forward filterbank,

$$\mathbf{P}_j = -\widetilde{\mathbf{G}}_j^\top\widetilde{\mathbf{J}}_j \tag{2.55}$$

$$\mathbf{U}_j = \widetilde{\mathbf{H}}_j^\top\widetilde{\mathbf{J}}_j^o. \tag{2.56}$$

The expressions (2.55) and (2.56) hold only if $\widetilde{\mathbf{G}}_j$ and $\widetilde{\mathbf{H}}_j$ are known to be derived from a lifting scheme with one prediction and one update step.

Exercise 2.2.3 *Consider the Haar transform at the coarsest level $L = 0$, with $n_{L+1} = 2$. Take $\widetilde{\mathbf{G}}_0^\top = [\ -1 \quad 1\ ]/2$ as in Exercise 2.2.1 and $\widetilde{\mathbf{H}}_0^\top = [\ 1 \quad 1\ ]/2$. Assuming that this transform can be factored as a lifting scheme with one prediction and one update, find the lifting matrices. Then check that the $\widetilde{\mathbf{G}}_0^\top$ following from the lifting matrices is not the one started from. This leads to the conclusion that the initial transform cannot be factored into a lifting scheme with one prediction and one update. This is confirmed by the scheme in Section 2.1.4, which contains an additional rescaling on the odd branch; see also Figures 2.9 and 2.10. Check that with an alternative normalisation, $\widetilde{\mathbf{G}}_0^\top = [\ -1 \quad 1\ ]$ and $\widetilde{\mathbf{H}}_0^\top = [\ 1 \quad 1\ ]/2$, the transform does admit a factoring into two lifting steps.*

For the reconstruction, identification of (2.26) and (2.25), further developed into $\boldsymbol{s}_{j+1,e} = \boldsymbol{s}_j - \mathbf{U}_j\boldsymbol{d}_j$, and $\boldsymbol{s}_{j+1,o} = \mathbf{P}_j\boldsymbol{s}_j + (\mathbf{I}-\mathbf{P}_j\mathbf{U}_j)\boldsymbol{d}_j$, with (2.46), i.e., $\boldsymbol{s}_{j+1,o} = \mathbf{H}_{j,o}\boldsymbol{s}_j + \mathbf{G}_{j,o}\boldsymbol{d}_j$, and $\boldsymbol{s}_{j+1,e} = \mathbf{H}_{j,e}\boldsymbol{s}_j + \mathbf{G}_{j,e}\boldsymbol{d}_j$ yields $\mathbf{H}_{j,e} = \mathbf{I}$; $\mathbf{G}_{j,e} = -\mathbf{U}_j$; $\mathbf{H}_{j,o} = \mathbf{P}_j$; and $\mathbf{G}_{j,o} = (\mathbf{I}-\mathbf{P}_j\mathbf{U}_j)$.
Using the identity in (2.54) it follows that

$$\boldsymbol{s}_{j+1} = \widetilde{\mathbf{J}}_j\boldsymbol{s}_{j+1,e} + \widetilde{\mathbf{J}}_j^o\boldsymbol{s}_{j+1,o} = \left[\widetilde{\mathbf{J}}_j + \widetilde{\mathbf{J}}_j^o\mathbf{P}_j\right]\boldsymbol{s}_j + \left[\widetilde{\mathbf{J}}_j^o(\mathbf{I}-\mathbf{P}_j\mathbf{U}_j) - \widetilde{\mathbf{J}}_j\mathbf{U}_j\right]\boldsymbol{d}_j. \tag{2.57}$$

Exercise 2.2.4 *Given the forward step of a Haar transform at level $j = 1$, with $n_j = n_1 = 4$, $\widetilde{\mathbf{G}}_0^\top = \begin{bmatrix} -1 & 1 & 0 & 0 \\ 0 & 0 & -1 & 1 \end{bmatrix}$ and $\widetilde{\mathbf{H}}_0^\top = \frac{1}{2}\begin{bmatrix} 1 & 1 & 0 & 0 \\ 0 & 0 & 1 & 1 \end{bmatrix}$, find the matrices of the inverse transform $\mathbf{H}_0$ and $\mathbf{G}_0$, knowing that this version of the Haar transform factors into one prediction, followed by one update step (see Exercise 2.2.3).*

Exercise 2.2.5 *Find the filterbank matrices $\widetilde{\mathbf{H}}_j$, $\widetilde{\mathbf{G}}_j$, $\mathbf{H}_j$, $\mathbf{G}_j$ in an update-first lifting scheme.*

2.2.3 Biorthogonality and perfect reconstruction

In some applications, especially for equispaced observations, such as images, it may be interesting to design appropriate filter matrices $\mathbf{H}_j$, $\mathbf{G}_j$, $\widetilde{\mathbf{H}}_j$ and $\widetilde{\mathbf{G}}_j$ without going through the lifting construction. One of the conditions to impose explicitly is then **perfect reconstruction**, listed as a key property in the construction of a wavelet analysis in Section 1.6.1. The perfect reconstruction condition (PR) states that $\mathbf{W}_j = \widetilde{\mathbf{W}}_j^{-1}$, that is

$$\mathbf{W}_j\widetilde{\mathbf{W}}_j = \mathbf{I}_{j+1} \Leftrightarrow \mathbf{H}_j\widetilde{\mathbf{H}}_j^\top + \mathbf{G}_j\widetilde{\mathbf{G}}_j^\top = \mathbf{I}_{j+1}. \tag{2.58}$$

Alternatively, the perfect reconstruction leads to

$$\widetilde{\mathbf{W}}_j\mathbf{W}_j = \mathbf{I}_{j+1} \Leftrightarrow \begin{bmatrix} \widetilde{\mathbf{H}}_j^\top\mathbf{H}_j & \widetilde{\mathbf{H}}_j^\top\mathbf{G}_j \\ \widetilde{\mathbf{G}}_j^\top\mathbf{H}_j & \widetilde{\mathbf{G}}_j^\top\mathbf{G}_j \end{bmatrix} = \mathbf{I}_{j+1} \tag{2.59}$$

In both equations, $\mathbf{I}_{j+1}$ stands for the identity matrix of size n_{j+1}. One of the benefits of a lifting scheme is that it leads automatically to a perfect reconstruction pair of filterbanks. This allows the designer to concentrate on other properties, such as sparsity, approximation power, smoothness, the bias-variance balance, aliasing.

Expression (2.59) can be decomposed into four separate expressions

$$\widetilde{\mathbf{H}}_j^\top\mathbf{H}_j = \mathbf{I}_{j+1,e}, \tag{2.60}$$

$$\widetilde{\mathbf{H}}_j^\top\mathbf{G}_j = \mathbf{0}_{j+1,e,o}, \tag{2.61}$$

$$\widetilde{\mathbf{G}}_j^\top\mathbf{H}_j = \mathbf{0}_{j+1,o,e}, \tag{2.62}$$

$$\widetilde{\mathbf{G}}_j^\top\mathbf{G}_j = \mathbf{I}_{j+1,o}, \tag{2.63}$$

where $\mathbf{I}_{j+1,e}$ is an identity matrix whose size equals the number of even indices at scale $j+1$. This is also the size of the vector $\boldsymbol{s}_j$. The matrix $\mathbf{0}_{j,e,o}$ has the same number of rows, while its columns are numbered by the odds at scale $j+1$, or, equivalently, by the size of the vector $\boldsymbol{d}_j$. Expressions (2.61) and (2.62) state that all columns of $\widetilde{\mathbf{H}}_j$ are orthogonal to all columns of $\mathbf{G}_j$ and all columns of $\widetilde{\mathbf{G}}_j$ are orthogonal to all columns of $\mathbf{H}_j$. The pairs $(\widetilde{\mathbf{H}}_j, \mathbf{G}_j)$ and $(\widetilde{\mathbf{G}}_j, \mathbf{H}_j)$ are said to be **biorthogonal**. Biorthogonality is basically nothing else but saying that $\widetilde{\mathbf{W}}_j$ and $\mathbf{W}_j$ are each other's inverses.

Whichever form we use to express biorthogonality or perfect reconstruction, it should be noted that the design of sparse matrices satisfying this condition is a nontrivial task. Indeed, the inverse of a sparse matrix is not at all guaranteed to be sparse. Hence, either the inverse or the forward transform (or both) can be computationally expensive. The lifting scheme can be seen as a tool in the design of wavelet transforms. It automatically constructs pairs of forward and inverse transforms that are both sparse.

Exercise 2.2.6 *(This exercise requires the use of software for matrix calculations.) Given the Haar refinement matrix where level* j *has* $n_{j+1} = 8$*, generate a random matrix* $\mathbf{G}_j$ *and find from there the corresponding matrices* $\widetilde{\mathbf{H}}_j$ *and* $\widetilde{\mathbf{G}}_j$*.*

Exercise 2.2.7 *In practice, the design of a wavelet transform from a given refinement matrix starts typically with the dual refinement matrix* $\widetilde{\mathbf{H}}_j$*, thus defining the projection onto the refinable basis. The primal and dual detail matrices* $\mathbf{G}_j$ *and* $\widetilde{\mathbf{G}}_j$ *then follow from the perfect reconstruction condition. The columns of* $\mathbf{G}_j$ *and* $\widetilde{\mathbf{G}}_j$ *span the left null spaces (cokernels) of* $\mathbf{H}_j$ *and* $\widetilde{\mathbf{H}}_j$*, while satisfying other requirements too, such as having zero elements for compact support of the corresponding basis functions. Let* $\mathbf{G}_j^{[1]} = \mathbf{G}_j\mathbf{A}_j$*, with* $\mathbf{A}_j$ *a* $n'_j \times n'_j$ *invertible matrix. Show that* $\mathbf{G}_j^{[1]}$ *is also a valid detail matrix*[2]*. What is* $\widetilde{\mathbf{G}}_j^{[1]}$ *in this case?*

2.2.4 The two-scale equation in a filterbank

We can also write the two-scale and wavelet equations, in the same way as we did for Expressions (2.28) and (2.29). More precisely, the combination of (1.29), i.e., $\Phi_{j+1}(x)\boldsymbol{s}_{j+1} = \Phi_j(x)\boldsymbol{s}_j + \Psi_j(x)\boldsymbol{d}_j$, with (2.46), i.e., $\boldsymbol{s}_{j+1} = \mathbf{H}_j\boldsymbol{s}_j + \mathbf{G}_j\boldsymbol{d}_j$, yields

$$\Phi_{j+1}(x)\mathbf{H}_j\boldsymbol{s}_j + \Phi_{j+1}(x)\mathbf{G}_j\boldsymbol{d}_j = \Phi_j(x)\boldsymbol{s}_j + \Psi_j(x)\boldsymbol{d}_j.$$

A necessary condition for this expression to hold for any $\boldsymbol{s}_j$ and any $\boldsymbol{d}_j$ is the **two-scale equation** or refinement equation, obtained by taking $\boldsymbol{s}_j = \mathbf{I}_{n_j}$ and $\boldsymbol{d}_j = \mathbf{0}$,

$$\Phi_j(x) = \Phi_{j+1}(x)\mathbf{H}_j. \tag{2.64}$$

This two-scale equation is a nonequispaced and matrix form of the two-scale equation in (1.35). The **wavelet equation** becomes

$$\Psi_j(x) = \Phi_{j+1}(x)\mathbf{G}_j. \tag{2.65}$$

Exercise 2.2.8 *Consider the binary refinement equation with* $\mathbf{H}_j$ *defined by its nonzero elements,* $H_{j;2k,k} = 1 = H_{j;2k+3,k}$*. (Compare with the Haar refinement matrix which has* $H_{j;2k,k} = 1 = H_{j;2k+1,k}$*). Check that a solution to this two-scale equation is given by* $\varphi_{j,k} = \chi_{[x_{j,k},x_{j,k+3})}(x)$ *with* $\chi_I(x)$ *the characteristic function on the interval* I*.*

Exercise 2.2.9 *What happens to the basis* $\Psi_j(x)$ *when the detail matrix* $\mathbf{G}_j$ *is replaced by* $\mathbf{G}_j^{[1]} = \mathbf{G}_j\mathbf{A}_j$*, as proposed in Exercise 2.2.7?*

---

[2] A procedure factoring the wavelet transform into lifting steps, described in Chapter 4, yields one possible $\mathbf{G}_j$ for a given $\widetilde{\mathbf{H}}_j$ through (4.35). Any other admissible $\mathbf{G}_j^{[1]} = \mathbf{G}_j\mathbf{A}_j$ requires a matrix multiplication with $\mathbf{A}_j$ in the reconstruction, before the rest of the inverse lifting scheme.

The two-scale equation can be iterated into a **repeated refinement** or **subdivision** process

$$\Phi_j(x) = \Phi_J(x)\mathbf{H}_J\mathbf{H}_{J-1}\ldots\mathbf{H}_j. \tag{2.66}$$

Multiplication of both sides in (2.66) with a canonical vector $e_{j,k}$ on the right leads to an expression for $\varphi_{j,k}(x)$ as a linear combination of fine scaling functions $\Phi_J(x)$. In theory, subdivision goes on up to an infinitely fine scale by letting $J \to \infty$. As mentioned already in Section 1.6.3, subdivision thus determines the infinitesimal properties of the scaling basis, in particular its smoothness. In practice, subdivision is a numerical technique for evaluation of the basis functions at finite resolution. This has been illustrated in Sections 1.3.8 and 2.1.2 for the Haar basis and in Section 2.1.5 for the triangular hat basis. Subdivision is further discussed in Section 2.2.5.

Exercise 2.2.10 *Consider the refinement matrix of Exercise 2.2.8. Let φ_j be the minimal subvector of the refined sequence at resolution level j, containing all the nonzeros in the sequence. Starting from a Kronecker delta at level $L = 0$, what is φ_j at level j? Does φ_j converge to a vector of function values of $\varphi_{0,0}(x_{j,k})$?*

> There is an important difference between the scopes and impact of two-scale and wavelet equations. The refinement equation (2.64), being the key to the subdivision procedure (2.66), fixes the scaling basis, and the smoothness of both the scaling and the wavelet bases. The wavelet equation (2.65) plays no role in the smoothness of the wavelet basis. Indeed, this equation is used at one scale only. It expresses wavelet functions in terms of scaling functions, further refinement always proceeds through the refinement equation. The wavelet equation is used mainly to impose vanishing moments. Unlike smoothness, the vanishing moment conditions are not an infinitesimal property.

Both refinement and wavelet equations can be integrated, using the definitions of the moments in (2.30) and (2.31), to find the **dual wavelet transform** for the moments

$$\boldsymbol{M}_j^{(q)} = \mathbf{H}_j^\top \boldsymbol{M}_{j+1}^{(q)}, \tag{2.67}$$

$$\boldsymbol{O}_j^{(q)} = \mathbf{G}_j^\top \boldsymbol{M}_{j+1}^{(q)}, \tag{2.68}$$

from which we conclude that moments are found by a forward wavelet transform with inverse transform matrices $\mathbf{H}_j$ and $\mathbf{G}_j$.

For those readers who like a graphical representation, Figure 2.11 depicts a flow chart for the filterbank implementation of a wavelet transform.

Exercise 2.2.11 *1. Let $\mathbf{W}_j$ be one step of the inverse transform,*

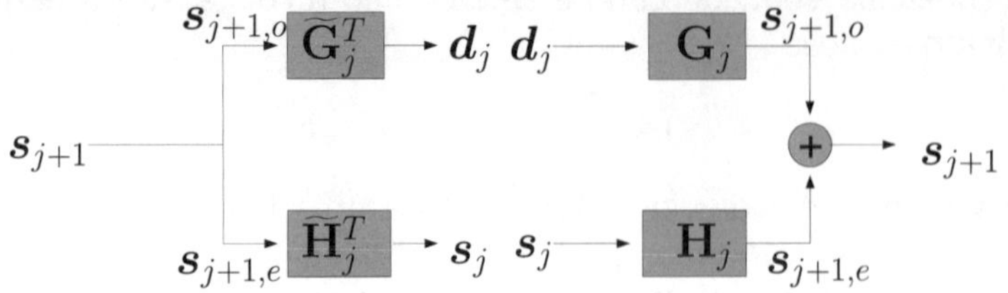

Figure 2.11
Forward and inverse filterbank for the implementation of one scale transition in a wavelet decomposition and reconstruction. The matrices $\mathbf{H}_j$, $\mathbf{G}_j$, $\widetilde{\mathbf{G}}_j$ and $\widetilde{\mathbf{H}}_j$ are rectangular, not squared.

defined in (2.48) and let $\widetilde{\mathbf{W}}_j$ *be one step of the forward transform defined in (2.44). Then check that the dual transform matrices are found by the inverse transposes of the given transform.*

2. From Exercise 2.2.5, we know that a lifting scheme with update $\mathbf{U}_j$ *first, prediction* $\mathbf{P}_j$ *second in the forward transform, has forward transform matrices*

$$\begin{aligned}\widetilde{\mathbf{H}}_j^\top &= \widetilde{\mathbf{J}}_j^\top + \mathbf{U}_j \widetilde{\mathbf{J}}_j^{o\top} \\ \widetilde{\mathbf{G}}_j^\top &= (\mathbf{I} - \mathbf{P}_j\mathbf{U}_j)\widetilde{\mathbf{J}}_j^{o\top} - \mathbf{P}_j\widetilde{\mathbf{J}}_j^\top.\end{aligned}$$

Use these expressions to identify the prediction and update steps in the dual transform of a prediction-first lifting scheme.

2.2.5 Subdivision

Let $f_j(x) = \Phi_j(x)\boldsymbol{s}_j$ be a function spanned by the scaling functions at scale j. In particular, taking a canonical vector as $\boldsymbol{s}_j$ yields a scaling basis function $\varphi_{j,k}(x)$ at scale j. Then the repeated refinement in (2.66) provides a routine to find the fine scaling coefficients $\boldsymbol{s}_J$ so that

$$\Phi_j(x)\boldsymbol{s}_j = \Phi_J(x)\mathbf{H}_J\mathbf{H}_{J-1}\ldots\mathbf{H}_j\boldsymbol{s}_j = \Phi_J(x)\boldsymbol{s}_J. \tag{2.69}$$

The calculation $\boldsymbol{s}_J = \mathbf{H}_J\mathbf{H}_{J-1}\ldots\mathbf{H}_j\boldsymbol{s}_j$ is nothing but an inverse transform where the detail coefficients $\boldsymbol{d}_i$ at all intermediate scales are zero. The numerical implementation thus runs the subdivision product in (2.69) from the right to the left, from coarse to fine scales, that is. The subdivision procedure has been developed for the case of triangular hat functions in Section 2.1.5.

The expression (2.69) does not fix the fine scaling functions $\Phi_J(x)$. As a matter of fact, these functions are free to choose. In the Haar transform, we proposed to take characteristic functions, while a scheme with linear interpolating prediction is better off with triangular hats, because the triangular hat family is preserved throughout scales by the two-scale equation; see Figure 2.3. In a general lifting scheme, it is not clear which choice at finest scale best fits with the two-scale equation. If the lifting scheme is designed to preserve a class of basis functions at all scales, then it is a natural choice to take these basis functions at the finest scale. For instance, the polylines generated by triangular hats can be generalised to piecewise polynomials, i.e., **splines**. Section 4.2 will develop a lifting scheme for this sort of basis. Other lifting schemes are designed not from or for the scaling basis, but rather from the prediction operation. For instance, the linear interpolating prediction associated with the triangular hats can be extended to higher order polynomial prediction. This prediction, further developed in Section 3.2.1, cannot be associated with a piecewise polynomial basis. In the absence of a naturally corresponding scaling basis, the finest scaling functions are often taken to be characteristic functions.

In order to reduce the artificial, staircase look of the coarse scaling functions, subdivision may be continued beyond the resolution of the observations, by inserting artificial knots. Let $J^* \to \infty$, an arbitrarily fine **superresolution**; then the scaling functions at level J^* can be taken to be characteristic functions, $\varphi_{J^*,k}(x) = \chi_{I_{J^*,k}}(x)$, where $\sum_{k=0}^{n_{J^*}} \varphi_{J^*,k}(x) = \chi_{[0,1]}(x)$ and $x_{J^*,k} \in I_{J,k}$. From there the two-scale equations (2.28) or (2.39) can be applied for $j = J^* - 1, J^* - 2, \ldots, J, J-1, \ldots, L$. The scaling basis obtained at level j can be denoted as $\Phi_j^{[J^*]}(x)$, with a superscript referring to the superresolution parameter of the numerical solution.

It is clear that this procedure involves issues of convergence when the superresolution J^* tends to infinity. The infinite refinement fixes the scaling basis and its smoothness properties. In most schemes, the outcome depends on the location of the artificial refinement knots. Through the wavelet equation (2.65), developed in specific cases as (2.29) or (2.40), the same infinite refinement also determines the smoothness of the wavelet basis. Given the multitude of possible refinement schemes beyond the observational scale, smoothness through subdivision is typically hard to analyse.

2.3 Dual basis functions

As a follow-up to the discussion in Section 2.2.4, let J^* be a superresolution at which the scaling functions are approximated as indicator functions. This superresolution may be the resolution of the observations, J, or finer. The coefficients $\boldsymbol{s}_{J^*}$ in the expression $s_{J^*}(x) = \Phi_{J^*}(x)\boldsymbol{s}_{J^*}$ are then given by

$$\boldsymbol{s}_{J^*} = \int_0^1 s_{J^*}(x)\widetilde{\Phi}_{J^*}^\top(x)dx, \tag{2.70}$$

where

$$\widetilde{\varphi}_{J^*,k}(x) = \varphi_{J^*,k}(x)/\Delta_{J^*,k}. \tag{2.71}$$

Here we take

$$\Delta_{J^*,k} = \int_0^1 \varphi_{J^*,k}(x)dx = \int_0^1 |\varphi_{J^*,k}(x)|^2 \, dx. \tag{2.72}$$

It then holds that

$$\int_0^1 \Phi_{J^*}^\top(x)\widetilde{\Phi}_{J^*}(x)dx = \mathbf{I}_{J^*} = \int_0^1 \widetilde{\Phi}_{J^*}^\top(x)\Phi_{J^*}(x)dx. \tag{2.73}$$

This is referred to by stating that $\widetilde{\Phi}_{J^*}(x)$ is the dual basis of $\Phi_{J^*}(x)$, which is then termed the primal basis. From the forward transform (2.43), it follows for

$$\widetilde{\Phi}_j(x) = \widetilde{\Phi}_{j+1}(x)\widetilde{\mathbf{H}}_j, \tag{2.74}$$

with $j = J^*, J^* - 1, \ldots$, that

$$\boldsymbol{s}_j = \int_0^1 s_{J^*}(x)\widetilde{\Phi}_j^\top(x)dx. \tag{2.75}$$

Expression (2.74) is the **dual refinement equation** or dual two-scale equation. In a similar way, there exist dual wavelet functions $\widetilde{\Psi}_j(x)$, defined by the dual wavelet equation,

$$\widetilde{\Psi}_j(x) = \widetilde{\Phi}_{j+1}(x)\widetilde{\mathbf{G}}_j, \tag{2.76}$$

so that

$$\boldsymbol{d}_j = \int_0^1 s_{J^*}(x)\widetilde{\Psi}_j^\top(x)dx. \tag{2.77}$$

Taking $s_{J^*}(x) = x^q$ in the two-step lifting scheme from Section 2.1.7, it follows that

$$\int_0^1 x^q \widetilde{\Psi}_j^\top(x)dx = 0, \tag{2.78}$$

for $q = 0, 1, \ldots, \widetilde{p} - 1$ if the prediction step $\mathbf{P}_j$ perfectly reconstructs polynomials of degree $\widetilde{p} - 1$. From (2.78), the parameter $\widetilde{p}$ can be interpreted as the number of **dual vanishing moments**. In a general wavelet transform, (2.78) implies that the functions x^q are spanned by the nested scaling bases; i.e., there exist vectors $\boldsymbol{s}_j$, so that $x^q = \Phi_j(x)\boldsymbol{s}_j$ at any resolution level j.

Exercise 2.3.1 *Use the perfect reconstruction properties and (2.73) to show*

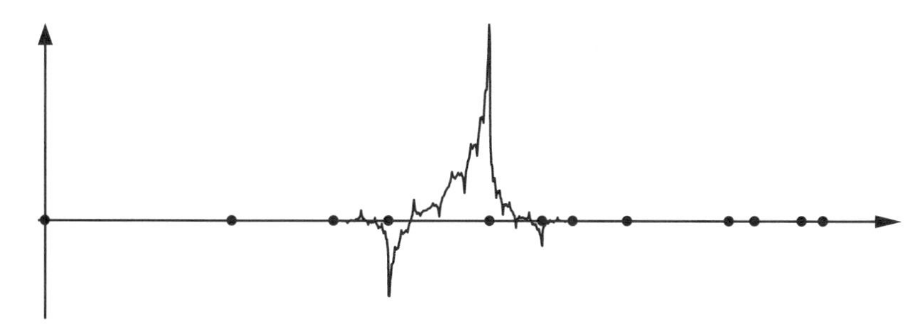

Figure 2.12
One element of one of the possible dual bases for the triangular hat basis in Figure 2.4.

that for all $j < J^*$,

$$\begin{aligned}
\int_0^1 \Phi_j^\top(x)\widetilde{\Phi}_j(x)dx &= \mathbf{I}_j = \int_0^1 \widetilde{\Phi}_j^\top(x)\Phi_j(x)dx,\\
\int_0^1 \Phi_j^\top(x)\widetilde{\Psi}_j(x)dx &= \mathbf{0} = \int_0^1 \widetilde{\Psi}_j^\top(x)\Phi_j(x)dx,\\
\int_0^1 \Psi_j^\top(x)\widetilde{\Phi}_j(x)dx &= \mathbf{0} = \int_0^1 \widetilde{\Phi}_j^\top(x)\Psi_j(x)dx,\\
\int_0^1 \Psi_j^\top(x)\widetilde{\Psi}_j(x)dx &= \mathbf{I}_j = \int_0^1 \widetilde{\Psi}_j^\top(x)\Psi_j(x)dx.
\end{aligned}$$

Remark 2.3.2 *In a way similar to (2.78), assuming that there exist $\widetilde{\boldsymbol{s}}_{j+1}$ so that $x^q = \widetilde{\Phi}_{j+1}(x)\widetilde{\boldsymbol{s}}_{j+1}$, the detail coefficients at level j are given by $\widetilde{\boldsymbol{d}}_j = \int_0^1 \Psi_j^\top(x)x^q dx$. The statement that $x^q = \widetilde{\Phi}_{j+1}(x)\widetilde{\boldsymbol{s}}_{j+1}$ can also be reproduced at level j, i.e., $x^q = \widetilde{\Phi}_j(x)\widetilde{\boldsymbol{s}}_j$ is then equivalent to imposing primal vanishing moments, $\int_0^1 \Psi_j^\top(x)x^q dx = \mathbf{0}$.*

The dual basis functions can also be used to define a projection of a function $f(x) \in L_2([0,1])$ onto the primal basis $\Phi_j(x)$ or $\Psi_j(x)$,

$$\boldsymbol{s}_j = \int_0^1 f(x)\widetilde{\Phi}_j^\top(x)dx, \tag{2.79}$$

and similar for $\boldsymbol{d}_j$. In general, the projection $\Phi_j(x)\boldsymbol{s}_j$ is not the orthogonal projection.

In biorthogonal wavelet transforms, the dual basis functions may have very different properties than the primal basis functions. In particular, while primal basis functions are designed to be smooth for smooth reconstructions $\Phi_j(x)\boldsymbol{s}_j$, no such property is needed for the dual functions. An example of a possible dual scaling function for the triangular hat basis in Figure 2.4 can be found in Figure 2.12. This dual basis function comes from a lifting scheme with linear interpolating prediction and a bidiagonal update for two primal vanishing moments as in (2.41).

Exercise 2.3.3 *How can we find a refinement scheme with three dual vanishing moments that admits the Haar basis as dual scaling basis?*

3

Using lifting for the design of a wavelet transform

In this chapter, wavelet transforms are designed by straightforwardly incorporating interesting properties into the lifting steps. As an important example in statistical applications, the wavelet decomposition should behave well when its coefficients are thresholded before reconstruction. A good behaviour under thresholding implies that both the bias and the variance after thresholding should be kept under control during the reconstruction phase. Orthogonal transforms perform optimally in preserving the bias-variance balance, but, unfortunately, orthogonality is often unfeasible, as it cannot be combined with other desired properties, especially for data on irregular knots. Therefore, this chapter studies how bias and variance can be controlled through lifting steps.

3.1 Key properties in the design of a lifting scheme

When lifting is used to construct a wavelet transform from the ground up, all desired properties should somehow be incorporated into the building blocks, the lifting steps, that is. The discussion starts off with the most important property defining a wavelet function, namely the zero integral.

3.1.1 Wavelets or hierarchical bases: zero integrals

In Sections 1.4, 2.1.6, and 2.1.7, it was suggested that the design of a wavelet transform should include an update step for basis function with zero integrals. Using componentwise integration of the row vector of functions $\Psi(x)$, defined in (1.29), this is $\int_0^1 \Psi_j^\top(x)dx = \mathbf{0}_{j'}$, with $\mathbf{0}_{j'}$ the zero vector of length $n'_j = n_{j+1} - n_j$. As mentioned in Section 1.3.6, functions with zero integrals fluctuate above and below the axis. The zero integral makes the wavelet have its wave-like shape. This section lists a couple of reasons why zero integrals are important.

DOI: 10.1201/9781003265375-3

A first argument for basis functions with zero integrals has been the control of **bias** or **approximation errors**, as discussed in Section 2.1.2 in the context of balanced versus unbalanced Haar transforms.

A second argument holds in applications where we need to control the integral of the reconstruction. This is the case, for instance, in **density estimation**, where the outcome is supposed to be a positive function with integral one. Let $\widehat{\boldsymbol{d}}_j$ be a processed version of $\boldsymbol{d}_j$; then the reconstruction

$$\widehat{f}_J(x) = \Phi_L(x)\boldsymbol{s}_L + \sum_{j=L}^{J-1} \Psi_j(x)\widehat{\boldsymbol{d}}_j$$

has the same integral as the original decomposition in (1.30).

The final and most important argument for zero integrals is that a multiscale basis with nonzero integrals is often not appropriate for approximations of general square integrable functions. Indeed, if the basis functions have nonzero integrals, then it may well be possible to construct **nontrivial decompositions of the zero function**, converging in $L_2([0,1])$.

In order to illustrate the claim, we first consider the case where $\mathbf{G}_j = \widetilde{\mathbf{J}}_j^o$, meaning that the wavelet equation reduces to $\Psi_j(x) = \Phi_{j+1,o}(x)$. In other words, the wavelets at scale j are just the odd indexed scaling functions at scale $j+1$. These detail functions have a nonzero integral. The resulting multiscale basis is termed a **hierarchical basis** (Yserentant, 1986). A simple construction of a hierarchical basis[1] follows from a lifting scheme with just one prediction and no update. This is evident from taking $\mathbf{U}_j = \mathbf{0}$ in (2.29), leading to

$$\mathbf{G}_j = \widetilde{\mathbf{J}}_j^o(\mathbf{I} - \mathbf{P}_j\mathbf{U}_j) - \widetilde{\mathbf{J}}_j\mathbf{U}_j = \widetilde{\mathbf{J}}_j^o.$$

Figure 3.1 depicts the example of the Haar transform without an update. Hierarchical bases can be used for L_2-functions in certain smoothness classes, such as Sobolev spaces[2] (Lorentz and Oswald, 2000). Thanks to the additional smoothness imposed by membership of these function classes, it is possible to apply nonlinear processing in a hierarchical basis decomposition. For general L_2-functions with discontinuities, a decomposition in a hierarchical basis will still lead to **multiscale, sparse** data representation. For nonlinear processing on the data, however, the zero integral is mostly indispensable.

> Hierarchical bases lead to sparse and multiscale representations of piecewise smooth data, yet, in general, they are not the appropriate tool for nonlinear processing.

[1] In general, the lifting scheme of a hierarchical basis contains several prediction and update steps. See Exercise 4.1.9.

[2] Sobolev spaces are further discussed in Section 8.3.1.

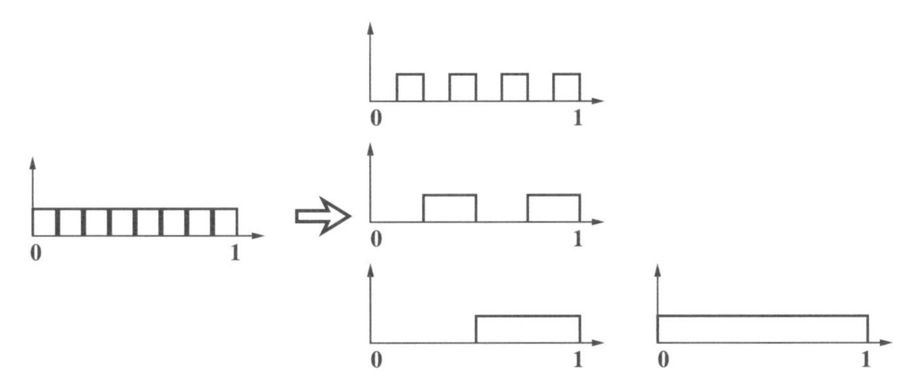

Figure 3.1
A Haar transform without update, i.e., where $d_{j,k} = s_{j+1,2k+1} - s_{j+1,2k}$ and $s_{j,k} = s_{j+1,2k}$, would lead to a multiscale basis where all wavelet functions are odd scaling functions at the next, finer scale, i.e., $\psi_{j,k}(x) = \varphi_{j+1,2k+1}$. This basis is not well conditioned in $L_2([0,1]$.

As an example, consider the Haar transform *without update*. The multiscale basis is depicted in Figure 3.1. Comparing the basis in Figures 3.1 and 1.18, we observe that both decompositions adopt the same formula for computing the detail coefficients, namely $d_{j,k} = s_{j+1,2k+1} - s_{j+1,2k}$. In a scheme without update step, or otherwise *before* the update $s_{j,k} = s_{j+1,2k} + (\Delta_{j+1,2k+1}/\Delta_{j,k})d_{j,k}$ has taken place, the detail $d_{j,k}$ indicates how far the odd input $s_{j+1,2k+1}$ is above the even input $s_{j+1,2k}$. *After* the update, the *same* offset tells us how far the odd input is above the average, and at the same time, how far the even input is below the average. The interpretation of a double offset against a common average is reflected in the waveform of the basis function.

Without update, it is easy to construct a nontrivial decomposition of the zero function that converges in $L_2([0,1])$ for an increasing finest scale J. Indeed, consider at each scale the first detail basis function

$$\psi_{j,0}(x) = \varphi_{j+1,1}(x) = \chi_{[x_{j+1,1},x_{j+1,2}[}(x).$$

In this definition, $x_{j,k}$ are the locations of nodes of the Haar basis functions. These nodes do not necessarily coincide with the points where the observations take place, as explained in Section 2.1.2. We also assume here that at each scale, the last node $x_{j,2^j} = 1$ coincides with the end point. Starting from $\varphi_{0,0}(x) = \chi_{[0,1[}(x)$, a simple subtraction of each of these detail basis functions leads to the function

$$\begin{aligned} f_J(x) &= \varphi_{0,0}(x) - \sum_{j=0}^{J-1} \psi_{j,0}(x) \\ &= \chi_{[0,1[}(x) - \sum_{j=0}^{J-1} \chi_{[x_{j+1,1},x_{j+1,2}[}(x) \\ &= \chi_{[0,1[}(x) - \sum_{j=0}^{J-1} \chi_{[x_{J,2^j},x_{J,2^j+1}[}(x) = \chi_{[0,x_{J,1}[}(x). \end{aligned}$$

It is clear that $\int_0^1 f_J^2(x)dx = x_{J,1}^2 \to 0$, as $J \to \infty$, at least if further refinement leaves no gap between $0 = x_{J,0}$ and $x_{J,1}$.

The converging expansion above is an example of a more general construction. Let $f_L(x) = \Phi_L(x)\boldsymbol{s}_L$ be a coarse scale function in $L_2([0,1])$. By orthogonal projection of $f_L(x)$ onto $\Psi_L(x)$,

$$\boldsymbol{d}_L = -\left[\int_0^1 \Psi_L(x)^\top \Psi_L(x)dx\right]^{-1} \left[\int_0^1 \Psi_L(x)^\top \Phi_L(x)dx\right] \boldsymbol{s}_L, \tag{3.1}$$

the norm of the function[3] $f_{L+1}(x) = \Phi_L(x)\boldsymbol{s}_L + \Psi_L(x)\boldsymbol{d}_L$ is minimised.

Exercise 3.1.1 *Consider a hierarchical basis of triangular hat functions, so* $\Phi_L(x)$ *is a triangular hat scaling basis at level* L, *and* $\Psi_L(x)$ *are the odd elements of the scaling basis at level* $L+1$. *Then find the coefficients* $\boldsymbol{d}_L$ *in the nontrivial decomposition of the zero function (3.1).*

If the wavelet functions are designed to make $\int_0^1 \Psi_L(x)^\top \Phi_L(x)dx$ the null matrix, then $\boldsymbol{d}_L = \boldsymbol{0}_{L'}$, the zero vector of length $n_{L+1} - n_L$, from which it follows that $\|f_{L+1}\|_2 = \|f_L\|_2$. Hence, the norm of

$$f_J(x) = \Phi_L(x)\boldsymbol{s}_L + \Psi_L(x)\boldsymbol{d}_L + \Psi_{L+1}(x)\boldsymbol{d}_{L+1} + \ldots + \Psi_J(x)\boldsymbol{d}_J$$

cannot possibly converge to zero. This design of a wavelet transform is termed **semi-orthogonal**. Semi-orthogonality is quite an attractive property, for various reasons, as will become clear in the next sections. It has, however, one main drawback: most semi-orthogonal wavelets have infinite support, leading to increased computational efforts for less localised basis functions. As a simpler alternative, the zero integral condition $\int_0^1 \Psi_L(x)^\top dx = \boldsymbol{0}_{L'}$ prevents $\|f_J\|_2$ from converging to zero. Indeed, writing $\overline{f} = \int_0^1 f_L(x)dx$, and $\widetilde{f}_L(x) = f_L(x) - \overline{f}$, it holds that $\|f_L\|_2^2 = \|\widetilde{f}_L\|_2^2 + \overline{f}^2$. Then, whatever the value

[3] Note that this construction is not a forward wavelet transform. A forward wavelet transform would involve the dual wavelet basis and a fine scale approximation to start from, i.e., $\boldsymbol{d}_L = \int_0^1 \widetilde{\Psi}_L(x)^\top f_{L+1}(x)dx$.

of $\boldsymbol{d}_L$, it holds that $\int_0^1 f_{L+1}(x)dx = \int_0^1 f_L(x)dx + 0 = \overline{f}$; hence $\|f_{L+1}\|_2^2 \geq \overline{f}^2$, and also $\|f_{L+1}\|_2^2 \geq \overline{f}^2$ for any $J \geq L$.

A further example of a nontrivial decomposition of the zero function in a hierarchical B-spline basis is developed in Section 8.3.2.

Exercise 3.1.2 *Consider a hierarchical basis of interpolating scaling functions.*

1. *Show that the refinement equation is given by* $\mathbf{H}_j = \widetilde{\mathbf{J}}_j + \widetilde{\mathbf{J}}_j^o \mathbf{P}_j$
2. *Replace* $\mathbf{G}_j = \widetilde{\mathbf{J}}_j^o$ *by* $\mathbf{G}_j = \widetilde{\mathbf{J}}_j^o \mathbf{A}_j$*, as in Exercise 2.2.7. This amounts to* $\widetilde{\mathbf{G}}_j^\top = \mathbf{A}_j^{-1}$*, as found in Exercise 2.2.7, meaning that the detail coefficient vector is multiplied by* $\mathbf{A}_j^{-1}$ *(and by* $\mathbf{A}_j$ *upon reconstruction). Can this construction help to create a wavelet basis with vanishing moments or to prevent nontrivial decompositions of the zero function?*

3.1.2 The role of the final update step

The lifting schemes discussed so far contained at most one update step, which was used to make sure that the detail basis functions $\Psi_j(x)$ were actually wavelets by imposing them to have **zero integrals**. In schemes with more than one update step, discussed further on in Section 4.1.5, the zero integral condition is satisfied by the last update in the forward scheme. This final update has no impact on the two-scale equation, and thus it does not affect the scaling basis $\Phi_j(x)$ and its infinitesimal properties.

From the primal vanishing moment conditions (2.34) and (2.41), it follows that one final update is enough to annihilate the integrals and higher order moments of the functions in $\Psi_j(x)$. For zero integrals alone, the update can be a diagonal matrix, while for p **primal vanishing moments**, the update must have p nonzeros in each column. The number of nonzeros in the update should not be unnecessarily large, as sparse update matrices lead to faster computations and basis functions $\Psi_j(x)$ with smaller supports, c.q., functions that are more precisely located.

As mentioned already in Section 2.1.7, an alternative way to describe the effect of the update concentrates on the coefficients in the transform (2.24) itself, instead of reasoning in terms of the basis functions. Taking away the update from a scheme (2.24) with only one update, the vector s_j would simply be the subsampled fine scale coefficients $s_{j+1,e}$. For data with noise, this means that the fluctuation on a single fine scale coefficient proceeds all the way up to the coarse scales, whereas in a scheme with updates, the data are smoothed throughout the process. The case of Haar is an illustrative example: with an update we get $(s_{j+1,2k}+s_{j+1,2k+1})/2$ instead of $s_{j+1,2k}$. The averaged value has clearly a lower **variance** than the unfiltered alternative.

Another effect of the update, well understood in signal processing, is

Figure 3.2
Aliasing. The left panel displays an image of 699×512 pixels. The right panel displays the same image after mere subsampling, reducing the size to 350×256 pixels. Note that the size of the display does not reflect the number of pixels. In the subsampled image, the roofs show spurious patterns, known as Moiré patterns.

anti-aliasing. Subsampling without any update has the effect that fluctuations with a frequency higher than half the subsampling rate[4] are misinterpreted as an oscillation with a lower frequency. For instance, binary subsampling on the alternation $(\ldots, -1, 1, -1, 1, \ldots)$ leads to a constant signal, either $(\ldots, 1, 1, 1, 1, \ldots)$ or $(\ldots, -1, -1, -1, -1, \ldots)$. The misinterpretation is known as **aliasing**, further discussed in Section 6.2.5. Aliasing is the reason why turning wheels in a moving picture sometimes apparently turn backwards: aliasing occurs, for instance, when in the time between two frames the wheel has made between a half and a full cycle: by the viewer, this is interpreted as less than one half in the opposite direction. Figure 3.2 illustrates the effect on images, where the aliasing results in false patterns, termed **Moiré patterns**. Aliasing can be controlled by filtering the data upon subsampling. The filtering does not prevent the fine details from being lost, but it prevents the creation of spurious patterns, otherwise leading to possibly false interpretations or conclusions at coarse scales.

[4]This is known as the Nyquist rate, see also Section 6.2.5.

As a summary, the final update step serves three main purposes.

1. In terms of the wavelet **basis**, it can be designed to add **primal vanishing moments** or other characteristics to the wavelet functions, thereby making the basis numerically well conditioned within the limits of the given scaling basis. Primal vanishing moments are also linked to a reduction of the **bias** or **approximation error**, as seen in Section 2.1.2 on the unbalanced Haar transform.
2. In terms of **coefficients**, the final update tends to **reduce the variance**, especially of coarse scaling coefficients.
3. In a scheme with just one update step, it has a smoothing **anti-aliasing** effect on the coefficients at coarse scales.

The final update has no effect on the two-scale equation, thus leaving the scaling basis $\Phi_j(x)$ unchanged.

Obviously, in many cases, including the **Haar transform**, the update has a simultaneous effect: it reduces the variance and at the same time introduces a vanishing moment in the basis. The choice between unbalanced and balanced Haar transform also illustrates that, for instance, variance reduction and vanishing moments are somehow conflicting objectives. Compromises can be made by giving up higher vanishing moments for the sake of some explicit variance reduction. This is further developed in Sections 3.1.3, 3.1.4, and 3.1.5.

Exercise 3.1.3 *Check that the update in a prediction-first two-step lifting scheme of Section 2.1.7 has no impact on the refinement matrix $\mathbf{H}_j$. When the update comes before the prediction in the forward transform, the refinement does depend on the update.*

Exercise 3.1.4 *Consider a triangular hat decomposition on an equidistant grid, with a bidiagonal update matrix. According to (2.15), the prediction coefficients are $P_{j;k,k} = 1/2 = P_{j;k,k+1}$. For reasons of symmetry, we propose $U_{j;k,k} = u = U_{j;k,k-1}$. From (2.15) and (2.16), it then follows that $s_{j,k} = (1-u)s_{j+1,2k} + us_{j+1,2k-1} + us_{j+1,2k+1} - (u/2)s_{j+1,2k-2} - (u/2)s_{j+1,2k+2}$. Assuming uncorrelated coefficients at level $j+1$, find the value of α minimising the variance at level j. Compare to the update for vanishing moments in Exercise 2.1.7. Explain why, at least on a regular grid, no compromise is possible between the two objectives, keeping one vanishing moment, while using one degree of freedom for minimising the variance. In practice, scaling coefficients in a triangular hat decomposition with updates will be uncorrelated at the finest level $j+1 = J$ at best. At all further scales, correlations will have to be taken into account. As a result, updates for minimum variances will be level dependent.*

3.1.3 Variance propagation in wavelet projections

The design of a wavelet transform for use with noisy observations requires some specific attention. Indeed, as discussed in Section 3.1.2, the final update tends to reduce the variance of coarse scaling coefficients. Without special attention to the variance, it may, however, not prevent undesired variance reinflation upon reconstruction.

3.1.3.1 A motivating experiment

The problem is illustrated by an experiment shown in Figure 3.3. In the first

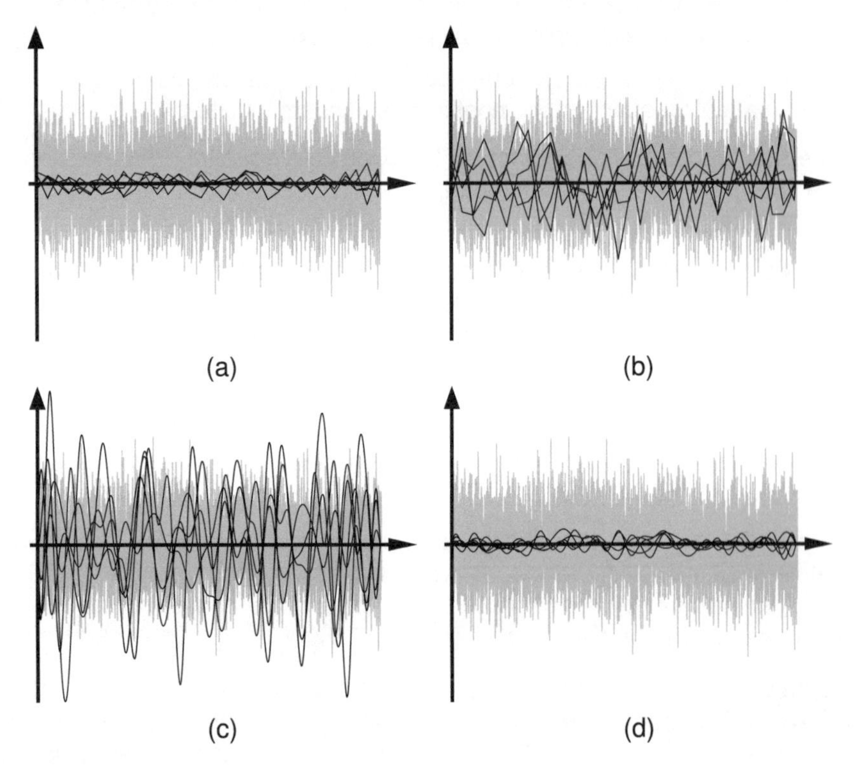

Figure 3.3
Reconstruction from 1000 observations of pure uncorrelated, equidistant, normal noise. (a) Using the triangular hat refinement of Section 2.1.5 and the update for two primal moments of Section 2.1.6. (b) Using the triangular hat refinement with no update. (c) Using a cubic B-spline refinement scheme as presented later, in Section 4.2.7, again with an update for two primal moments. (d) The same cubic B-spline refinement, now with an update for semi-orthogonal wavelets (see text).

experiment, represented in the left panel of the figure, a triangular hat wavelet transform equipped with an update step for two primal vanishing moments is applied to five samples of 1000 i.i.d. normal observations, associated to an equispaced grid of knots. As an oracle reveals to the algorithm that the observations contain only noise, the algorithm decides to take away all detail coefficients, so that the reconstruction is based on 32 coarse scaling coefficients only. This is a linear operation. The figure depicts these five reconstructions in black solid line against a background of the normal observations in grey colour. The reconstructions can be seen to take the form of polylines, as could be expected from a triangular hat basis. It is also clear that the variance in the reconstruction is much smaller than in the input.

In Figure 3.3(b) the same experiment is repeated using a triangular hat wavelet transform without update step. The variance is far less reduced in this setting, which illustrates the claim that the final update indeed may contribute to variance reduction. In Figure 3.3(c), the update for two vanishing moments is put back in place, while the triangular hat is replaced by a more sophisticated cubic B-spline refinement, introduced later[5], in Chapter 4. Surprisingly, the update does not seem able to prevent an unpleasant variance inflation, making the combination of cubic spline refinement and an update for two vanishing moments unsuitable for use in practice. In Figure 3.3(d), the update for vanishing moments is replaced by an update for maximum variance control. In Sections 3.1.4 and 3.1.5 it is found that maximum variance control leads to the same semi-orthogonal wavelets, already encountered in Section 3.1.1. The following sections develop updates for semi-orthogonal wavelets, as well as updates that combine variance control and primal vanishing moments.

3.1.3.2 Wavelet reconstructions as projections

In order to describe the outcome of the experiment, the observations are supposed to be available at fine scale, following the model

$$Y(x) = \Phi_J(x)(\boldsymbol{s}_J + \boldsymbol{\varepsilon}_J),$$

where $\boldsymbol{s}_J$ is the unknown parameter vector and where $\boldsymbol{\varepsilon}_J$ is a vector of zero mean random variables with covariance matrix $\boldsymbol{\Sigma}_J$. Then the wavelet transform that maps $Y(x)$ onto the multiscale basis

$$\Psi(x) = \{\Phi_L(x), \Psi_L(x), \Psi_{L+1}(x), \ldots, \Psi_{J-1}(x)\}$$

can be represented by the matrix $\widetilde{\mathbf{W}}$, introduced in (1.24) and again in (2.45). By linearity, the contribution of the noise to the multiscale decomposition is given by the vector $\boldsymbol{\eta}_L = \widetilde{\mathbf{W}}\boldsymbol{\varepsilon}_J$, consisting of

$$\boldsymbol{\eta}_L^\top = \left[\begin{array}{ccccc} \boldsymbol{\varepsilon}_L^\top & \boldsymbol{\delta}_L^\top & \boldsymbol{\delta}_{L+1}^\top & \ldots & \boldsymbol{\delta}_{J-1}^\top \end{array}\right].$$

[5]The triangular hat can be seen as the linear representative of the same family of B-spline refinement, as discussed in Chapter 4

Most wavelet processing methods then operate on the detail coefficients, thus affecting the random detail vectors δ_j, for $j = L, L+1, \ldots, J-1$. For these operations to work without unexpected effects, at least the reconstruction with all details replaced by zero should not lead to variance inflation. The experiment illustrated in Figure 3.3 reveals that this is far from trivial.

In particular, let $\widetilde{\mathbf{V}}_L$ be defined by backward recursion on $L = J-1, J-2, \ldots$

$$\widetilde{\mathbf{V}}_L = \widetilde{\mathbf{H}}_L^\top \widetilde{\mathbf{V}}_{L+1} = \widetilde{\mathbf{H}}_L^\top \widetilde{\mathbf{H}}_{L+1}^\top \ldots \widetilde{\mathbf{H}}_{J-1}^\top; \tag{3.2}$$

then $\boldsymbol{\varepsilon}_L = \widetilde{\mathbf{V}}_L \boldsymbol{\varepsilon}_J$. The reconstruction from $\boldsymbol{\varepsilon}_L$ with all detail offsets set to zero is given by $\mathbf{V}_L \boldsymbol{\varepsilon}_L$, also defined by a recursion,

$$\mathbf{V}_L = \mathbf{V}_{L+1} \mathbf{H}_L = \mathbf{H}_{J-1} \mathbf{H}_{J-2} \ldots \mathbf{H}_{L+1} \mathbf{H}_L. \tag{3.3}$$

Since all operations are linear, the projection $\widetilde{\mathbf{P}}_L = \mathbf{V}_L \widetilde{\mathbf{V}}_L$, applied to $Y(x)$, provides an unbiased estimator of $\widetilde{\mathbf{P}}_L \Phi_J(x) \boldsymbol{s}_J$, i.e., $E(\widetilde{\mathbf{P}}_L \boldsymbol{\varepsilon}_J) = \mathbf{0}_J$. The only conribution to the bias through the matrices $\mathbf{V}_L$ and $\widetilde{\mathbf{V}}_L$ lies in the approximation error of $\widetilde{\mathbf{P}}_L \Phi_J(x) \boldsymbol{s}_J$. The impact of the noise $\boldsymbol{\varepsilon}_J$ is a mere variance propagation.

In order to design wavelet transforms with good variance propagation properties, we first note that from the perfect reconstruction condition (2.60) and the definitions (3.2) and (3.3) it follows that $\widetilde{\mathbf{V}}_L \mathbf{V}_L = \mathbf{I}_L$, with $\mathbf{I}_L$ the identity matrix of size n_L. By multiplication on both sides of the identity $\widetilde{\mathbf{V}}_{L+1} \mathbf{V}_{L+1} = \mathbf{I}_{L+1}$ with $\widetilde{\mathbf{H}}_L^\top$, it also holds that

$$\widetilde{\mathbf{H}}_L^\top = \widetilde{\mathbf{V}}_L \mathbf{V}_{L+1}. \tag{3.4}$$

Expression (3.4) can be used to construct the scale by scale realisation of a multiresolution decomposition whose projection has a desired behaviour in terms of variance propagation.

The discussion of wavelet transforms as projections in the subsequent sections will rest on the following well-known results from linear algebra.

Lemma 3.1.5 *Any projection matrix $\widetilde{\mathbf{P}}$, i.e., any idempotent matrix, has the following properties.*

1. *$\widetilde{\mathbf{P}}$ has eigenvalues 0 and/or 1. The number of linearly independent eigenvectors equals the dimension of the matrix.*
2. *There exists an orthogonal $\mathbf{Q}$ so that $\widetilde{\mathbf{P}} = \mathbf{Q} \begin{bmatrix} \mathbf{I} & \mathbf{0} \\ \mathbf{A} & \mathbf{0} \end{bmatrix} \mathbf{Q}^\top$.*
3. *Singular values of $\widetilde{\mathbf{P}}$ are either 0 or larger than 1.*

Proof.

1. Let $\widetilde{\mathbf{P}} x = \lambda x$ with x nontrivial; then $\lambda^2 x = \widetilde{\mathbf{P}} \widetilde{\mathbf{P}} x = \widetilde{\mathbf{P}} x = \lambda x$; hence $\lambda^2 = \lambda$.

The number of linearly independent eigenvectors with eigenvalue 0 equals the dimension of the null space of $\widetilde{\mathbf{P}}$. The number of linearly independent eigenvectors with eigenvalue 1 is given by the dimension of the null space of $\mathbf{I} - \widetilde{\mathbf{P}}$, which is also a projection, complementary to $\widetilde{\mathbf{P}}$.

2. Let $\mathbf{E} = \mathbf{QL}$ be a variant of the QR decomposition with a lower triangular matrix $\mathbf{L}$ and an orthogonal matrix $\mathbf{Q}$. Applied within the eigenvalue decomposition of $\widetilde{\mathbf{P}}$, this yields

$$\begin{aligned}\widetilde{\mathbf{P}} &= \mathbf{E}\begin{bmatrix} \mathbf{I} & \mathbf{0} \\ \mathbf{0} & \mathbf{0} \end{bmatrix}\mathbf{E}^{-1} = \mathbf{QL}\begin{bmatrix} \mathbf{I} & \mathbf{0} \\ \mathbf{0} & \mathbf{0} \end{bmatrix}\mathbf{L}^{-1}\mathbf{Q}^\top \\ &= \mathbf{Q}\begin{bmatrix} \mathbf{I} & \mathbf{0} \\ \mathbf{L}_{21}(\mathbf{L}^{-1})_{11} & \mathbf{0} \end{bmatrix}\mathbf{Q}^\top,\end{aligned}$$

where $\mathbf{L}_{21}$ is a submatrix of $\mathbf{L} = \begin{bmatrix} \mathbf{L}_{11} & \mathbf{0} \\ \mathbf{L}_{21} & \mathbf{L}_{22} \end{bmatrix}$, while in a similar way, $(\mathbf{L}^{-1})_{11}$ is a submatrix of $\mathbf{L}^{-1}$.

3. Singular values of $\widetilde{\mathbf{P}}$ are eigenvalues of

$$\widetilde{\mathbf{P}}^\top\widetilde{\mathbf{P}} = \mathbf{Q}\begin{bmatrix} \mathbf{I} + \mathbf{A}^\top\mathbf{A} & \mathbf{0} \\ \mathbf{0} & \mathbf{0} \end{bmatrix}\mathbf{Q}^\top,$$

which can be found[6] to be either 0 or larger than 1.

□

Section 3.1.4 sets a benchmark for the assessment of the variance propagation throughout a wavelet analysis and reconstruction. Then, in Section 3.1.5, the final update is used to approximate the ideal case as closely as possible under constraints of compact support and vanishing moments.

Exercise 3.1.6 *Following Exercise 3.1.4, consider a triangular hat decomposition on an equidistant grid, with a bidiagonal update matrix with* $U_{j;k,k} = u = U_{j;k,k-1}$. *From the forward transform* $s_{j,k} = (1-u)s_{j+1,2k} + us_{j+1,2k-1} + us_{j+1,2k+1} - (u/2)s_{j+1,2k-2} - (u/2)s_{j+1,2k+2}$, *and the reconstruction in (2.17) and (2.18), page 43, find the projection* $\widetilde{\mathbf{P}}_L = \mathbf{V}_L\widetilde{\mathbf{V}}_L$, *for* $L = J-1$ *and the value of* u *that minimises the variances of the odd coefficients in* $\widehat{s}_J = \widetilde{\mathbf{P}}_L s_J$.

Exercise 3.1.7 *Replace* $\mathbf{G}_j$ *by* $\mathbf{G}_j^{[1]} = \mathbf{G}_j\mathbf{A}_j$, *as in Exercise 2.2.7. Show that the choice of* $\mathbf{A}_j$ *has no impact on the multiscale variance propagation of the wavelet transform.*

---

[6]Indeed, $(\mathbf{I} + \mathbf{A}^\top\mathbf{A})e = \lambda e \Leftrightarrow \mathbf{A}^\top\mathbf{A}e = (\lambda - 1)e$. But the eigenvalues of $\mathbf{A}^\top\mathbf{A}$ are nonnegative (because the matrix is symmetric positive semi-definite), leading to the conclusion that $\lambda - 1 \geq 0 \Leftrightarrow \lambda \geq 1$.

3.1.4 Semi-orthogonal wavelets

An interesting choice for use in a multiscale analysis is the **orthogonal projection** onto the columns of the reconstruction $\mathbf{V}_L$.

Definition 3.1.8 *A projection $\widetilde{\mathbf{P}}_\perp$ is orthogonal if for any vector x, the residual $x - \widetilde{\mathbf{P}}_\perp x$ is either zero or perpendicular to the projection $\widetilde{\mathbf{P}}_\perp x$. This means that $(\mathbf{I} - \widetilde{\mathbf{P}}_\perp)^\top \widetilde{\mathbf{P}}_\perp = \mathbf{0}$.*

The matrix of an orthogonal projection is symmetric[7] and all symmetric, idempotent matrices represent orthogonal projections. Indeed, $(\mathbf{I} - \widetilde{\mathbf{P}}_\perp)^\top \widetilde{\mathbf{P}}_\perp = \mathbf{0} \Leftrightarrow \widetilde{\mathbf{P}}_\perp^\top \widetilde{\mathbf{P}}_\perp = \widetilde{\mathbf{P}}_\perp$. The left hand side of this equality is a symmetric matrix (taking the transpose leads to the same matrix), and so must be the right hand side, $\widetilde{\mathbf{P}}_\perp$. On the other hand, if an idempotent matrix is symmetric, then $\widetilde{\mathbf{P}}_\perp = \widetilde{\mathbf{P}}_\perp \widetilde{\mathbf{P}}_\perp = \widetilde{\mathbf{P}}_\perp^\top \widetilde{\mathbf{P}}_\perp$, from which the orthogonality follows.

An orthogonal projection onto the columns of the reconstruction $\mathbf{V}_L$ is obtained by taking

$$\widetilde{\mathbf{V}}_L = (\mathbf{V}_L^\top \mathbf{V}_L)^{-1} \mathbf{V}_L^\top. \tag{3.5}$$

Using (3.4) the orthogonal projection can be realised step by step, from fine to coarse scales,

$$\widetilde{\mathbf{H}}_L^\top = \widetilde{\mathbf{V}}_L \mathbf{V}_{L+1} = (\mathbf{V}_L^\top \mathbf{V}_L)^{-1} \mathbf{V}_L^\top \mathbf{V}_{L+1}. \tag{3.6}$$

As the columns of $\mathbf{V}_L$ themselves are not (necessarily) orthogonal, this wavelet transform is termed **semi-orthogonal**.

Definition 3.1.9 *A semi-orthogonal wavelet transform is a wavelet transform where the analysis $\widetilde{\mathbf{H}}_L^\top$ performs orthogonal projections onto the columns of the reconstruction matrix $\mathbf{V}_L$, as given by (3.6).*

In a wavelet analysis derived from a lifting scheme with one prediction and one update[8], the update realises a semi-orthogonal wavelet transform if, following (2.56),

$$\mathbf{U}_L = \widetilde{\mathbf{H}}_L^\top \widetilde{\mathbf{J}}_L^o = (\mathbf{V}_L^\top \mathbf{V}_L)^{-1} \mathbf{V}_L^\top \mathbf{V}_{L+1} \widetilde{\mathbf{J}}_L^o.$$

From the perfect reconstruction condition in (2.61), we find an orthogonality expression involving the synthesis matrices $\mathbf{H}_L$ and $\mathbf{G}_L$ only,

$$\begin{aligned} & \mathbf{0}_{L+1,e,o} = \widetilde{\mathbf{H}}_L^\top \mathbf{G}_L = (\mathbf{V}_L^\top \mathbf{V}_L)^{-1} \mathbf{V}_L^\top \mathbf{V}_{L+1} \mathbf{G}_L \\ \Leftrightarrow \quad & \mathbf{0}_{L+1,e,o} = \mathbf{V}_L^\top \mathbf{V}_{L+1} \mathbf{G}_L = (\mathbf{V}_{L+1} \mathbf{H}_L)^\top (\mathbf{V}_{L+1} \mathbf{G}_L). \end{aligned} \tag{3.7}$$

[7] It is *not* an orthogonal matrix, i.e., $\widetilde{\mathbf{P}}_\perp^\top \widetilde{\mathbf{P}}_\perp \neq \mathbf{I}$.

[8] In more general wavelet transforms, an update for semi-orthogonal wavelet transforms is provided by the expression (4.35).

3.1.4.1 Semi-orthogonal wavelet basis functions

The recursive definition of $\mathbf{V}_L$ in (3.3) proceeds from fine to coarse resolution levels, which is the opposite direction of subdivision. Subdivision can be taken into account by looking at the refinable basis functions. Combining the recursive definition in (3.3) with the refinement equation in (2.64) yields $\Phi_L(x) = \Phi_J(x)\mathbf{V}_L = \Phi_J(x)\mathbf{V}_{L+1}\mathbf{H}_L$. Define the fine scale Gramian matrix,

$$\mathbf{\Pi}_J = \int_{-\infty}^{\infty} \Phi_J^\top(x)\Phi_J(x)dx; \tag{3.8}$$

then refinement yields

$$\mathbf{\Pi}_L = \mathbf{H}_L^\top \mathbf{\Pi}_{L+1}\mathbf{H}_L = \mathbf{V}_L^\top \mathbf{\Pi}_J \mathbf{V}_L. \tag{3.9}$$

Furthermore, let $\mathbf{\Upsilon}_L$ be the Gramian submatrix

$$\mathbf{\Upsilon}_L = \int_{-\infty}^{\infty} \Phi_L^\top(x)\Psi_L(x)dx; \tag{3.10}$$

then this Gramian submatrix can be refined as

$$\begin{aligned}\mathbf{\Upsilon}_L &= \int_{-\infty}^{\infty} \mathbf{H}_L^\top \Phi_{L+1}^\top(x)\Phi_{L+1}(x)\mathbf{G}_L\, dx \\ &= \mathbf{H}_L^\top \mathbf{\Pi}_{L+1}\mathbf{G}_L = (\mathbf{V}_{L+1}\mathbf{H}_L)^\top \mathbf{\Pi}_J(\mathbf{V}_{L+1}\mathbf{G}_L). \end{aligned} \tag{3.11}$$

By Expression (3.7), a semi-orthogonal wavelet transform has $\mathbf{\Upsilon}_j = \mathbf{0}_{L+1,e,o}$ at all resolution levels if the finest scaling functions satisfy $\mathbf{\Pi}_J = \mathbf{I}_J$. In other words, if the finest scaling functions are orthogonal, then all detail bases in a semi-orthogonal wavelet transform are orthogonal to the scaling bases at the same level. The subspace spanned by the detail basis is then the orthogonal complement of the subspace spanned by the coarse scaling basis within the space spanned by the fine scaling basis. In applications where the finest scaling functions are free to choose, they can be taken to be characteristic functions, as proposed in Section 2.2.5. Characteristic functions on disjoint intervals are obviously orthogonal, thus satisfying $\mathbf{\Pi}_J = \mathbf{I}_J$. If $\mathbf{\Pi}_J \neq \mathbf{I}_J$; then the orthogonal projection in (3.5) is replaced by $\widetilde{\mathbf{V}}_L = (\mathbf{V}_L^\top \mathbf{\Pi}_J \mathbf{V}_L)^{-1}\mathbf{V}_L^\top \mathbf{\Pi}_J$, to construct a semi-orthogonal multiresolution analysis replacing (3.6),

$$\widetilde{\mathbf{H}}_L^\top = \widetilde{\mathbf{V}}_L \mathbf{V}_{L+1} = (\mathbf{V}_L^\top \mathbf{\Pi}_J \mathbf{V}_L)^{-1}\mathbf{V}_L^\top \mathbf{\Pi}_J \mathbf{V}_{L+1} = \mathbf{\Pi}_L^{-1}\mathbf{H}_L^\top \mathbf{\Pi}_{L+1}. \tag{3.12}$$

The last equality follows from the refinement of $\mathbf{\Pi}_L$ in (3.9). Substitution of (3.12) in the perfect reconstruction condition $\widetilde{\mathbf{H}}_L^\top \mathbf{G}_L = \mathbf{0}_{L+1,e,o}$ then leads to the conclusion that $\mathbf{\Upsilon}_L = \mathbf{0}_{L+1,e,o}$.

On an equidistant grid, the Gramian matrices $\mathbf{\Pi}_L$ and $\mathbf{\Pi}_{L+1}$, showing in Expression (3.12), depend on the normalisation of the wavelet basis only, not on the grid $\boldsymbol{x}_{L+1}$. Moreover, the normalisation in $\mathbf{\Pi}_L^{-1}$ undoes the normalisation in $\mathbf{\Pi}_{L+1}$. As a result, the elements of the dual refinement matrix in an equispaced semi-orthogonal wavelet transform are level independent, $\widetilde{\mathbf{H}}_L^\top = \widetilde{\mathbf{H}}^\top$, at least if the primal refinement matrix is taken to be the same at each level.

Defining the detail Gramian matrix

$$\begin{aligned}\boldsymbol{\Xi}_L = \int_{-\infty}^{\infty} \Psi_L^\top(x)\Psi_L(x)\,dx &= \int_{-\infty}^{\infty} \mathbf{G}_L^\top \Phi_{L+1}^\top(x)\Phi_{L+1}(x)\mathbf{G}_L\,dx \\ &= \mathbf{G}_L^\top \boldsymbol{\Pi}_{L+1}\mathbf{G}_L, \end{aligned} \tag{3.13}$$

the dual detail matrix $\widetilde{\mathbf{H}}_L^\top$ is given by

$$\widetilde{\mathbf{G}}_L^\top = \boldsymbol{\Xi}_L^{-1}\mathbf{G}_L^\top \boldsymbol{\Pi}_{L+1}. \tag{3.14}$$

This is verified by checking the two perfect reconstruction conditions involving $\widetilde{\mathbf{G}}_L^\top$, namely $\widetilde{\mathbf{G}}_L^\top \mathbf{H}_L = \boldsymbol{\Xi}_L^{-1}\mathbf{G}_L^\top \boldsymbol{\Pi}_{L+1}\mathbf{H}_L = \boldsymbol{\Xi}_L^{-1}\boldsymbol{\Upsilon}_L^\top = \boldsymbol{\Xi}_L^{-1}\mathbf{0}_{L+1,o,e}$, and $\widetilde{\mathbf{G}}_L^\top \mathbf{G}_L = \boldsymbol{\Xi}_L^{-1}\mathbf{G}_L^\top \boldsymbol{\Pi}_{L+1}\mathbf{G}_L = \boldsymbol{\Xi}_L^{-1}\boldsymbol{\Xi}_L = \mathbf{I}_L$.

A semi-orthogonal multiresolution basis has $\boldsymbol{\Upsilon}_j = \mathbf{0}_{j+1,e,o}$ at all levels, meaning that at each level the scaling and detail bases span orthogonal complements. The detail basis is then also orthogonal to any linear combination of the scaling basis. In particular, the semi-orthogonality implies that detail bases at different levels span orthogonal subspaces,

$$\boldsymbol{\Upsilon}_j = \mathbf{0}_{j+1,e,o} \Rightarrow \int_{-\infty}^{\infty} \Psi_j^\top(x)\Psi_i(x)dx = \mathbf{0}, \tag{3.15}$$

for any $i < j$. Obviously, for the case $j > i$, the orthogonality of detail functions at different scales follows from $\boldsymbol{\Upsilon}_i = \mathbf{0}_{i+1,e,o}$.

In a semi-orthogonal multiresolution decomposition, the primal wavelets inherit the dual **vanishing moments**. Indeed, let $\widetilde{\boldsymbol{x}}_j^{[q]}$ denote the coefficients in the power function expansion $x^q = \Phi_j(x)\widetilde{\boldsymbol{x}}_j^{[q]}$. Then, the semi-orthogonality $\boldsymbol{\Upsilon}_j = \mathbf{0}_{j+1,e,o}$ implies

$$\int_{-\infty}^{\infty} \Psi_j^\top(x)x^q\,dx = \int_{-\infty}^{\infty} \Psi_j^\top(x)\Phi_j(x)\widetilde{\boldsymbol{x}}_j^{[q]}\,dx = \mathbf{0}_{j'}.$$

In practice, boundary effects conflict with the unbounded support of the primal wavelet functions, leading to approximate primal vanishing moments.

3.1.4.2 Dual basis functions in a semi-orthogonal transform

Figure 3.4 depicts a dual scaling basis function in the semi-orthogonal wavelet transform for a triangular hat scaling basis. The dual basis function has **no compact support**. On the positive side, a semi-orthogonal transform shows optimal behaviour when it comes to **variance propagation**, as will be discussed in Section 3.1.5.

The construction perfectly controls the **smoothness of the dual functions**, according to the following result.

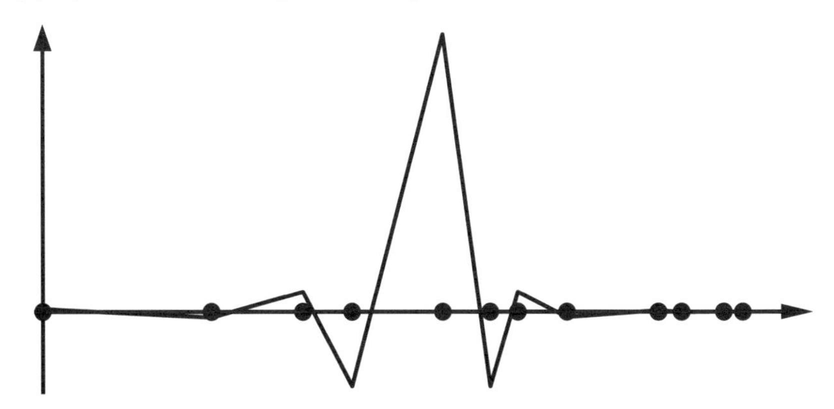

Figure 3.4
One element of the dual scaling basis in the semi-orthogonal wavelet transform operating on the triangular hat scaling basis in Figure 2.4. The semi-orthogonal dual basis has no compact support. In theory, the associated primal wavelet basis inherits the vanishing moments from the dual wavelets. The dual functions also inherit the smoothness of the primal basis. The dual scaling basis can be compared with the dual basis in Figure 2.12, which has two primal vanishing moments.

Lemma 3.1.10 *In a semi-orthogonal wavelet transform, the dual scaling functions are linear combinations of the primal scaling basis, more precisely*

$$\widetilde{\Phi}_j(x) = \Phi_j(x)\mathbf{\Pi}_j^{-1} \tag{3.16}$$

with $\mathbf{\Pi}_J$ *the Gramian matrix as defined in (3.8).*

Proof. Using (3.12), the dual multiscale refinement is given by

$$\widetilde{\Phi}_j(x) = \widetilde{\Phi}_{j+1}(x)\widetilde{\mathbf{H}}_j^\top = \widetilde{\Phi}_{j+1}(x)\mathbf{\Pi}_{j+1}\mathbf{H}_j\mathbf{\Pi}_j^{-1},$$

or, equivalently, $\widetilde{\Phi}_j(x)\mathbf{\Pi}_j = \widetilde{\Phi}_{j+1}(x)\mathbf{\Pi}_{j+1}\mathbf{H}_j$. This is nothing but the primal two-scale equation, $\Phi_j(x) = \Phi_{j+1}(x)\mathbf{H}_j$. The result then follows by identification of the two members of the equation. □

Since the dual basis is a linear combination of the primal basis, the two bases share the smoothness characteristics. The illustration in Figure 3.4 shows that both the primal and the dual bases in a semi-orthogonal triangular hat transform consist of polylines, i.e., continuous piecewise linear functions. This is in contrast to the dual basis with two primal vanishing moments, depicted in Figure 2.12. In a semi-orthogonal transform, the primal and dual bases can easily switch roles. Switching roles can be interesting in applications that run more forward transforms than syntheses. Since the dual basis has no compact support, the corresponding transformation is computationally more complex than that of the primal basis.

Because of Lemma 3.1.10, the dual Gramian matrix can be easily found as

$$\widetilde{\mathbf{\Pi}}_j = \int_{-\infty}^{\infty} \widetilde{\Phi}_j^\top(x)\widetilde{\Phi}_j(x)\,dx = \mathbf{\Pi}_j^{-1}\int_{-\infty}^{\infty} \Phi_j^\top(x)\Phi_j(x)\,dx\mathbf{\Pi}_j^{-1} = \mathbf{\Pi}_j^{-1}.$$

This result in its turn can be used in the proof of the subsequent result.

Lemma 3.1.11 *In a semi-orthogonal multiresolution scheme, the dual wavelet functions are linear combinations of the primal wavelet functions at the same scale,*

$$\widetilde{\Psi}_j(x) = \Psi_j(x)\mathbf{\Xi}_j^{-1}. \tag{3.17}$$

Proof. The proof develops the dual wavelet equation at level j, using Lemma 3.1.10 and inserting the perfect reconstruction of (2.58).

$$\begin{aligned}
\widetilde{\Psi}_j(x) &= \widetilde{\Phi}_{j+1}(x)\widetilde{\mathbf{G}}_j = \Phi_{j+1}(x)\mathbf{\Pi}_{j+1}^{-1}\widetilde{\mathbf{G}}_j \\
&= \Phi_{j+1}(x)(\mathbf{H}_j\widetilde{\mathbf{H}}_j^\top + \mathbf{G}_j\widetilde{\mathbf{G}}_j^\top)\widetilde{\mathbf{\Pi}}_{j+1}^{-1}\widetilde{\mathbf{G}}_j \\
&= \Phi_{j+1}(x)\mathbf{H}_j \cdot \widetilde{\mathbf{H}}_j^\top\widetilde{\mathbf{\Pi}}_{j+1}^{-1}\widetilde{\mathbf{G}}_j + \Phi_{j+1}(x)\mathbf{G}_j \cdot \widetilde{\mathbf{G}}_j^\top\widetilde{\mathbf{\Pi}}_{j+1}^{-1}\widetilde{\mathbf{G}}_j \\
&= \Phi_j(x)\widetilde{\mathbf{\Upsilon}}_j + \Psi_j(x)\widetilde{\mathbf{\Xi}}_j = \mathbf{0}_{j'} + \Psi_j(x)\mathbf{\Xi}_j^{-1}.
\end{aligned}$$

The last line uses the fact that $\widetilde{\mathbf{\Upsilon}}_j = \mathbf{0}_{j+1,e,o}$, which holds in a semi-orthogonal multiresolution analysis because

$$\widetilde{\mathbf{\Upsilon}}_j = \widetilde{\mathbf{H}}_j^\top\widetilde{\mathbf{\Pi}}_{j+1}\widetilde{\mathbf{G}}_j = \mathbf{\Pi}_j^{-1}\mathbf{H}_j^\top\mathbf{\Pi}_{j+1}\widetilde{\mathbf{\Pi}}_{j+1}\mathbf{\Pi}_{j+1}\mathbf{G}_j\mathbf{\Xi}_j^{-1} = \mathbf{\Pi}_j^{-1}\mathbf{\Upsilon}_j\mathbf{\Xi}_j^{-1},$$

where $\mathbf{\Upsilon}_j = \mathbf{0}_{L+1,e,o}$ in a semi-orthogonal setting. □

Exercise 3.1.12 *Construct an alternative proof, working from $\Psi_j(x)$ towards $\widetilde{\Psi}_j(x)\mathbf{\Xi}_j$.*

As a result of Lemma 3.1.11, the set of detail coefficients at a given level in a semi-orthogonal multiresolution scheme, given by $\boldsymbol{d}_j = \int_{-\infty}^{\infty}\widetilde{\Psi}_j^\top(x)f_J(x)\,dx$, can be interpreted as an orthogonal projection onto the detail basis at that resolution level.

3.1.5 Variance control in the final update step

In order to compare semi-orthogonal and other wavelet transforms, we first formulate a few results on orthogonal projections.

Lemma 3.1.13 *1. The orthogonal projection matrix*[9]

$$\widetilde{\mathbf{P}}_{L\perp} = \mathbf{V}_L(\mathbf{V}_L^\top\mathbf{V}_L)^{-1}\mathbf{V}_L^\top$$

is symmetric and has no singular values but zeros or ones.

[9] also known as the Hat matrix, e.g. Ruppert et al. (2003, p.21).

2. When the input noise is uncorrelated and homoscedastic, then, among all projections on the columns of a given matrix $\mathbf{V}_L$, the orthogonal projection minimises the variance of any component in the reconstruction[10].

Proof. The covariance matrix after orthogonal projection is given by $\boldsymbol{\Sigma}_{\widetilde{\mathbf{P}}_L\boldsymbol{\varepsilon}} = \sigma^2\widetilde{\mathbf{P}}_L\widetilde{\mathbf{P}}_L^\top$. For any other projection $\widetilde{\mathbf{P}}_L$, we have $\widetilde{\mathbf{P}}_L\widetilde{\mathbf{P}}_{L\perp} = \widetilde{\mathbf{P}}_{L\perp}$. As the orthogonal projection is symmetric and idempotent, $\widetilde{\mathbf{P}}_L\widetilde{\mathbf{P}}_L^\top - \widetilde{\mathbf{P}}_{L\perp}\widetilde{\mathbf{P}}_{L\perp}^\top = \widetilde{\mathbf{P}}_L(I - \widetilde{\mathbf{P}}_{L\perp})\widetilde{\mathbf{P}}_L^\top$. The matrix $I - \widetilde{\mathbf{P}}_{L\perp}$ is positive semi-definite; hence all diagonal elements of $\widetilde{\mathbf{P}}_L\widetilde{\mathbf{P}}_L^\top - \widetilde{\mathbf{P}}_{L\perp}\widetilde{\mathbf{P}}_{L\perp}^\top$ must be non-negative, from which it follows that for all components i, $\mathrm{var}((\widetilde{\mathbf{P}}_L\boldsymbol{\varepsilon})_i) \geq \mathrm{var}((\widetilde{\mathbf{P}}_{L\perp}\boldsymbol{\varepsilon})_i)$. □

Unfortunately, in most cases the forward transform $\widetilde{\mathbf{V}}_L = (\mathbf{V}_L^\top\mathbf{V}_L)^{-1}\mathbf{V}_L^\top$ in the semi-orthogonal decomposition is a full matrix. The dual scaling functions thus have no compact support, making the coefficients to carry more diffuse, less localised, spatial information[11]. Without compact support, the computational complexity of the analysis increases from linear to that of a full matrix-vector multiplication, which is quadratic.

Therefore, a final update matrix with a limited number of nonzero elements near the diagonal can be designed to approach the optimal variance propagation property of the orthogonal projection, and at the same time provide some vanishing moments.

The variance propagation of a reconstruction from a wavelet analysis is described by the matrix $\widetilde{\mathbf{P}}_L\widetilde{\mathbf{P}}_L^\top$, and so by its eigenvalue decomposition. The eigenvalues of $\widetilde{\mathbf{P}}_L\widetilde{\mathbf{P}}_L^\top$ are the singular values of $\widetilde{\mathbf{P}}_L$, thus motivating the following formal definition of the variance propagation.

Definition 3.1.14 *(multiscale variance propagation)*[12] *Given a wavelet transform defined by the sequences of forward and inverse refinement matrices $\mathbf{H}_j$ and $\widetilde{\mathbf{H}}_j$, where $j = L, \ldots, J-1$, let $\mathbf{V}_L$ and $\widetilde{\mathbf{V}}_L$ be defined as in (3.2) and (3.3), and let $\widetilde{\mathbf{P}}_L = \mathbf{V}_L\widetilde{\mathbf{V}}_L$. Then the multiscale variance propagation equals*

$$\kappa_F(\mathbf{V}_L, \widetilde{\mathbf{V}}_L) = \|\boldsymbol{\sigma}(\widetilde{\mathbf{P}}_L)\|_2/\sqrt{n_L} = \|\widetilde{\mathbf{P}}_L\|_F/\sqrt{n_L}, \tag{3.18}$$

where $\boldsymbol{\sigma}(\widetilde{\mathbf{P}}_L)$ is the vector of singular values of $\widetilde{\mathbf{P}}_L$, while n_L is the number of columns in $\mathbf{V}_L$, which is also the number of nonzero singular values of

[10] The orthogonal projection onto the scaling basis *functions*, which involves the Gramian matrix $\boldsymbol{\Pi}_J$, is slightly suboptimal from the purely statistical perspective of variance control. In practice, it does not seem to make much of a difference.

[11] See also Section 2.1.7.

[12] This notion is related to the numerical condition, defined in Section 5.4.6. The classical definition of numerical condition is, however, insufficient as a measure for variance inflation. Well conditioned transforms may still inflate the variance. The problem is that the notion of numerical condition cannot describe the effect of a projection within a forward-inverse transform scheme, as numerical condition assumes the operation to be invertible. Non-invertible operations have infinite condition, by definition.

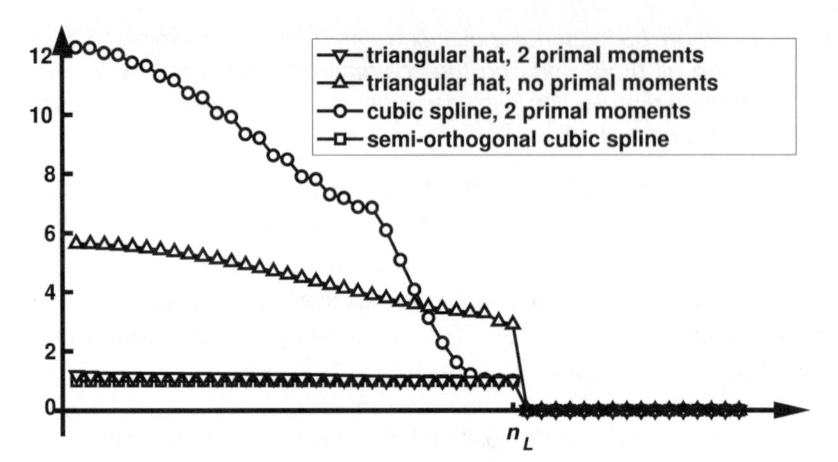

Figure 3.5
Singular values in descending order for projections $\widetilde{\mathbf{P}}_L = \mathbf{V}_L\widetilde{\mathbf{V}}_L$ corresponding to the reconstructions on equispaced point sets in Figure 3.3. The semi-orthogonal cubic B-spline projection has optimal variance reduction, because its nonzero singular values all have the minimal value, one. Although variance propagation was no issue in the design of the triangular hat projection with two primal vanishing moments, its nonzero singular values are close to one, explaining the excellent variance reduction in Figure 3.3(a). The triangular hat without update has large nonzero singular values, and this holds even more for the cubic B-spline with an update for two vanishing moments.

$\widetilde{\mathbf{P}}_L$. *The notation* $\|\widetilde{\mathbf{P}}_L\|_F$ *stands for the Frobenius norm*[13] *of* $\widetilde{\mathbf{P}}_L$, *whereas* $\|\boldsymbol{\sigma}(\widetilde{\mathbf{P}}_L)\|_2 = \sqrt{\boldsymbol{\sigma}(\widetilde{\mathbf{P}}_L)^\top\boldsymbol{\sigma}(\widetilde{\mathbf{P}}_L)}$ *is the classical Euclidean vector norm.*

Figure 3.5 has plots of the singular values of the projections used in the reconstructions of Figure 3.3. It illustrates the use of Definition 3.1.14 in quantifying the variance propagation throughout a wavelet transform. All projections have n_L nonzero singular values. The projections with small nonzero singular values are those reducing the variance in Figures 3.3 (a) and (d), while projections with large nonzero singular values tend to blow up the variance, as in Figures 3.3 (b) and (c).

As an alternative to (3.18), the multiscale variance propagation can also

[13]The Frobenius norm of a real $n \times m$ matrix $\mathbf{A}$ is defined as $\|\mathbf{A}\|_F^2 = \sum_{i=1}^{n}\sum_{j=1}^{m} a_{ij}^2 = \mathrm{tr}(\mathbf{A}^\top\mathbf{A})$ Let $\mathbf{B} = \mathbf{AQ}$, with $\mathbf{Q}$ an orthogonal transform; then $\mathrm{tr}(\mathbf{B}^\top\mathbf{B}) = \mathrm{tr}(\mathbf{Q}^\top\mathbf{A}^\top\mathbf{A}\mathbf{Q}) = \mathrm{tr}(\mathbf{A}^\top\mathbf{A}\mathbf{Q}\mathbf{Q}^\top) = \mathrm{tr}(\mathbf{A}^\top\mathbf{A})$, where we use the fact that a trace is invariant under cyclic permutation of the factors, i.e., $\mathrm{tr}(\mathbf{KL}) = \mathrm{tr}(\mathbf{LK})$, which is easy to check by writing out the elements of the matrix product. Let $\mathbf{A} = \mathbf{USV}^\top$ be the singular value decomposition of $\mathbf{A}$, where $\mathbf{S}$ is a diagonal matrix and $\mathbf{U}$ and $\mathbf{V}$ are orthogonal; then it follows that $\|\mathbf{A}\|_F = \|\mathbf{S}\|_F = \|\boldsymbol{\sigma}(\mathbf{A})\|_2$.

be defined as

$$\kappa_2(\mathbf{V}_L, \widetilde{\mathbf{V}}_L) = \max(\boldsymbol{\sigma}(\widetilde{\mathbf{P}}_L)) = \|\widetilde{\mathbf{P}}_L\|_2, \tag{3.19}$$

where $\|\widetilde{\mathbf{P}}_L\|_2$ denotes the matrix norm induced from the Euclidean vector norm, which is equal to the largest singular value of the matrix.

Since the singular values of $\widetilde{\mathbf{P}}_L$ are either zero or greater than one, we have that $\kappa_F(\mathbf{V}_L, \widetilde{\mathbf{V}}_L) \geq 1$, and $\kappa_F(\mathbf{V}_L, \widetilde{\mathbf{V}}_L) = 1$ if and only if $\widetilde{\mathbf{P}}_L$ is an orthogonal projection. Let $\mathbf{U}_L$ be the final update matrix at resolution level L, which we impose to have a limited number of nonzero elements only. Let $U_{L;r,s}$ be one of these elements; then putting

$$\begin{aligned} 0 &= \frac{\partial \kappa_F(\mathbf{V}_L, \widetilde{\mathbf{V}}_L)}{\partial U_{L;r,s}} = \frac{\partial \|\mathbf{V}_L \widetilde{\mathbf{V}}_L\|_F^2}{\partial U_{L;r,s}} = \frac{\partial \sum_{i=1}^n \sum_{j=1}^n (\mathbf{V}_L \widetilde{\mathbf{V}}_L)_{i,j}^2}{\partial U_{L;r,s}} \\ &= \sum_{i=1}^n \sum_{j=1}^n 2(\mathbf{V}_L \widetilde{\mathbf{V}}_L)_{i,j} \frac{\partial (\mathbf{V}_L \widetilde{\mathbf{V}}_L)_{i,j}}{\partial U_{L;r,s}}, \end{aligned}$$

subject to the constraint of primal vanishing moments as in (2.34) or (2.41), leads to a set of equations for the nonzero elements in $\mathbf{U}_L$ (Jansen, 2016). In particular, the partial derivative is further developed as

$$\frac{\partial (\mathbf{V}_L \widetilde{\mathbf{V}}_L)}{\partial U_{L;r,s}} = \mathbf{V}_L \frac{\partial \widetilde{\mathbf{V}}_L}{\partial U_{L;r,s}},$$

reflecting the fact that the subdivision from coarse scaling coefficients, $\mathbf{V}_L$, does not depend on the update $\mathbf{U}_L$. In other words, the bases of scaling functions $\Phi_j(x)$ at subsequent resolution levels j do not depend on the final update. As a result, the approach in this section cannot possibly lead to fully orthogonal bases. Orthogonal scaling bases and orthogonal wavelet transforms are the topic of Section 3.1.6.

3.1.6 Orthogonal wavelets

Let $\Phi_j(x) = \Phi_{j+1}(x)\mathbf{H}_j$ be a refinement equation as in (2.64), and define a transformed multiscale basis $\Phi_{A,j}(x) = \Phi_j(x)\mathbf{A}_j$ with at each level j an invertible $n_j \times n_j$ matrix $\mathbf{A}_j$. Then it is straightforward to check that the transformed basis satisfies the refinement $\Phi_{A,j}(x) = \Phi_{A,j+1}(x)\mathbf{H}_{A,j}$, where $\mathbf{H}_{A,j} = \mathbf{A}_{j+1}^{-1}\mathbf{H}_j\mathbf{A}_j$. If the bases $\Phi_j(x)$ have compact support and if the transform matrices $\mathbf{A}_j$ are band limited, then the transformed bases will have compact support too. But even if the refinement matrix $\mathbf{H}_j$ has a band structure, the transformed refinement $\mathbf{H}_{A,j}$ is expected to be a full matrix, because the inverse $\mathbf{A}_{j+1}^{-1}$ is a full matrix, and in general the band structure is not restored by multiplication with $\mathbf{H}_j\mathbf{A}_j$.

In cases where $\mathbf{A}_j$ is a full matrix, both the refinement and the refinable basis can be expected to be nonzero all over as well. This holds in particular when $\mathbf{A}_j$ represents an orthogonalisation of the scaling basis, which is when

$\mathbf{A}_j^\top \mathbf{A}_j = \mathbf{\Pi}_j^{-1}$, where $\mathbf{\Pi}_j$ is the Gramian matrix from (3.8). The orthogonalised refinable basis $\Phi_{A,j}(x)$ can be combined with orthogonal projections in the forward transform. Since the Gramian matrix of the orthogonalised basis is the identity matrix, application of (3.12) leads to $\widetilde{\mathbf{H}}_j^\top = \mathbf{H}_j$, while by Lemma 3.1.10, it holds that $\widetilde{\Phi}_j(x) = \Phi_j(x)$. The wavelet transform is then **fully orthogonal**.

In a fully orthogonal wavelet transform, all dual filters and bases equal their primal counterparts.

It should be noted that orthogonalisation of the scaling basis leaves other options for the transform than the fully orthogonal one. For instance, it is possible to combine a full refinement matrix $\mathbf{H}_j$ in an orthogonal basis of scaling functions with band limited dual refinements $\widetilde{\mathbf{H}}_j$. The transform as a whole is then not orthogonal, while the scaling functions within each level are orthogonal.

In terms of variance propagation, fully orthogonal wavelet transforms add little benefit to semi-orthogonal transforms. One minor advantage is that an orthogonal transform of homoscedastic, uncorrelated noise is again homoscedastic and uncorrelated. When all wavelet coefficients have the same variance, a single threshold on the magnitudes of the coefficients can be used to select the significant coefficients. Fine-tuning the single threshold can be based on all coefficients, which may help in making a well balanced decision.

On the downside, orthogonalised scaling bases combined with orthogonal projections in the forward transform lead to wavelet and scaling basis functions with infinite support. Both the forward and the inverse transforms operate with full (non-structured) matrices, leading to quadratic instead of linear computational complexity.

Examples of orthogonal wavelet transforms include the Battle-Lemarié wavelets, which come from orthogonalised spline basis functions (Battle, 1987; Lemarié, 1988), see also Strang and Nguyen (1996, p. 254) and Daubechies (1992, p. 146).

On equispaced grids, it is possible to construct orthogonal wavelet transforms with compactly supported basis functions and fast wavelet transforms. This is the topic of Section 7.1.

3.1.7 The role of the prediction in a two-step lifting scheme

In a scheme with two steps, all properties not provided by the update step should come from the prediction step. In particular, the two-scale equations (2.28) and (2.39) depend on the prediction. The prediction thus plays a central role in refinement, which defines the scaling basis $\Phi_j(x)$ and its continuity or **smoothness**.

Just like for the update, the effect of the prediction step can be studied on the coefficient level, rather than on the basis function level. A common way to design the prediction is to impose that if $\Phi_{j+1}(x)\boldsymbol{s}_{j+1}$ is a polynomial of degree $\widetilde{p}-1$, then all wavelet coefficients at level j should be zero, i.e., $\boldsymbol{d}_j = \boldsymbol{0}_{j'}$. As a result, functions that are well approximated by polynomials will have small wavelet coefficients. In statistical applications, good **approximation** means small **bias** after reconstruction.

Based on an interpretation explained in Section 2.3 The parameter $\widetilde{p}$ is termed the number of **dual vanishing moments**. In association with this term, a prediction step is also known as **dual lifting**, while an update step is called a **primal lifting** step, even if it does not create any primal vanishing moment.

Just as at least one primal vanishing moment is necessary in practice, the first dual primal vanishing moment is indispensable. A refinement should therefore reproduce constant vectors. Refinement schemes preserving constants are termed **affine**. Non-affine refinement is of little use in practice. Even more, a refinement scheme involving a grid dependent prediction[14] should have at least two dual vanishing moments, in order to reproduce not just constants but also the identical function $f(x) = x$. Otherwise an approximation of any other smooth function $f(x)$ in the wavelet basis will reflect the irregularity of the knot spacing, doing no better than the grid independent refinement of the Haar decomposition.

As a summary, the prediction in a two-step scheme has the following effects.

1. In terms of the wavelet **basis**, it controls the **smoothness** of the scaling and wavelet functions through the subdivision or refinement scheme.
2. In terms of the wavelet **coefficients**, it controls the **sparsity** of the coefficient vectors through the **dual vanishing moments**. The sparsity in its turn determines the **error rate** of a linear or nonlinear **approximation** in the wavelet basis. From there, the prediction has an important effect on the **bias** in the reconstruction from estimated wavelet coefficients.

In a lifting scheme with more than one final update, the origin of the smoothness and the sparsity is more diffuse among all lifting steps. In Section 4.2, this will be illustrated for B-spline wavelets. In a B-spline lifting scheme, there is not a single prediction or update that *adds* dual vanishing moments or smoothness to the bias. Instead, all steps are designed to *preserve* the spline properties throughout the decomposition.

[14] The Haar prediction does not depend on the grid. It is an acceptable exception to the rule.

3.2 Interpolating scaling bases

Sections 2.1.2 and 2.1.5 have introduced two examples of a wavelet scheme with one interpolating prediction step. The constant extrapolation in the Haar transform has led to piecewise constant scaling functions, while the linear interpolation has led to piecewise linear scaling functions. As mentioned in Section 2.2.5, subdivision with higher order prediction no longer leads to piecewise polynomial scaling functions. At this point, one has to choose either for the closed formula of a piecewise polynomial scaling function, or for the straightforward implementation of the lifting scheme with a single polynomial interpolating prediction. A prototypical basis of piecewise polynomial scaling functions will be developed in Section 4.2. These will be the **B-spline** scaling functions. First, in Section 3.2.1, we explain why a prediction with a higher order interpolating polynomial does not lead to a piecewise polynomial basis function.

3.2.1 Lagrange interpolating prediction, Deslauriers-Dubuc

Figure 3.6 illustrates the first two steps of the refinement process, $\boldsymbol{s}_{j+1,e} = \boldsymbol{s}_j$ and $\boldsymbol{s}_{j+1,o} = \mathbf{P}_j \boldsymbol{s}_{j+1,e}$, where $j = L, L+1$. As always, refinement boils down to the inverse wavelet transform (2.25) and (2.26) with all details $\boldsymbol{d}_j$ equal to zero. In the top figure, the initial coarse vector $\boldsymbol{s}_L$ is the kth canonical vector $\boldsymbol{e}_{L,k}$, i.e., a Kronecker delta, or, equivalently, the kth column of the $n_L \times n_L$ identity matrix. As a result, the refined vector $\boldsymbol{s}_{L+1}$ contains the kth column of $\mathbf{P}_L$.

In the Deslauriers-Dubuc scheme (Deslauriers and Dubuc, 1987; Donoho and Yu, 1999), the prediction $\mathbf{P}_j$ is constructed by polynomial interpolation at the nearest coarse scale knots. The formalisation of this idea writes the prediction $\mathbf{P}_j$ as a matrix of evaluations of n_j elementary refinement functions $\mathrm{P}_{j,l}(x; \boldsymbol{x}_j)$ at the $n'_j = n_{j+1} - n_j$ refinement points $x_{j+1,2k+1}$,

$$P_{j;k,l} = \mathrm{P}_{j,l}(x_{j+1,2k+1}; \boldsymbol{x}_j). \tag{3.20}$$

The elementary prediction functions $\mathrm{P}_{j,l}(x; \boldsymbol{x}_j)$ are interpolating piecewise polynomials of degree $\widetilde{p} - 1$ with kinks (discontinuous derivatives, that is) at the knots. Each polynomial segment is either zero, or a $\widetilde{p}-1$ degree Lagrange basis polynomial. The full expression is given by

$$\mathrm{P}_{j,l}(x; \boldsymbol{x}_j) = \sum_{m=0}^{n_j-1} \chi_{[x_{j,m}, x_{j,m+1}]}(x) \prod_{r \in \partial m \setminus \{l\}} \frac{x - x_{j,r}}{x_{j,l} - x_{j,r}}, \tag{3.21}$$

where $\partial m = \{m - \lfloor (\widetilde{p}-1)/2 \rfloor, m - \lfloor (\widetilde{p}-1)/2 \rfloor + 1, \ldots, m + \lceil (\widetilde{p}-1)/2 \rceil\}$ are the $\widetilde{p}$ knots used in the interpolating polynomial for $x \in [x_{j,m}, x_{j,m+1}]$.

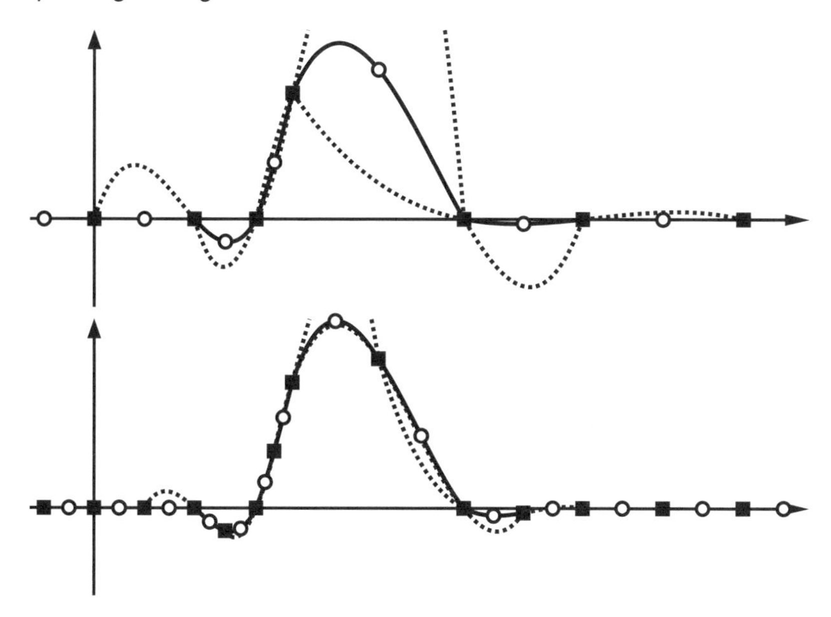

Figure 3.6
Two consecutive steps of the cubic Deslauriers-Dubuc refinement scheme. The coarse scale knots are displayed as squares, the refinement knots are displayed as bullets. In each step, the solid line represents the refinement function of (3.22). The prediction function consists of segments from polynomials, whose other segments are depicted as dotted lines.

In Figure 3.6(top), one of the functions $\mathrm{P}_{j,l}(x;\boldsymbol{x}_j)$ is depicted as a solid line. It consists of segments of interpolating polynomials. The segments of the interpolating polynomials that are not part of $\mathrm{P}_{j,l}(x;\boldsymbol{x}_j)$ are depicted as dotted lines.

From the construction of the prediction matrix, it follows that the refinement can also be described as the evaluation of an interpolating prediction function at the refinement knots, $s_{j+1,2k+1} = s_{j+1}(x_{j+1,2k+1})$ where $s_{j+1}(x)$ is defined as

$$s_{j+1}(x) = \sum_{l=0}^{n_j-1} \mathrm{P}_{j,l}(x;\boldsymbol{x}_j)s_{j,l}. \tag{3.22}$$

The prediction function satisfies the interpolation property $s_{j+1}(x_{j,l}) = s_{j,l}$.

In the example of cubic polynomial interpolation, the value of $s_{j+1,2k+1}$ is found by evaluation at the refinement point $x_{j+1,2k+1}$ of the cubic polynomial interpolating the even indexed points $(x_{j+1,2k-2}, x_{j+1,2k}, x_{j+1,2k+2}, x_{j+1,2k+4})$ with function values $(s_{j+1,2k-2}, s_{j+1,2k}, s_{j+1,2k+2}, s_{j+1,2k+4})$. The refinement thus proceeds by interpolation with two coarse scale knots on the left and two on the right. Near the end knots, there may not be enough coarse scale

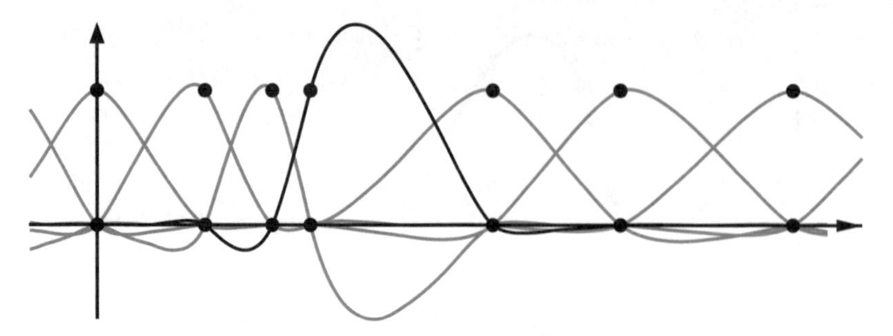

Figure 3.7
Deslauriers-Dubuc scaling basis for cubic polynomial interpolating prediction (i.e., $\widetilde{p} = 4$) showing 7 knots. The basis function in the black line follows from the subdivision process in Figure 3.6.

neighbours on each side, so the prediction may use three or even four neighbours on a single side.

Unlike the linear prediction scheme in Figure 2.2, the cubic prediction scheme never recycles the same polynomial for further refinement at a next, finer scale. As a result, the limiting curve cannot possibly be a piecewise polynomial.

3.2.2 Interpolating scaling functions

Figure 3.7 depicts all scaling functions at level j and on an internal subinterval where no boundary effects are visible. The basis functions are generated by refinement or subdivision up to an arbitrarily fine superresolution, using interpolating prediction as refinement operation. At the superresolution the basis functions $\varphi_{J^*,k}(x)$ are taken to be triangular hat functions, introduced in Section 2.1.5. These functions are interpolating, meaning that $\varphi_{J^*,k}(x_{J^*,m}) = \delta_{km}$. Then, by construction, the basis functions $\varphi_{j,k}(x)$ are polylines, but with the superresolution arbitrarily fine, this is hardly or not at all visible in Figure 3.7. Moreover, the basis functions at all scales satisfy the **interpolation** property

$$\varphi_{j,k}(x_{j,m}) = \delta_{km}, \tag{3.23}$$

where $\delta_{k\ell}$ is the Kronecker delta. As a consequence, the function

$$f_j(x) = \sum_{k=0}^{n_j-1} f(x_{j,k})\varphi_{j,k}(x) \tag{3.24}$$

interpolates the function $f(x)$ in $\boldsymbol{x}_j$, i.e., $f_j(x_{j,k}) = f(x_{j,k})$. As discussed later, in Section 8.4.2, the function $f_j(x)$ has good approximation properties with respect to $f(x)$. Therefore the interpolating polynomial prediction is not just an intuitive way to construct a multiscale transform; it can also take the observations as initial fine scale coefficients, i.e., $s_{J,k} = f(x_{J,k})$ is a legitimate choice to initialise the forward Deslauriers-Dubuc wavelet transform.

Another motivation for the use of function values as fine scaling coefficients in interpolating scaling bases follows from the next result.

Theorem 3.2.1 *Let $\Phi_j(x)$ be a scaling basis with $\widetilde{p}$ dual vanishing moments; then for $q < \widetilde{p}$, it holds that*

$$x^q = \sum_{k=0}^{n_j-1} x_{j,k}^q \varphi_{j,k}(x). \tag{3.25}$$

Proof. This result follows immediately from the fact that the decomposition $x^q = \Phi_j(x)\boldsymbol{s}_j$ exists, followed by an evaluation of that decomposition in the knots, leading to the conclusion that $s_{j,k} = x_{j,k}^q$. □

As a conclusion from Theorem 3.2.1, we find that polynomials up to degree $\widetilde{p} - 1$ are reconstructed exactly from their function values, plugged in as scaling coefficients in an interpolating scaling basis. It can be anticipated that functions that are well approximated by polynomials are also well approximated by a reconstruction with their function values as coefficients in such a basis. This is confirmed by the analysis in Section 8.4.2.

The B-spline wavelets, discussed in Section 4.2, do not satisfy any interpolation property. Finding appropriate finest scaling coefficients $s_{J,k}$ from the observations $f(x_{J,k})$ is then a nontrivial task. Simply taking the observations as input is only possible in exceptional cases, to be discussed in Section 8.4.

The relationship between interpolating prediction and interpolating scaling bases also holds in the opposite direction, as is pointed out by the following result.

Theorem 3.2.2 *Interpolating scaling bases are refined with a single prediction lifting step. In other words, if a scaling basis $\Phi_j(x)$ satisfies the interpolating property (3.23) at all levels j, then there exists a single prediction matrix $\mathbf{P}_j$ so that the refinement equation (2.64) takes the form (2.28).*

Proof. With $h_{j,k,l}$ the element on row k, column l of $\mathbf{H}_j$, the two-scale equation can be written as

$$\varphi_{j,l}(x) = \sum_{k=0}^{n_{j+1}-1} h_{j,k,l}\varphi_{j+1,k}(x).$$

From the interpolation property (3.23), we find for $m \neq l$ that

$$\begin{aligned}\varphi_{j,l}(x_{j,m}) = 0 &= \sum_{k=0}^{n_{j+1}-1} h_{j,k,l}\varphi_{j+1,k}(x_{j,m}) = \sum_{k=0}^{n_{j+1}-1} h_{j,k,l}\varphi_{j+1,k}(x_{j+1,2m}) \\ &= h_{j,2m,l} \cdot 1 + 0 = h_{j,2m,l}.\end{aligned}$$

Likewise, for $m = l$, we have $h_{j,2l,l} = 1$. We can write $\Phi_j(x) = \Phi_{j+1,e}(x)\mathbf{H}_{j,e} + \Phi_{j+1,o}(x)\mathbf{H}_{j,o}$, where $\mathbf{H}_{j,e} = \mathbf{I}_{n_j}$. Identification of $\mathbf{H}_{j,o} = \mathbf{P}_j$ completes the proof. □

The interpolating property and the intuitive prediction operation provide good arguments in favour of the Deslauriers-Dubuc wavelets. On the downside is the lack of a closed formula for the scaling functions. Such a formula exists for B-splines, which are piecewise polynomial functions, providing not just an easy evaluation but also full information about the smoothness of the function. The smoothness of the Deslauriers-Dubuc basis functions has to be analysed through the subdivision scheme, which is a difficult task, and for which only partial results are known (Daubechies et al., 1999). One of the problems is that the basis functions in $\Phi_j(x)$ depend not only on the values in $\boldsymbol{x}_j$, but also on all further refinement steps and on the order of refinement. From Figure 3.7, it is can be seen that on irregular knots, the basis functions may have heavy side lobes and as such they are unbounded. This is in contrast to B-spline functions, which are bounded between 0 and 1. In general, unbounded scaling functions are more difficult for application in the construction of wavelet basis functions.

Exercise 3.2.3 *Let $\Phi_{j+1}(x)$ be an interpolating scaling basis and $f_{j+1}(x) = \Phi_{j+1}\boldsymbol{f}_{j+1}$, where $\boldsymbol{f}_{j+1}$ is the vector of function values $f(x_{j+1,k})$. Then $\boldsymbol{f}_{j+1} = \Phi_j(x)\boldsymbol{s}_j + \Psi_j(x)\boldsymbol{d}_j$. Prove that $\boldsymbol{s}_j = \boldsymbol{f}_j$ if and only if all details are zero or the forward transform consists of one prediction step only.*

3.2.3 Using the interpolation at superresolution

The Deslauriers-Dubuc scheme is an instance of a class of refinement schemes where the refinement matrix $\mathbf{H}_j$ comes from a vector of refinement functions, parametrically depending on the coarse scale knots, and evaluated at the fine scale knots,

$$H_{j;k,l} = \mathrm{P}_{j,l}(x_{j+1,k}; \boldsymbol{x}_j). \tag{3.26}$$

For a Deslauriers-Dubuc scheme, the functions $\mathrm{P}_{j,l}(x; \boldsymbol{x}_j)$ are given by the interpolating piecewise polynomials (3.21).

While in subdivision, the prediction functions $\mathrm{P}_{j,l}(x; \boldsymbol{x}_j)$ are used for evaluation at the refinement points only, it is possible to use the same functions in the choice of the scaling functions at the finest resolution. So, for $i = J$ or even for $i = J^*$ at some superresolution, we can define

$$\overline{\varphi}_{i,l}^{[i]}(x) = \mathrm{P}_{i,l}(x; \boldsymbol{x}_J), \tag{3.27}$$

replacing the characteristic or indicator functions $\varphi_{i,l}(x) = \chi_{I_{i,l}}(x)$ of Section 2.2.5. From there the scaling basis at level j can be found through the usual subdivision of the two-scale equation (2.64),

$$\overline{\Phi}_j^{[i]} = \overline{\Phi}_{j+1}^{[i]}\mathbf{H}_j, \tag{3.28}$$

thereby iterating over $j = i-1, i-2, \ldots$.

The following lemma provides an alternative to subdivision as a numerical procedure for solving the two-scale systems.

Lemma 3.2.4 *If the refinement matrix $\mathbf{H}_j$ comes from a vector of refinement functions $\mathrm{P}_{i,l}(x;\boldsymbol{x}_i)$ by (3.26), then the scaling basis at level j in (3.28) is obtained by recursion on $i = j+1, j+2, \ldots$ of*

$$\overline{\Phi}_j^{[i]}(x) = \sum_{l=0}^{n_i-1} \overline{\Phi}_j^{[i-1]}(x_{i,l})\mathrm{P}_{i,l}(x;\boldsymbol{x}_i), \tag{3.29}$$

initialised by (3.27) for $i = j$.

Moreover, we have the following result, stating that subdivision and the recursion in (3.29) lead to similar numerical results.

Theorem 3.2.5 *The set of continuous, linearly independent functions $\overline{\Phi}_j^{[i]}$, defined recursively by (3.29), interpolates the set of piecewise constant, linearly independent functions $\Phi_j^{[i]}(x)$ obtained by subdivision from characteristic functions,*

$$\overline{\Phi}_j^{[i]}(\boldsymbol{x}_J) = \Phi_j^{[i]}(\boldsymbol{x}_J) = \mathbf{H}_{i-1}\mathbf{H}_{i-2}\ldots\mathbf{H}_j. \tag{3.30}$$

In Equation (3.30), the expression $\Phi_j^{[i]}(\boldsymbol{x}_J)$ stands for the $n_i \times n_j$ matrix with elements $\varphi_{j,k}^{[i]}(x_{i,l})$ on row k, column l, while $\overline{\Phi}_j^{[i]}(\boldsymbol{x}_i)$ is of course the same matrix, now using the functions in $\overline{\Phi}_j^{[i]}(x)$.

The sequence of bases $\overline{\Phi}_j^{[i]}(x)$, for $i = j, j+1, \ldots$ can be used for the analysis of the smoothness of the basis obtained by letting $i \to \infty$.

3.3 Average interpolating prediction

The elementary prediction functions $\mathrm{P}_{j,l}(x;\boldsymbol{x}_j)$ in the interpolating polynomial subdivision scheme of Deslauriers-Dubuc, given by (3.21), are composed from segments of interpolating polynomials. On the interval $[x_{j,m}, x_{j,m+1}]$, the prediction function coincides with the polynomial interpolating the pairs $(x_{j,r}, \delta_{lr})$, where $r \in \partial m = \{m - \lfloor(\widetilde{p}-1)/2\rfloor, m - \lfloor(\widetilde{p}-1)/2\rfloor + 1, \ldots, m + \lceil(\widetilde{p}-1)/2\rceil\}$ and δ_{lr} the Kronecker delta.

As an alternative, the elementary prediction functions can be composed as segments from polynomials satisfying an average interpolating condition. In particular, we fix the elementary prediction functions $\mathrm{P}_{j,l}(x;\boldsymbol{x}_j)$ as

$$\mathrm{P}_{j,l}(x;\boldsymbol{x}_j) = \sum_{m=0}^{n_j-1} \chi_{[x_{j,m},x_{j,m+1}]}(x)\,\frac{1}{\Delta_j(x)}\int_x^{x_{j,m+1}} p_{j,l,m}(u)du, \tag{3.31}$$

where $\Delta_j(x) = x_{j,m+1} - x$. The construction in (3.31) means that in each interval $[x_{j,m}, x_{j,m+1}]$, the prediction is the average over the subinterval $[x, x_{j,m+1}]$ of a polynomial $p_{j,l,m}(x)$. This polynomial is chosen so that its averages over the adjacent intervals are given by the coarse scaling coefficients, i.e., for $r \in \partial m = \{m - \lceil(\widetilde{p}-1)/2\rceil, m - \lceil(\widetilde{p}-1)/2\rceil + 1, \ldots, m + \lfloor\widetilde{p}-1)/2\rfloor\}$, we have

$$\frac{1}{\Delta_{j,r}} \int_{x_{j,r}}^{x_{j,r+1}} p_{j,l,m}(x)dx = \delta_{lr}, \tag{3.32}$$

with $\Delta_{j,r} = x_{j,r+1} - x_{j,r}$.

The prediction function $s_{j+1}(x)$ in (3.22) is then a piecewise polynomial function so that

$$\frac{1}{\Delta_{j,l}} \int_{x_{j,l}}^{x_{j,l+1}} s_{j+1}(x)dx = s_{j,l},$$

consisting of segments from the polynomials $s_{j+1,m}(x)$ that are defined by the average interpolating conditions,

$$\frac{1}{\Delta_{j,r}} \int_{x_{j,r}}^{x_{j,r+1}} s_{j+1,m}(x)dx = s_{j,r},$$

for $r \in \partial m$. This refinement is termed **average interpolation**, as it interprets the coarse scale coefficients $s_{j,r}$ as averages of a function, say $f(x)$, over the interval $[x_{j,r}, x_{j,r+1}]$. The function $f(x)$ is then locally approximated by a polynomial in order to produce refinements.

Once the refinement procedure has been fixed, a forward, fine-to-coarse that is, wavelet transform can be defined on top of it. Indeed, let $\boldsymbol{d}_j = \boldsymbol{s}_{j+1,o} - \mathbf{P}_j \boldsymbol{s}_j$, where $\mathbf{P}_j$ is the prediction matrix obtained from the prediction functions as in (3.20). The refinement does not fix the coarse scaling coefficients, $\boldsymbol{s}_j$. We could take $\boldsymbol{s}_j = \boldsymbol{s}_{j+1,e}$, as in the interpolating subdivision case. This choice is in conflict with the interpretation of the scaling functions as averages. Indeed, the same value would then represent the averages on both the intervals $[x_{j,k}, x_{j,k+1}]$ and $[x_{j+1,2k}, x_{j+1,2k+1}]$. We therefore propose a transformation from $\boldsymbol{s}_{j+1,e}$to $\boldsymbol{s}_j$ that allows us to interpret the coefficients at both levels as averages over an interval. More precisely, let $f_j(x)$ be a resolution level j approximation of a function $f(x)$, constructed by

$$f_j(x) = \sum_{k=0}^{n_j-1} s_{j,k}\varphi_{j,k}(x), \tag{3.33}$$

where we take as scaling coefficients $\overline{f}_{j,k}$ the averages

$$s_{j,k} = \overline{f}_{j,k} = \frac{1}{\Delta_{j,k}} \int_{x_{j,k}}^{x_{j,k+1}} f(x)dx; \tag{3.34}$$

then it holds that

$$s_{j,k} = \left[\Delta_{j+1,2k} s_{j+1,2k} + \Delta_{j+1,2k+1} s_{j+1,2k+1}\right] / \Delta_{j,k}. \tag{3.35}$$

The operation (3.35) is nothing but the scale and update part (2.6) of a forward unbalanced Haar transform. Hence, if the prediction step in the forward transform is preceded by the unbalanced Haar update $\boldsymbol{s}_j = \mathbf{D}_j \boldsymbol{s}_{j+1,e} + \mathbf{U}_j \boldsymbol{s}_{j+1,o}$, then it is possible to interpret the scaling coefficients as in (3.34). Writing $\widetilde{\varphi}_{H;j,k}(x) = \chi_{j,k}(x)/\Delta_{j,k}$ the dual Haar scaling basis (as the Haar transform is orthogonal, the dual basis equals the primal basis up to constant), the interpretation in (3.34) becomes

$$s_{j,k} = \int_{-\infty}^{\infty} \widetilde{\varphi}_{H;j,k}(x) f(x) dx.$$

Defining the scaling coefficients as averages is not the only possible choice for the $\boldsymbol{s}_j$ in an average interpolating scheme. Indeed, the unbalanced Haar update and the average interpolating prediction can be followed by another update of the values in $\boldsymbol{s}_j$. These values are then no longer the averages over the interval; neither are the dual scaling functions those from the Haar basis. The primal basis functions, $\varphi_{j,k}(x)$ are, however, untouched by a second update. These basis functions have the average interpolating property, as follows from the next result.

Theorem 3.3.1 *Let $\Phi_j(x)$ be a scaling basis associated with the refinement defined by an unbalanced Haar update and an average interpolating prediction. If the scaling basis functions are average interpolating at the finest scale J, then they are average interpolating at all scales $j = J, J-1, \ldots, L$. Formally, if $\Phi_J(x)$ is chosen so that*

$$\int_{x_{j,l}}^{x_{j,l+1}} \varphi_{j,k}(x) dx = \delta_{kl} \Delta_{j,k} \tag{3.36}$$

holds for $j = J$, then the same expression holds for $j = J, J-1, \ldots, L$.

Average interpolating basis functions (3.3.1) should be seen in contrast to interpolating basis functions (3.23), with the Haar scaling basis at the intersection between the two classes. The Haar scaling basis is also a valid choice at the finest scale (or at a superresolution) for both classes of basis functions (although triangular hat functions lead to visually more pleasant results in the interpolating case).

As a result, the approximation (3.33) satisfies

$$s_{j,k} = \frac{1}{\Delta_{j,k}} \int_{x_{j,k}}^{x_{j,k+1}} f_j(x) dx. \tag{3.37}$$

The difference between (3.37) and (3.34) lies in the subscript j in the integrand $f_j(x)$. Indeed, (3.37) is a property of an approximation to $f(x)$. This property is thus about a synthesis of a function from its scaling coefficients, whereas (3.34) describes a possible criterion in the design of the analysis of $f(x)$.

Proof. (of Theorem 3.3.1) The two-scale equation is a slight variant of (2.39), now involving the rescale matrix $\mathbf{D}_j$,

$$\Phi_j(x) = \Phi_{j+1,e}(x)\mathbf{D}_j^{-1}(\mathbf{I} - \mathbf{U}_j\mathbf{P}_j) + \Phi_{j+1,o}(x)\mathbf{P}_j,$$

where $\mathbf{U}_j$ is a diagonal matrix. Denote by $\mathbf{A}_j$ the matrix with columns

$$\boldsymbol{a}_{j;l} = \int_{x_{j,l}}^{x_{j,l+1}} \Phi_j^\top(x)dx,$$

i.e., with elements

$$A_{j;k,l} = \int_{x_{j,l}}^{x_{j,l+1}} \varphi_{j,k}(x)dx;$$

then we have

$$\begin{aligned}\int_{x_{j,l}}^{x_{j,l+1}} \varphi_{j+1,k}(x)dx &= \int_{x_{j+1,2l}}^{x_{j+1,2l+1}} \varphi_{j+1,k}(x)dx + \int_{x_{j+1,2l+1}}^{x_{j+1,2l+2}} \varphi_{j+1,k}(x)dx \\ &= A_{j+1;k,2l} + A_{j+1;k,2l+1}.\end{aligned}$$

Integration of the transposed two-scale equation then yields

$$\mathbf{A}_j = (\mathbf{I} - \mathbf{P}_j^\top\mathbf{U}_j^\top)\mathbf{D}_j^{-1}(\mathbf{A}_{j+1;e,e} + \mathbf{A}_{j+1;e,o}) + \mathbf{P}_j^\top(\mathbf{A}_{j+1;o,e} + \mathbf{A}_{j+1;o,o}),$$

where $\mathbf{A}_{j+1;e,o}$ denotes the submatrix of $\mathbf{A}_{j+1}$ containing the even rows and odd columns, with the other submatrices defined in a similar way. The submatrices $\mathbf{A}_{j+1;e,o}$ and $\mathbf{A}_{j+1;o,o}$ have an additional zero column if the number of evens exceeds the number of odds by one. Assuming that (3.3.1) holds at level $j+1$, we have that $\mathbf{A}_{j+1}$ is a diagonal matrix with $A_{j+1,i,i} = \Delta_{j+1,i}$. Then $\mathbf{A}_{j+1;e,o}$ and $\mathbf{A}_{j+1;o,e}$ are zero matrices, while $\mathbf{U}_j^\top\mathbf{D}_j^{-1}\mathbf{A}_{j+1;e,e} = \mathbf{A}_{j+1;o,o}$, because $\mathbf{U}_j$ and $\mathbf{D}_j$ are diagonal matrices with the elements of the forward unbalanced Haar scaling filter. As a result, the transposed two-scale equation becomes

$$\mathbf{A}_j = \mathbf{D}_j^{-1}\mathbf{A}_{j+1;e,e} + \mathbf{P}_j^\top\mathbf{0}_{j+1;o,e},$$

which is the diagonal matrix corresponding to (3.3.1) at scale j. □

The flow chart corresponding to this wavelet scheme has the same form as that of a Haar transform in Figure 2.7, replacing the identical operation in the last lifting step by the prediction $\mathbf{P}_j$.

3.4 Nonlinear and adaptive wavelet transforms

Although wavelets are popular for use in nonlinear data processing, most wavelet transforms themselves are linear. The nonlinearity occurs in the processing between forward and inverse wavelet transforms. Nevertheless, lifting steps can be nonlinear as well. An important class of nonlinear lifting schemes are **integer wavelet transforms** (Calderbank et al., 1998), where each lifting step ends by a rounding operation. If the input consists of integer values, then so do all intermediate and output coefficients. In applications such as signal and image compression, working with integers rather than with floating points may increase lossless compression rates.

Other forms of nonlinear lifting steps involve median filtering or the computation of maxima (Claypoole et al., 1998; Goutsias and Heijmans, 2000a,b; Jansen and Ooninx, 2005).

Nonlinearity may also follow from **adaptive lifting**, where the lifting operation is chosen from two or more options, depending on the input scaling coefficients. For instance, it is possible to switch locally between high- and low-order interpolation as a prediction (Claypoole et al., 2003). An important issue in adaptive lifting is the inverse transform. In order to reconstruct the original data from an adaptive transform, we need to know which choices the forward transform had made. In signal or image compression, storage of the adaptive choice may undo the gain in disk space obtained from the adaptivity. Carefully designed adaptive transforms produce detail coefficients from which the choices made in the forward transform can be recovered (Piella and Heijmans, 2002; Heijmans et al., 2006; Ooninx and de Zeeuw, 2003; Jansen and Ooninx, 2005). It is crucial, of course, not to lose that crucial, hidden information during the compression stage following the decomposition.

For that reason, adaptive transforms in denoising applications use a different strategy (Nunes et al., 2006; Knight and Nason, 2006). In denoising, the manipulation of the coefficients is typically much more thorough. On the other hand, as economic use of external memory is not an a goal as such, the adaptive choices can easily be stored for use upon the reconstruction.

Choosing between alternatives in an adaptive transform amounts to choosing between several simple, linear wavelet transforms. Since every linear wavelet transform corresponds to a wavelet basis, an adaptive lifting scheme can be seen as a basis (or model) selection procedure. More examples of adaptive basis selection will follow in Section 9.4.

4

Wavelet transforms from factored refinement schemes

The main risk in constructing your wavelet from the ground up lies in the refinement part. Characteristics added to the final update, such as zero integrals and good variance control, are safe. Other properties that are appealing in one refinement step may have undesired effects after recursive refining. In particular, it is notoriously difficult to control the smoothness of the basis functions from the design of a single refinement step through lifting. This holds even if the same approach, such as polynomial prediction, is applied in every refinement step. One of the reasons is that the limiting basis function may heavily depend on the location of the intermediate refinement points.

As an alternative to design of the transform through the lifting steps, this chapter proposes to start from existing refinable bases. Adding a final update step as in Chapter 3 to the refinement scheme of such a basis then yields a wavelet decomposition. For this approach to work, we need to identify the refinement scheme for a given refinable basis. In the first instance, sketched in Section 4.1, the refinement scheme is assumed to be known. The scheme is then factored into elementary lifting steps. Secondly, when the exact refinement scheme is unknown, the general form of the factoring may still be established. This situation is illustrated by the case of **B-spline bases**. Section 4.2 explains how the B-spline refinement can be found from its factoring into lifting steps.

As a result, this chapter puts B-spline refinement in contrast to the Deslauriers-Dubuc refinement of Chapter 3. One of the main advantages of B-spline refinement is that the limiting basis functions are known: they are B-splines. The smoothness of these functions is of course well understood. Moreover, the basis is independent from the order of refinement, i.e., the location of the intermediate refinement points.

4.1 Factoring of wavelet transforms into lifting steps

Section 2.1 has introduced the lifting scheme as a tool for the design and construction of wavelet transforms on arbitrary settings. This section

DOI: 10.1201/9781003265375-4

analyses a given wavelet transform, using the same lifting scheme, but now in breaking the wavelet transform down again, i.e., decomposing it into its elementary building blocks. The factoring into elementary lifting steps will be used in Section 4.2 for the assembly of B-splines and B-spline wavelets.

4.1.1 The polyphase representation of a wavelet transform

Obviously, the wavelet transforms in a lifting scheme fit into the filterbank framework of Section 2.2. As developed in Section 4.1.3, it is also true, though a bit less trivial, that all filterbanks can be factored into a sequence of alternating lifting steps. In order to bring both implementations of a wavelet transform in accordance to each other, we first insert the even-odd partition from the lifting scheme into the filterbank version. The resulting implementation of the filterbank is termed the **polyphase representation**.

As in Section 2.2.2, denote by $\mathbf{H}_{j,e}$ the submatrix of $\mathbf{H}_j$ containing the even rows of $\mathbf{H}_j$, and let $\mathbf{H}_{j,o}$ be the submatrix with the odd rows. Similar definitions are of course possible for the other matrices in the filterbank. Then the inverse transform (2.46) can be rewritten as

$$\boldsymbol{s}_{j+1,e} = \mathbf{H}_{j,e}\boldsymbol{s}_j + \mathbf{G}_{j,e}\boldsymbol{d}_j, \tag{4.1}$$

$$\boldsymbol{s}_{j+1,o} = \mathbf{H}_{j,o}\boldsymbol{s}_j + \mathbf{G}_{j,o}\boldsymbol{d}_j. \tag{4.2}$$

When we ignore the reordering in the merging of even and odd entries of $\boldsymbol{s}_{j+1}$, we can write the inverse transform matrix at level j as

$$\mathbf{W}_j = \begin{bmatrix} \mathbf{H}_{j,e} & \mathbf{G}_{j,e} \\ \mathbf{H}_{j,o} & \mathbf{G}_{j,o} \end{bmatrix}, \tag{4.3}$$

so that

$$\begin{bmatrix} \boldsymbol{s}_{j+1,e} \\ \boldsymbol{s}_{j+1,o} \end{bmatrix} = \mathbf{W}_j \begin{bmatrix} \boldsymbol{s}_j \\ \boldsymbol{d}_j \end{bmatrix}. \tag{4.4}$$

The two-scale equation becomes

$$\Phi_j(x) = \Phi_{j+1,e}(x)\mathbf{H}_{j,e} + \Phi_{j+1,o}(x)\mathbf{H}_{j,o}. \tag{4.5}$$

Although a bit trivial at this point, the forward transform is written here, for further reference, with even and odd submatrices of $\widetilde{\mathbf{H}}$ and $\widetilde{\mathbf{G}}$ as well,

$$\boldsymbol{s}_j = \widetilde{\mathbf{H}}_{j,e}^\top \boldsymbol{s}_{j+1,e} + \widetilde{\mathbf{H}}_{j,o}^\top \boldsymbol{s}_{j+1,o}, \tag{4.6}$$

$$\boldsymbol{s}_j = \widetilde{\mathbf{G}}_{j,e}^\top \boldsymbol{s}_{j+1,e} + \widetilde{\mathbf{G}}_{j,o}^\top \boldsymbol{s}_{j+1,o}. \tag{4.7}$$

4.1.2 Adding lifting steps to existing wavelet transforms

We now investigate how to add a lifting step to an existing filterbank. This allows us, in Section 4.1.3, to go the other way around, and factor an existing filterbank into a simpler version and a lifting step. Suppose that we are given a filterbank $\mathbf{W}_j^{[s+1]}$ with filters $\mathbf{H}_j^{[s+1]}$ and $\mathbf{G}_j^{[s+1]}$. We wish to add further lifting to this filterbank. The objective of further lifting is to add more properties to the filterbank. By convention, we let the superscript $[s]$ go down when adding a new lifting step. We make sure that the last lifting step brings us from $\mathbf{W}_j^{[0]}$ to the final filterbank $\mathbf{W}_j$, from which the superscript, $[-1]$, has been omitted. In some applications, the number of lifting steps can be computed in advance. This holds for the lifting factoring of a spline refinement, to be developed in Section 4.2. In other applications, the number of lifting steps follows in the course of the calculations, and so the superscripts $[s]$ in each step are assigned a posteriori. In order to go from $\mathbf{W}_j^{[s+1]}$ to $\mathbf{W}_j^{[s]}$ we have four options.

1. **Add an update lifting step on the fine scale side.** This lifting step takes place after the operation $\mathbf{W}_j^{[s+1]}$ in the inverse transform. Let $\boldsymbol{s}_{j+1}^{[s+1]}$ be the fine scale coefficients before the update $\mathbf{U}_j^{[s+1]}$; then after the update, we have $\boldsymbol{s}_{j+1,o}^{[s]} = \boldsymbol{s}_{j+1,o}^{[s+1]}$ and $\boldsymbol{s}_{j+1,e}^{[s]} = \boldsymbol{s}_{j+1,e}^{[s+1]} - \mathbf{U}_j^{[s+1]}\boldsymbol{s}_{j+1,o}^{[s+1]}$. All together, the update can be written as

$$\begin{aligned}\mathbf{W}_j^{[s]} &= \begin{bmatrix} \mathbf{H}_{j,e}^{[s]} & \mathbf{G}_{j,e}^{[s]} \\ \mathbf{H}_{j,o}^{[s]} & \mathbf{G}_{j,o}^{[s]} \end{bmatrix} \\ &= \begin{bmatrix} \mathbf{I}_{j+1,e} & -\mathbf{U}_j^{[s+1]} \\ \mathbf{0}_{j,o,e} & \mathbf{I}_{j+1,o} \end{bmatrix} \begin{bmatrix} \mathbf{H}_{j,e}^{[s+1]} & \mathbf{G}_{j,e}^{[s+1]} \\ \mathbf{H}_{j,o}^{[s+1]} & \mathbf{G}_{j,o}^{[s+1]} \end{bmatrix}. \qquad (4.8)\end{aligned}$$

To find the effect on the basis functions, we start from the identity

$$\Phi_{j+1,e}^{[s]}(x)\boldsymbol{s}_{j+1,e}^{[s]}+\Phi_{j+1,o}^{[s]}(x)\boldsymbol{s}_{j+1,o}^{[s]} = \Phi_{j+1,e}^{[s+1]}(x)\boldsymbol{s}_{j+1,e}^{[s+1]}+\Phi_{j+1,o}^{[s+1]}(x)\boldsymbol{s}_{j+1,o}^{[s+1]}.$$

We rephrase the update for use in a forward wavelet transform as $\boldsymbol{s}_{j+1,o}^{[s+1]} = \boldsymbol{s}_{j+1,o}^{[s]}$ and $\boldsymbol{s}_{j+1,e}^{[s+1]} = \boldsymbol{s}_{j+1,e}^{[s]} + \mathbf{U}_j^{[s+1]}\boldsymbol{s}_{j+1,o}^{[s+1]}$. This we substitute to get

$$\begin{aligned}\Phi_{j+1,e}^{[s+1]}(x)(\boldsymbol{s}_{j+1,e}^{[s]} + \mathbf{U}_j^{[s+1]}\boldsymbol{s}_{j+1,o}^{[s]}) + \Phi_{j+1,o}^{[s+1]}(x)\boldsymbol{s}_{j+1,o}^{[s]} \\ = \Phi_{j+1,e}^{[s]}(x)\boldsymbol{s}_{j+1,e}^{[s]} + \Phi_{j+1,o}^{[s]}(x)\boldsymbol{s}_{j+1,o}^{[s]}.\end{aligned}$$

As this holds for arbitrary $\boldsymbol{s}_{j+1}^{[s]}$, we find $\Phi_{j+1,e}^{[s]}(x) = \Phi_{j+1,e}^{[s+1]}(x)$ and $\Phi_{j+1,o}^{[s]}(x) = \Phi_{j+1,o}^{[s+1]}(x) + \Phi_{j+1,e}^{[s+1]}(x)\mathbf{U}_j^{[s+1]}$. Like in Section 2.1.7, we conclude that an **update** step changes the **values of the even coefficients**. The odd values are left untouched, but the update changes the **odd basis functions**, the interpretation, that is, of the odd coefficients.

2. **Add a prediction step on the fine scale side.** A prediction step $\mathbf{P}_j^{[s+1]}$ at the end of the inverse transform leads to a similar result.

The coefficients before and after the prediction step are related by $s^{[s]}_{j+1,o} = s^{[s+1]}_{j+1,o} + \mathbf{P}^{[s+1]}_j s^{[s+1]}_{j+1,e}$. The inverse transformation matrices are given by

$$\begin{aligned}\mathbf{W}^{[s]}_j &= \begin{bmatrix} \mathbf{H}^{[s]}_{j,e} & \mathbf{G}^{[s]}_{j,e} \\ \mathbf{H}^{[s]}_{j,o} & \mathbf{G}^{[s]}_{j,o} \end{bmatrix} \\ &= \begin{bmatrix} \mathbf{I}_{j+1,e} & \mathbf{0}_{j,e,o} \\ \mathbf{P}^{[s+1]}_j & \mathbf{I}_{j+1,o} \end{bmatrix} \begin{bmatrix} \mathbf{H}^{[s+1]}_{j,e} & \mathbf{G}^{[s+1]}_{j,e} \\ \mathbf{H}^{[s+1]}_{j,o} & \mathbf{G}^{[s+1]}_{j,o} \end{bmatrix}. \end{aligned} \tag{4.9}$$

The scaling basis functions are updated as $\Phi^{[s]}_{j+1,o}(x) = \Phi^{[s+1]}_{j+1,o}(x)$ and $\Phi^{[s]}_{j+1,e}(x) = \Phi^{[s+1]}_{j+1,e}(x) - \Phi^{[s+1]}_{j+1,o}(x)\mathbf{P}^{[s+1]}_j$.
A lifting step at the fine scale side affects half of both matrices $\mathbf{H}^{[s+1]}_j$ and $\mathbf{G}^{[s+1]}_j$. For instance, with a prediction step, we have $\mathbf{H}^{[s]}_{j,e} = \mathbf{H}^{[s+1]}_{j,e}$, while $\mathbf{H}^{[s]}_{j,o} = \mathbf{H}^{[s+1]}_{j,o} + \mathbf{P}^{[s+1]}_j \mathbf{H}^{[s+1]}_{j,e}$. A similar statement holds for $\mathbf{G}^{[s+1]}_j$. Both matrices are lifted separately, i.e., using their own columns only. This is different in the next two lifting operations.

3. **Add an update lifting step on the coarse scale side.** As explained in Section 4.1.5, one single update step can be added on the coarse scale side at the very end of the factoring process, when $s+1=0$. In the inverse transform, this lifting step takes place before the operation $\mathbf{W}^{[0]}_j$. The outcome of the update step is a new coarse scaling coefficient vector $s_j = s^{[0]}_j - \mathbf{U}_j d^{[0]}_j$. The full transform at resolution level j becomes

$$\begin{aligned}\mathbf{W}_j &= \begin{bmatrix} \mathbf{H}_{j,e} & \mathbf{G}_{j,e} \\ \mathbf{H}_{j,o} & \mathbf{G}_{j,o} \end{bmatrix} \\ &= \begin{bmatrix} \mathbf{H}^{[0]}_{j,e} & \mathbf{G}^{[0]}_{j,e} \\ \mathbf{H}^{[0]}_{j,o} & \mathbf{G}^{[0]}_{j,o} \end{bmatrix} \begin{bmatrix} \mathbf{I}_{j+1,e} & -\mathbf{U}_j \\ \mathbf{0}_{j+1,o,e} & \mathbf{I}_{j+1,o} \end{bmatrix}. \end{aligned} \tag{4.10}$$

In this case, the effect of the lifting step can be described without the even-odd partitioning.

$$\mathbf{G}_j = \mathbf{G}^{[0]}_j - \mathbf{H}^{[0]}_j \mathbf{U}_j, \tag{4.11}$$

$$\mathbf{H}_j = \mathbf{H}^{[0]}_j. \tag{4.12}$$

The update affects both even and odd columns of $\mathbf{G}_j$, in a way depending on $\mathbf{H}_j$. In terms of basis functions, the effect of this update step can be written as $\Phi_j(x) = \Phi^{[0]}_j(x)$, while

$$\Psi_j(x) = \Psi^{[0]}_j(x) + \Phi^{[0]}_j(x)\mathbf{U}_j. \tag{4.13}$$

As explained in Section 3.1.2, Expression (4.13) can be used to design $\mathbf{U}_j$ in a way that $\Psi_j(x)$ has the desired properties, such as vanishing moments.

Exercise 4.1.1 *Let* $\begin{bmatrix} \mathbf{H}_j^\top \\ \mathbf{G}_j^\top \end{bmatrix}$ *be the forward, dual wavelet transform, as introduced in (2.67) and (2.68). The reconstruction is then given by* $[\widetilde{\mathbf{H}}_j\, \widetilde{\mathbf{G}}_j]$*. Prove that the effect of the lifting in (4.11) on the inverse dual transform is given by*

$$\widetilde{\mathbf{H}}_j = \widetilde{\mathbf{H}}_j^{[0]} + \widetilde{\mathbf{G}}_j^{[0]}\mathbf{U}_j^\top,$$

and give an interpretation of this expression.

4. **Add a prediction lifting step on the coarse scale side.** Although this step will not be used in the discussion below, we mention, for the sake of completeness, that a prediction step can be inserted on the coarse scale side. Again, we suppose that $s+1=0$. This lifting step has an effect on all columns of $\mathbf{H}_j$, in a way depending on $\mathbf{G}_j$.

$$\mathbf{H}_j = \mathbf{H}_j^{[0]} + \mathbf{G}_j^{[0]}\mathbf{P}_j, \tag{4.14}$$

$$\mathbf{G}_j = \mathbf{G}_j^{[0]}, \tag{4.15}$$

which is equivalent to

$$\mathbf{W}_j = \begin{bmatrix} \mathbf{H}_{j,e} & \mathbf{G}_{j,e} \\ \mathbf{H}_{j,o} & \mathbf{G}_{j,o} \end{bmatrix} = \begin{bmatrix} \mathbf{H}_{j,e}^{[0]} & \mathbf{G}_{j,e}^{[0]} \\ \mathbf{H}_{j,o}^{[0]} & \mathbf{G}_{j,o}^{[0]} \end{bmatrix} \begin{bmatrix} \mathbf{I}_{j+1,e} & \mathbf{0}_{j+1,e,o} \\ \mathbf{P}_j & \mathbf{I}_{j+1,o} \end{bmatrix}. \tag{4.16}$$

4.1.3 Factoring into lifting steps

Armed with the results of Section 4.1.2, we now factor an existing refinement matrix $\mathbf{H}_j^{[s]}$ into lifting steps. We want to do this independently from any possibly accompanying $\mathbf{G}_j$, whose lifting implementation will follow afterwards. We will use (4.8) and (4.9), which can be summarised as

$$\mathbf{H}_{j,o}^{[s]} = \mathbf{H}_{j,o}^{[s+1]} + \mathbf{P}_j^{[s+1]}\mathbf{H}_{j,e}^{[s]}, \tag{4.17}$$

for the prediction step, and

$$\mathbf{H}_{j,e}^{[s]} = \mathbf{H}_{j,e}^{[s+1]} - \mathbf{U}_j^{[s+1]}\mathbf{H}_{j,o}^{[s]}, \tag{4.18}$$

for the update step. In (4.17) the even columns of $\mathbf{H}_{j,e}^{[s+1]}$ remain unchanged, whereas in (4.18) the odd columns are left untouched. The objective is to find, for given $\mathbf{H}_{j,e}^{[s]}$ and $\mathbf{H}_{j,o}^{[s]}$, an update $\mathbf{U}_j^{[s+1]}$ or a prediction $\mathbf{P}_j^{[s+1]}$ so that the resulting $\mathbf{H}_{j,e}^{[s+1]}$ or $\mathbf{H}_{j,o}^{[s+1]}$ are simpler operations, i.e., closer to the identity matrix. In particular we want the matrices after factoring to have their nonzero entries to be more concentrated near the main diagonal. For the further development in this context we will use the following definition.

Definition 4.1.2 *(bandwidth of a refinement matrix) Let A be an $m \times n$ matrix; then its* ***bandwidth*** *is the smallest integer b for which there exists an increasing sequence of integers $k_1(l)$ so that*

$$A_{k,l} \neq 0 \Rightarrow k_1(l) \leq k \leq k_1(l) + b - 1. \tag{4.19}$$

The definition imposes that $k_1(l)$ is increasing, so that the inverse sequence $l_2(k)$ can defined by $l_2(k_1(l)) = l$. As a consequence, it may happen that $A_{k,l} = 0$, even inside the nonzero band $\{k_1(l), \ldots, k_1(l) + b - 1\}$. If a refinement matrix $\mathbf{H}_j$ contains zeros inside its nonzero band, its even submatrix $\mathbf{H}_{j,e}$ may have a different band structure than its odd submatrix $\mathbf{H}_{j,o}$. This occurs, for instance, in interpolating refinement matrices. A different band structure in the submatrices may also occur in the case of a non-binary even-odd partitioning, i.e., when evens and odds do not alternate as in the classical definition of even and odd numbers. The result stated below holds for the regular case, where the two submatrices have overlapping bands, with bandwidths that differ by one or zero. The bandwidth of $\mathbf{H}_j$ then equals the sum of the bandwidths of the submatrices.

Theorem 4.1.3 *(Factoring into bidiagonal lifting steps) Let $\mathbf{H}_j^{[s]}$ be a refinement matrix whose bandwidth equals the sum of the bandwidths of its even and odd submatrices $\mathbf{H}_{j,e}^{[s]}$ and $\mathbf{H}_{j,o}^{[s]}$.*

If the bandwidth of $\mathbf{H}_{j,e}^{[s]}$ is larger by one than the bandwidth of $\mathbf{H}_{j,o}^{[s]}$, and each column of $\mathbf{H}_{j,o}^{[s]}$ contains nonzeros at the endpoints of the bandwidth, then there exists a bidiagonal *matrix $\mathbf{U}_j^{[s+1]}$ and a matrix $\mathbf{H}_{j,e}^{[s+1]}$ with bandwidth smaller than that of $\mathbf{H}_{j,o}^{[s]}$ for use in (4.18).*

Conversely, if the bandwidth of $\mathbf{H}_{j,o}^{[s]}$ is larger by one than the bandwidth of $\mathbf{H}_{j,e}^{[s]}$, and each column of $\mathbf{H}_{j,e}^{[s]}$ contains nonzeros at the endpoints of the bandwidth, then there exists a bidiagonal *matrix $\mathbf{P}_j^{[s+1]}$ and a matrix $\mathbf{H}_{j,o}^{[s+1]}$ with bandwidth smaller than that of $\mathbf{H}_{j,e}^{[s]}$ for use in (4.17).*

If $\mathbf{H}_{j,e}^{[s]}$ and $\mathbf{H}_{j,o}^{[s]}$ have equal bandwidths, then both factorings (4.18) and (4.17) are possible.

Proof. The proof proceeds by construction on each column of the refinement matrix separately. In each column, the work amounts to the proof for refinement on a regular grid in Daubechies and Sweldens (1998). We develop the case where $\mathbf{H}_{j,e}^{[s]}$ has a larger bandwidth than $\mathbf{H}_{j,o}^{[s]}$. Denote by $k_1(l)$ and $k_2(l)$ the indices of the first and the last nonzero elements in column l of $\mathbf{H}_{j,e}^{[s]}$. Then all nonzeros in column l of $\mathbf{H}_{j,o}^{[s]}$ are in $\{k_1(l), \ldots, k_2(l) - 1\}$. We also assume that $k_1(l)$ and $k_2(l)$ are strictly increasing sequences. Otherwise, some of the zero elements in $\mathbf{H}_{j,e}^{[s]}$ can be counted as nonzeros in the definition of $k_1(l)$ and $k_2(l)$. For $k_1(l) < k < k_2(l)$, the lifting step (4.18) can be written as

$$H_{j,2k,l}^{[s]} = H_{j,2k,l}^{[s+1]} - \sum_{m=k_1(l)}^{k_2(l)-1} U_{j,k,m}^{[s+1]} H_{j,2m+1,l}^{[s]}. \tag{4.20}$$

For $k \in \{k_1(l), k_2(l)\}$, the same step results in two equations of the form

$$H_{j,2k,l}^{[s]} = - \sum_{m=k_1(l)}^{k_2(l)-1} U_{j,k,m}^{[s+1]} H_{j,2m+1,l}^{[s]}. \tag{4.21}$$

The equations in (4.21) contain as unknowns the elements of $\mathbf{U}_j^{[s+1]}$. From there, the equations in (4.20) can be solved for $\mathbf{H}_{j,e}^{[s+1]}$.

While the Equations (4.21) have been stated for fixed l, the solutions are found by considering the two equations for given k, i.e., by looking at $l = l_2(k)$ and $l = l_1(k)$, where $l_2(k)$ is the inverse of $k_1(l)$, i.e., $l_2(k_1(l)) = l$, and $l_1(k)$ denotes the inverse of $k_2(l)$.

For $k = k_1(l)$, i.e., $l = l_2(k)$, we set $U_{j,k,m}^{[s+1]} = 0$ for all $m = k + 1, \ldots, k_2(l_2(k)) - 1$, while for $m = k$, it then follows that

$$U_{j,k,k}^{[s+1]} = -H_{j,2k,l_2(k)}^{[s]} / H_{j,2k+1,l_2(k)}^{[s]}. \tag{4.22}$$

Indeed, this is needed in order to satisfy (4.21) for $k = k_1(l)$ or $l = l_2(k)$.

For $k = k_2(l)$, i.e., $l = l_1(k)$, we set $U_{j,k,m}^{[s+1]} = 0$ for all $m = k_1(l_1(k)), \ldots, k-2$, while for $m = k - 1$, it then follows that

$$U_{j,k,k-1}^{[s+1]} = -H_{j,2k,l_1(k)}^{[s]} / H_{j,2k-1,l_1(k)}^{[s]}. \tag{4.23}$$

Again this is required to satisfy (4.21), this time for $k = k_2(l)$ or $l = l_1(k)$.

Once the diagonal and the lower diagonal of $\mathbf{U}_j$ have been found, all the other entries of this matrix can be filled with zeros. This leaves us with $k_2(l) - k_1(l) - 1$ equations of the form (4.20). Each of these equations allows us to find exactly one element in column l of $\mathbf{H}_{j,e}^{[s+1]}$. □

The alternating factoring of $\mathbf{H}_{j,o}^{[s]}$ and $\mathbf{H}_{j,e}^{[s]}$ proceeds exactly as **Euclid's algorithm** for finding the greatest common divisor (Daubechies and Sweldens, 1998). Indeed, both factorings (4.17) and (4.18) are of the form

$$a = b \times q + r. \tag{4.24}$$

In the ring of integers, Euclid's algorithm is based on the property that if d is a common divisor of a and b, then it also divides r. Indeed if there exist quotients q_a and q_b so that $a = dq_a$ and $b = dq_b$, it follows that $r = d(q_a - q_b q)$.

In (4.17), the roles of dividend, divisor, quotient, and remainder are filled in with $a = \mathbf{H}_{j,o}^{[s]}$, $b = \mathbf{H}_{j,e}^{[s]}$, $q = \mathbf{P}_j^{[s+1]}$ and $r = \mathbf{H}_{j,o}^{[s+1]}$, while in (4.18), this is $a = \mathbf{H}_{j,e}^{[s]}$, $b = \mathbf{H}_{j,o}^{[s]}$, $q = -\mathbf{U}_j^{[s+1]}$ and $r = \mathbf{H}_{j,e}^{[s+1]}$. With $\mathbf{H}_{j,e}^{[s+1]}$ the remainder, and $\mathbf{U}_j^{[s]}$ the quotient when dividing $\mathbf{H}_{j,e}^{[s]}$ by $\mathbf{H}_{j,o}^{[s]}$, the bandwidths of these matrices is the analogue of the polynomial degree in a Euclidean algorithm applied in the ring[1] of polynomials. In particular, if the degree of 'polynomial' $\mathbf{H}_{j,e}^{[s]}$ is one more than that of $\mathbf{H}_{j,o}^{[s]}$, then the remainder $\mathbf{H}_{j,e}^{[s+1]}$ has a degree one less than that of $\mathbf{H}_{j,o}^{[s]}$. The quotient has degree 1, where the constant term corresponds to the diagonal of $\mathbf{U}_j^{[s]}$ and the linear term corresponds to the lower diagonal.

In the first instance, the Euclidean algorithm is used to find a lifting implementation for a given refinement matrix $\mathbf{H}_j$. In some applications, it can be used to even construct the matrix $\mathbf{H}_j$ from scratch. This is illustrated in Section 4.2 with the construction of B-spline (scaling) functions using a lifting scheme.

4.1.4 The final factoring step

As in Euclid's algorithm, the factoring in Theorem 4.1.3 can be repeated up to the point where the remainder matrix reaches a form that can be used in the construction of a multiscale transform. This form is a diagonal matrix, as explained below. The factoring then provides a fast forward and inverse wavelet transform. This transform includes a detail matrix $\mathbf{G}_j^{[0]}$, which follows implicitly from the factoring process. As $\mathbf{G}_j^{[0]}$ may not have the right properties yet, Section 4.1.5 will discuss how to further update the transform.

Theorem 4.1.4 *(Wavelet transform from lifting factoring) If the factoring $\mathbf{H}_{j,e}^{[s]} = \mathbf{H}_{j,e}^{[s+1]} - \mathbf{U}_j^{[s+1]}\mathbf{H}_{j,o}^{[s]}$, as in (4.18), has an invertible remainder $\mathbf{H}_{j,e}^{[s+1]} = \mathbf{D}_j$, then we can construct a wavelet transform containing $\mathbf{H}_j^{[s]}$ as the refinement matrix. In particular, we let $\mathbf{P}_j^{[s+1]} = \mathbf{H}_{j,o}^{[s]}\mathbf{D}_j^{-1}$, and take as forward transform*

$$\boldsymbol{s}_j = \mathbf{D}_j^{-1}(\boldsymbol{s}_{j+1,e} + \mathbf{U}_j^{[s+1]}\boldsymbol{s}_{j+1,o}), \tag{4.25}$$

$$\boldsymbol{d}_j = \boldsymbol{s}_{j+1,o} - \mathbf{P}_j^{[s+1]}\mathbf{D}_j\boldsymbol{s}_j. \tag{4.26}$$

[1] The matrices $\mathbf{H}_{j,e}^{[s]}$ and $\mathbf{H}_{j,o}^{[s]}$ do *not* live in a ring. These matrices may even have incompatible sizes for addition. On equidistant knots, these matrices have a Toeplitz structure, which can be associated one-to-one to polynomials. In that case, the lifting factoring operates within the framework of a ring, see Daubechies and Sweldens (1998). Otherwise, we are cutting the corners of the theoretical framework, forcing us to be careful about what we are doing.

The synthesis or inverse transform consists of

$$\boldsymbol{s}_{j+1,o} = \boldsymbol{d}_j + \mathbf{P}_j^{[s+1]}\mathbf{D}_j\boldsymbol{s}_j, \quad (4.27)$$

$$\boldsymbol{s}_{j+1,e} = \mathbf{D}_j\boldsymbol{s}_j - \mathbf{U}_j^{[s+1]}\boldsymbol{s}_{j+1,o}. \quad (4.28)$$

Proof. From the inverse transform, we can check that $\mathbf{H}_{j,o}^{[s]} = \mathbf{P}_j^{[s+1]}D_j$, while $\boldsymbol{s}_{j+1,e} = \mathbf{D}_j\boldsymbol{s}_j - \mathbf{U}_j^{[s+1]}(\boldsymbol{d}_j + \mathbf{P}_j^{[s+1]}\mathbf{D}_j\boldsymbol{s}_j)$, meaning that $\mathbf{H}_{j,e}^{[s]} = \mathbf{D}_j - \mathbf{U}_j^{[s+1]}\mathbf{P}_j^{[s+1]}\mathbf{D}_j = \mathbf{H}_{j,e}^{[s+1]} - \mathbf{U}_j^{[s+1]}\mathbf{H}_{j,o}^{[s]}$. □

In principle, the construction of the wavelet transform in Theorem 4.1.4 can be applied at any stage s of the factoring process, but the inverse $\mathbf{D}_j^{-1}$ is sparse only if $\mathbf{D}_j$ is a diagonal matrix.

A diagonal structure is not required for the odd rows in $\mathbf{H}_{j,o}^{[s+1]}$. In other words, the last step in a full factoring of a refinement filter must always be a prediction step. This prediction is also the first step in the inverse wavelet transform, and so it is the last step in the forward transform. It is followed by a diagonal rescaling $\mathbf{D}_j$.

If the matrix $\mathbf{D}_j$ turns out to have zeros on its diagonal, then the construction of a forward wavelet transform as in (4.25) is impossible. In the extreme case where $\mathbf{D}_j$ is the zero matrix, we can see that $\mathbf{H}_{j,e}^{[s]} = -\mathbf{U}_j^{[s+1]}\mathbf{H}_{j,o}^{[s]}$. A non-invertible diagonal after factoring can be interpreted as two matrices, $\mathbf{H}_{j,e}^{[s]}$ and $\mathbf{H}_{j,o}^{[s]}$, not being *coprime*.

Definition 4.1.5 *(coprime matrices) In a ring, where most elements do not have a multiplicative inverse, two elements a and b are said to be coprime (or relatively prime) if their greatest common divisor is an element with a multiplicative inverse within the ring. In the context of the factoring into lifting steps[2], the $l \times c$ matrix $\mathbf{A}_l$ and $r \times c$ matrix $\mathbf{A}_r$ are said to be coprime (or relatively prime) if repeated factoring of the matrix $\mathbf{A}_1 = \begin{bmatrix} \mathbf{A}_l^\top & \mathbf{A}_r^\top \end{bmatrix}^\top$ using the routine in Theorem 4.1.3 leads to an invertible diagonal matrix.*

The factoring step in Theorem 4.1.4 can be written as

$$\begin{bmatrix} \mathbf{H}_{j,e}^{[s]} \\ \mathbf{H}_{j,o}^{[s]} \end{bmatrix} = \begin{bmatrix} \mathbf{I}_{n_j} & -\mathbf{U}_j^{[s+1]} \\ \mathbf{0} & \mathbf{I}_{n'_j} \end{bmatrix} \begin{bmatrix} \mathbf{I}_{n_j} & \mathbf{0} \\ \mathbf{P}_j^{[s+1]} & \mathbf{I}_{n'_j} \end{bmatrix} \begin{bmatrix} \mathbf{D}_j \\ \mathbf{0} \end{bmatrix},$$

where $n'_j = n_{j+1} - n_j$.

Let $\mathbf{H}_j$ be a general refinement matrix, so that $\mathbf{H}_{j,e}$ has a larger bandwidth than $\mathbf{H}_{j,o}$; then a full factoring starts off by assigning $\mathbf{H}_j^{[0]} = \mathbf{H}_j$, which is then factored into

$$\mathbf{H}_j = \begin{bmatrix} \mathbf{H}_{j,e} \\ \mathbf{H}_{j,o} \end{bmatrix} = \left(\prod_{s=1}^{u} \begin{bmatrix} \mathbf{I}_{n_j} & -\mathbf{U}_j^{[s]} \\ \mathbf{0} & \mathbf{I}_{n'_j} \end{bmatrix} \begin{bmatrix} \mathbf{I}_{n_j} & \mathbf{0} \\ \mathbf{P}_j^{[s]} & \mathbf{I}_{n'_j} \end{bmatrix} \right) \begin{bmatrix} \mathbf{D}_j \\ \mathbf{0} \end{bmatrix}, \quad (4.29)$$

[2]this context is not a ring, see page 104.

where u is the number of update steps[3]. The refinement matrix can be expanded into a full, invertible two-scale transform by adding independent columns to the last factor, thus defining a detail matrix in

$$\begin{aligned} \begin{bmatrix} \mathbf{H}_j & \mathbf{G}_j^{[0]} \end{bmatrix} &= \begin{bmatrix} \mathbf{H}_{j,e} & \mathbf{G}_{j,e}^{[0]} \\ \mathbf{H}_{j,o} & \mathbf{G}_{j,o}^{[0]} \end{bmatrix} \\ &= \left(\prod_{s=1}^{u} \begin{bmatrix} \mathbf{I}_{n_j} & -\mathbf{U}_j^{[s]} \\ \mathbf{0} & \mathbf{I}_{n'_j} \end{bmatrix} \begin{bmatrix} \mathbf{I}_{n_j} & \mathbf{0} \\ \mathbf{P}_j^{[s]} & \mathbf{I}_{n'_j} \end{bmatrix} \right) \begin{bmatrix} \mathbf{D}_j & \mathbf{0} \\ \mathbf{0} & \mathbf{I}_{n'_j} \end{bmatrix}. \end{aligned} \tag{4.30}$$

The matrix $\mathbf{G}_j^{[0]}$ defines a **primitive wavelet** basis

$$\Psi_j^{[0]}(x) = \Phi_{j+1}(x)\mathbf{G}_j^{[0]}, \tag{4.31}$$

so that

$$\Phi_{j+1}(x)\boldsymbol{s}_{j+1} = \Phi_j(x)\boldsymbol{s}_j^{[0]} + \Psi_j^{[0]}(x)\boldsymbol{d}_j.$$

The primitive wavelet basis is not really a wavelet basis, because the basis functions have not been designed to be wavelets, to have zero integrals, that is. Section 4.1.5 will discuss the upgrading of $\mathbf{G}_j^{[0]}$ to a proper wavelet transform. The upgrade will affect the detail basis $\Psi_j^{[0]}(x)$ and the scaling coefficients $\boldsymbol{s}_j^{[0]}$.

Remark 4.1.6 *With $\mathbf{A}_j$ a non-singular $n'_j \times n'_j$ matrix, replacing*

$$\begin{bmatrix} \mathbf{D}_j & \mathbf{0} \\ \mathbf{0} & \mathbf{I}_{n'_j} \end{bmatrix} \text{ by } \begin{bmatrix} \mathbf{D}_j & \mathbf{0} \\ \mathbf{0} & \mathbf{A}_j \end{bmatrix}$$

leads to a primitive wavelet transform with detail matrix $\mathbf{G}_j^{[0]}\mathbf{A}_j$ and detail basis $\Psi_j^{[0]}(x)\mathbf{A}_j$ still spanning the same subspace.

Expression (4.30) is the factoring of a multiscale synthesis, i.e., the inverse wavelet transform $\boldsymbol{s}_{j+1} = \mathbf{H}_j\boldsymbol{s}_j^{[0]} + \mathbf{G}_j^{[0]}\boldsymbol{d}_j$. The forward wavelet transform, i.e., the wavelet analysis follows from inversion or perfect reconstruction,

$$\begin{aligned} \begin{bmatrix} \widetilde{\mathbf{H}}_j^{[0]\top} \\ \widetilde{\mathbf{G}}_j^{\top} \end{bmatrix} &= \begin{bmatrix} \mathbf{H}_j & \mathbf{G}_j^{[0]} \end{bmatrix}^{-1} \\ &= \begin{bmatrix} \mathbf{D}_j^{-1} & \mathbf{0} \\ \mathbf{0} & \mathbf{I}_{n'_j} \end{bmatrix} \left(\prod_{s=0}^{u-1} \begin{bmatrix} \mathbf{I}_{n_j} & \mathbf{0} \\ -\mathbf{P}_j^{[q-s]} & \mathbf{I}_{n'_j} \end{bmatrix} \begin{bmatrix} \mathbf{I}_{n_j} & \mathbf{U}_j^{[q-s]} \\ \mathbf{0} & \mathbf{I}_{n'_j} \end{bmatrix} \right). \end{aligned} \tag{4.32}$$

[3]Until now, the index s has been used to count the individual lifting steps. From here on, it runs over pairs of lifting steps, each involving one update and one prediction. This way, we avoid working with overloaded indices $2s$ and $2s+1$.

If $\mathbf{H}_{j,o}$ has a larger bandwidth than $\mathbf{H}_{j,e}$, then the full factoring of the refinement matrix is

$$\begin{aligned}\mathbf{H}_j &= \begin{bmatrix}\mathbf{H}_{j,e}\\ \mathbf{H}_{j,o}\end{bmatrix} \qquad (4.33)\\ &= \begin{bmatrix}\mathbf{I}_{n_j} & \mathbf{0}\\ \mathbf{P}_j^{[0]} & \mathbf{I}_{n'_j}\end{bmatrix}\left(\prod_{s=1}^{u}\begin{bmatrix}\mathbf{I}_{n_j} & -\mathbf{U}_j^{[s]}\\ \mathbf{0} & \mathbf{I}_{n'_j}\end{bmatrix}\begin{bmatrix}\mathbf{I}_{n_j} & \mathbf{0}\\ \mathbf{P}_j^{[s]} & \mathbf{I}_{n'_j}\end{bmatrix}\right)\begin{bmatrix}\mathbf{D}_j\\ \mathbf{0}\end{bmatrix}.\end{aligned}$$

4.1.5 Finding detail coefficients for a given subdivision scheme

The factoring of $\mathbf{H}_j$ inserts new lifting steps (4.8) and (4.9) between the existing steps on the **fine scale side** and the remaining refinement in polyphase form on the coarse scale side. Full factoring leads to a sequence of lifting steps always ending with a prediction step on the coarse scale side. This sequence defines a wavelet transform that has $\mathbf{H}_j$ as refinement matrix. It also fixes an associated wavelet equation matrix $\mathbf{G}_j^{[0]}$. This matrix has not been taken into account during the factoring process, and so it is unlikely to be equal to the matrix $\mathbf{G}_j$ that had been proposed in combination with $\mathbf{H}_j$, or the matrix $\mathbf{G}_j$ that satisfies the conditions imposed by the application. In particular, it is unlikely that the wavelet basis $\Psi^{[0]}$ already has a vanishing moment. Nevertheless, two wavelet transforms sharing the same refinement matrix $\mathbf{H}_j$ have factorings that differ only by one update step after the final prediction at **the coarse scale side**, as in (4.11). Indeed, we have the following result.

Theorem 4.1.7 *(Final update step) Let $\mathbf{W}_j = [\mathbf{H}_j\ \mathbf{G}_j]$ and $\mathbf{W}_j^{[0]} = [\mathbf{H}_j\ \mathbf{G}_j^{[0]}]$ be two wavelet transforms with common refinement matrix $\mathbf{H}_j$. The dual detail matrix $\widetilde{\mathbf{G}}_j^\top$, spanning the left null space of $\mathbf{H}_j$ according to the perfect reconstruction in (2.62) is assumed to be the same*[4] *in the two forward transforms. Then there exists an update operation $\mathbf{U}_j$ so that*

$$\mathbf{G}_j = \mathbf{G}_j^{[0]} - \mathbf{H}_j\mathbf{U}_j. \qquad (4.34)$$

More precisely, an explicit form of the update is given by

$$\mathbf{U}_j = \widetilde{\mathbf{H}}_j^\top \mathbf{G}_j^{[0]}, \qquad (4.35)$$

where $\widetilde{\mathbf{H}}_j^\top$ is a dual refinement matrix, satisfying the perfect reconstruction conditions in (2.60) and (2.61). In the forward transform, adding the update $\mathbf{U}_j$ amounts to a final update of the coarse scaling coefficients, before proceeding to the next scale:

$$\boldsymbol{s}_j = \boldsymbol{s}_j^{[0]} + \mathbf{U}_j\boldsymbol{d}_j. \qquad (4.36)$$

[4] A priori, it is to be expected that the dual detail matrices of $\mathbf{W}_j$ and $\mathbf{W}_j^{[0]}$, though spanning the same null space, are not the same. They follow from each other through a matrix multiplication, as in Exercise 2.2.7.

Proof. With $\mathbf{G}_j = \mathbf{G}_j^{[0]} - \mathbf{H}_j\mathbf{U}_j$ and $\mathbf{U}_j = \widetilde{\mathbf{H}}_j^\top \mathbf{G}_j^{[0]}$, the perfect reconstruction condition in (2.61) is satisfied, as indeed

$$\widetilde{\mathbf{H}}_j^\top \mathbf{G}_j = \widetilde{\mathbf{H}}_j^\top (\mathbf{G}_j^{[0]} - \mathbf{H}_j\mathbf{U}_j) = \mathbf{U}_j - \mathbf{I}_{j+1,e}\mathbf{U}_j = \mathbf{0}_{j+1,e,o},$$

while the condition in (2.63) becomes

$$\widetilde{\mathbf{G}}_j^\top \mathbf{G}_j = \widetilde{\mathbf{G}}_j^\top \mathbf{G}_j^{[0]} - \widetilde{\mathbf{G}}_j^\top \mathbf{H}_j\mathbf{U}_j = \mathbf{I}_{j+1,o} - \mathbf{0}_{j+1,o,e}\mathbf{U}_j = \mathbf{I}_{j+1,o}.$$

This is because $\widetilde{\mathbf{G}}_j$, $\mathbf{G}_j^{[0]}$ and $\mathbf{H}_j$ are already three out of four matrices in a perfect reconstruction filterbank. The other two perfect reconstruction conditions do not involve $\mathbf{G}_j$. □

As a consequence, after factoring of $\mathbf{H}_j$ using bidiagonal lifting steps on the fine scale side, the detail matrix $\mathbf{G}_j$ is achieved from $\mathbf{G}_j^{[0]}$ by adding a single update step $\mathbf{U}_j$ at the coarse scale side. The update $\mathbf{U}_j$ may have more than two nonzeros in each column. If $\mathbf{U}_j$ is used for vanishing moments in $\Psi_j(x)$, then the design of $\mathbf{U}_j$ follows from integration of (4.13).

Exercise 4.1.8 *How can the factoring of the refinement $\mathbf{H}_j$ be used for the construction of a full semi-orthogonal wavelet transform?*

Exercise 4.1.9 *Using the expression $\mathbf{H}_j = \widetilde{\mathbf{J}}_j + \widetilde{\mathbf{J}}_j^o\mathbf{P}_j$, derived in Exercise 3.1.2, check that $\widetilde{\mathbf{H}}_j = \widetilde{\mathbf{J}}_j$ is an admissible projection onto an interpolating scaling basis. Find the general form of the final update in a hierarchical interpolating basis transform from the factoring of the refinement equations.*

Exercise 4.1.10 *Let both $\mathbf{H}_j$ and $\mathbf{G}_j^{[1]}$ be given. Develop a wavelet transform using the factoring of $\mathbf{H}_j$ into lifting steps. Use the results of Exercise 2.2.7 to deal with the fact that the $\widetilde{\mathbf{G}}_j$ from the factoring algorithm spans the same null space without necessarily being the same matrix as the $\widetilde{\mathbf{G}}_j^{[1]}$ satisfying the perfect reconstruction property with $\mathbf{H}_j$ and $\mathbf{G}_j^{[1]}$. Apply to a non-interpolating, hierarchical basis, defined by $\Psi_j(x) = \Phi_{j+1,e}(x)$.*

4.2 B-splines and B-spline wavelets

4.2.1 Situation: piecewise polynomials as scaling functions

The lifting scheme itself, developed in Section 2.1, provides the tools to construct interpolating wavelet transforms in Section 3.2 or the average interpolating transforms in Section 3.3. The design of these wavelet transforms is based on the properties of the lifting steps. This section starts from a different viewpoint by imposing properties not on the lifting operations in the wavelet transform, but rather on the wavelet basis that can be associated to the transform. The factoring of the transform into lifting steps is used to translate the imposed properties into properties of the lifting steps. This section develops the important class of wavelets with B-splines as scaling functions.

Spline functions are piecewise polynomials, parametrised by a polynomial degree but also by a set of knots in which the polynomial segments meet in a continuous way. Unlike genuine polynomials, the space of splines of a certain degree and with a given set of knots can be generated by a set of compactly supported basis functions, termed **B-splines**. The B-splines are now used as the scaling basis in a multiscale decomposition. Unlike the interpolating wavelet transforms of Section 3.2, the B-spline wavelet transforms in this section start from scaling functions with a closed formula, not depending on the refinement relation. Otherwise stated, in contrast to the Deslauriers-Dubuc wavelets, the expression for the B-spline scaling functions depends only on the current scale, not on further refinement. This way, we have full control over the smoothness and other properties of the basis.

The current section provides straightforward arguments for the existence of refinement and lifting schemes for splines and spline wavelets. The design of these schemes, however, requires the use of the factoring procedure of Section 4.1. The factoring procedure, developed as a way to break down a given wavelet transform into lifting steps, is thus used here in the design of a new transform.

In the setting of equidistant knots, the wavelets constructed in this section are well known as the **Cohen-Daubechies-Feauveau** (CDF) family (Cohen et al., 1992). One of the most important fields of applications of the CDF family is their use in image processing and compression.

4.2.2 Splines and B-splines

A spline is a real function, denoted here by $f_j(x)$. The index j refers to the vector of knots $\boldsymbol{x}_j$ that appears as a parameter in the definition beneath of a spline. The elements of the knot vector are denoted by $x_{j,k}$, with $k \in \{0, \ldots, n_j - 1\}$. In a later stage, this grid of knots will be refined or it will be subsampled, as in Definition 2.1.1. The index j then refers to the resolution level of the grid. Unless otherwise specified, the knots are supposed to be ordered and non-coincident, i.e., $x_{j,0} < x_{j,1} < \ldots < x_{j,n_j-1}$. Another convention, tacitly adopted throughout this section, is that $f_j(x)$ is defined on the interval $\left[x_{j,0}, x_{j,n_j-1}\right]$.

Definition 4.2.1 *(splines) A spline (de Boor, 2001) of order $\widetilde{p}$, defined on the ordered vector of non-coincident knots $\boldsymbol{x}_j$, is a $\widetilde{p} - 2$ times continuously differentiable function that takes the form of a polynomial of degree $\widetilde{p} - 1$ on each open interval $(x_{j,k-1}, x_{j,k})$ between two successive knots.*

In other words, a spline of **order** $\widetilde{p}$, sometimes referred to as a spline of

degree $\widetilde{p}-1$, is a piecewise polynomial of degree $\widetilde{p}-1$ with $\widetilde{p}-2$ continuous derivatives at the interior knots $x_{j,k}$, with $k \in \{1,\ldots,n_j-2\}$.

Let $\mathcal{S}(\boldsymbol{x}_j,\widetilde{p})$ be the vector space of spline functions of order $\widetilde{p}$ defined on the vector of knots $\boldsymbol{x}_j$; then B-splines will serve as a basis for $\mathcal{S}(\boldsymbol{x}_j,\widetilde{p})$.

The basis functions will be denoted by $\varphi_{j,k}^{[\widetilde{p}]}(x)$. There are several ways to defined the B-spline basis functions. In the first instance, we adopt the following recursion (Qu and Gregory, 1992; Daubechies et al., 2001, page 497).

Definition 4.2.2 *(B-splines) B-splines of order 1 defined on the n_j-tuple of knots $\boldsymbol{x}_j$ are the n_j-1 characteristic functions $\varphi_{j,k}^{[0]}(x)=\chi_{j,k}(x)$, where $k=0,\ldots,n_j-2$ and $\chi_{j,k}(x)=1 \Leftrightarrow x\in[x_{j,k},x_{j,k+1})$ while $\chi_{j,k}(x)=0$ otherwise. B-splines of order 1 are also known as B-splines of degree 0. If $k\notin\{0,\ldots,n_j-2\}$, we formally define $\varphi_{j,k}^{[0]}(x)=0$ for all values of x.*

B-splines of order $\widetilde{p}$, i.e., degree $\widetilde{p}-1$, for $\widetilde{p}>0$, are defined recursively as

$$\begin{aligned}\varphi_{j,k}^{[\widetilde{p}]}(x) &= \frac{x-x_{j,k-\lfloor\widetilde{p}/2\rfloor}}{x_{j,k+\lceil\widetilde{p}/2\rceil-1}-x_{j,k-\lfloor\widetilde{p}/2\rfloor}}\varphi_{j,k-1+\mathrm{rem}(\widetilde{p}/2)}^{[\widetilde{p}-1]}(x)\\ &+ \frac{x_{j,k+\lceil\widetilde{p}/2\rceil}-x}{x_{j,k+\lceil\widetilde{p}/2\rceil}-x_{j,k-\lfloor\widetilde{p}/2\rfloor+1}}\varphi_{j,k+\mathrm{rem}(\widetilde{p}/2)}^{[\widetilde{p}-1]}(x).\end{aligned} \tag{4.37}$$

In this equation $\mathrm{rem}(p/q)=p-q\lfloor p/q\rfloor$ denotes the remainder from an integer division. Furthermore, the recursion includes knots with indices outside $\{0,\ldots,n_j-1\}$. These knots are defined to coincide with one of the end knots. In particular, $x_{j,l}=x_{j,0}$ for $l<0$ and $x_{j,r}=x_{j,n_j-1}$ for $r>n_j-1$. In all cases where two coincident knots show up together in one of the denominators of (4.37), the accompanying $\varphi_{j,\ell}^{[\widetilde{p}-1]}(x)$ turns out to be the zero function. By convention, the resulting term of (4.37) is set to zero.

In Section 4.2.5, the recursion will be replaced by a refinement scheme as an equivalent definition of B-splines.

As suggested by the notation, there is a link between basis function $\varphi_{j,k}^{[\widetilde{p}]}(x)$ and knot $x_{j,k}$: the basis function is more or less centred around the corresponding knot, as follows from the theorem below. The main result of the theorem is that B-splines are indeed splines.

Theorem 4.2.3 *(Piecewise polynomials on bounded intervals)*
For $k\in\{\lfloor\widetilde{p}/2\rfloor,\ldots,n_j-1-\lceil\widetilde{p}/2\rceil\}$ the function $\varphi_{j,k}^{[\widetilde{p}]}(x)$ is zero outside the interval $S_{j,k}=[x_{j,l_k},x_{j,r_k})$, where $l_k=k-\lfloor\widetilde{p}/2\rfloor$ and $r_k=k+\lceil\widetilde{p}/2\rceil$. Inside this interval, $\varphi_{j,k}^{[\widetilde{p}]}(x)$ is a polynomial of degree $\widetilde{p}-1$ between two knots $x_{j,k}$ and $x_{j,k+1}$, while at the knots, the function and its first $\widetilde{p}-2$ derivatives are continuous.

The proof of Theorem 4.2.3 follows by induction, using Definition 4.2.2.

Remark 4.2.4 *The result in Theorem 4.2.3 should be amended for basis functions near the boundaries, more precisely when the left knot index l_k is negative or when the right knot index r_k is beyond the number of knots. Therefore, Theorem 4.2.3, as well as some of the subsequent results, holds on the interior interval I_j, defined as*

$$I_j = \left[x_{j,\widetilde{p}-1}, x_{j,n_j-\widetilde{p}}\right]. \tag{4.38}$$

It then holds that $x \in I_j$ if and only if all basis functions $\varphi_{j,k}^{[\widetilde{p}]}(x)$ with nonzero value in x have a left knot index $l_k \geq 0$ and a right knot index $r_k \leq n_j - 1$.

Remark 4.2.5 *An alternative notation, far more popular in the spline literature than the $\varphi_{j,k}^{[\widetilde{p}]}(x)$, uses a shifted indexing, defined by*

$$\varphi_{j,k}^{[\widetilde{p}]}(x) = N_{j,k-\lfloor\widetilde{p}/2\rfloor}^{[\widetilde{p}]}(x).$$

In the notation $N_{j,k}^{[\widetilde{p}]}(x)$, the index refers to the left end point of the support of the B-spline, rather than to the knot in the middle. In general, the end point indexing leads to more elegant expressions in the recursion expressions of B-splines and in the proofs of the results below. In wavelet literature, however, it is more common to have a basis function indexed according to its centre. This leads to more elegant expressions for bases and coefficients at different scales, but linked to one, given location.

The following lemma states that B-splines of order $\widetilde{p}$ are the only splines with support bounded by $\widetilde{p}+1$ knots.

Lemma 4.2.6 *Using the notations l_k and r_k as in Theorem 4.2.3, let $s_{j,k}(x)$ be a spline of order $\widetilde{p}$, taking the form*

$$\sum_{q=0}^{\widetilde{p}-1} a_{l,q} x^q$$

on the interval $[x_{j,l}, x_{j,l+1}]$ for $l \in \mathcal{I}_k = \{l_k, l_k+1, \ldots, r_k\}$, and zero otherwise. If $a_{l,\widetilde{p}-1} \neq 0$ for at least one $l \in \mathcal{I}_k$, then $s_{j,k}(x) = a\varphi_{j,k}^{[\widetilde{p}]}(x)$, is a multiple of the B-spline defined in Definition 4.2.2.

The proof follows from establishing the set of equations for the coefficients $a_{l,q}$, leading to a unique solution.

Exercise 4.2.7 *Count all conditions that lead to the equations in $a_{l,q}$ defining the B-spline and check that these fix the spline up to a constant.*

An induction argument, similar to the one used for Theorem 4.2.3, leads to an expression for the derivatives of a B-spline, which we mention here without detailed proof.

Lemma 4.2.8 *The derivative of a B-spline is given by*

$$\frac{d}{dx}\varphi_{j,k}^{[\widetilde{p}]}(x) = (\widetilde{p}-1)\left[\frac{\varphi_{j,k-1+\text{rem}(\widetilde{p}/2)}^{[\widetilde{p}-1]}(x)}{x_{j,k+\lceil\widetilde{p}/2\rceil-1} - x_{j,k-\lfloor\widetilde{p}/2\rfloor}} - \frac{\varphi_{j,k+\text{rem}(\widetilde{p}/2)}^{[\widetilde{p}-1]}(x)}{x_{j,k+\lceil\widetilde{p}/2\rceil} - x_{j,k-\lfloor\widetilde{p}/2\rfloor+1}}\right]. \tag{4.39}$$

4.2.3 B-splines as basis functions

From Theorem 4.2.3 we know that B-splines are indeed splines. We now argue that B-splines may serve as a basis. A straightforward argument by induction reveals the dimension of the vector space of spline functions $\mathcal{S}(\boldsymbol{x}_j, \widetilde{p})$.

Lemma 4.2.9 *On n_j knots, we have $n_j + \widetilde{p} - 2$ nonzero B-splines of order $\widetilde{p}$.*

This result is in accordance with the dimension of $\mathcal{S}(\boldsymbol{x}_j, \widetilde{p})$. Indeed, on $n_j - 1$ intervals between n_j knots, the definition of polynomials of degree $\widetilde{p} - 1$ involves $(n_j - 1)\widetilde{p}$ coefficients. The $(n_j - 2)(\widetilde{p} - 1)$ continuity conditions for the function values and derivatives at the interior knots reduce the number of degrees of freedom to $n_j + \widetilde{p} - 2$.

From Theorem 4.2.3, it follows that all linear combinations of B-splines

$$f_j(x) = \sum_{k=-\lfloor\widetilde{p}/2\rfloor}^{n_j-2+\lceil\widetilde{p}/2\rceil} s_{j,k}\varphi_{j,k}^{[\widetilde{p}]}(x), \tag{4.40}$$

are spline functions of order $\widetilde{p}$ on the given set of knots. Conversely, it can be verified that any $\widetilde{p}$th order spline on $\boldsymbol{x}_j$ can be decomposed into a linear combination of B-splines.

Theorem 4.2.10 *(B-spline basis) Let $f_j(x)$ be a $\widetilde{p}$th order spline on n_j-tuple of knots $\boldsymbol{x}_j$; then there exist constants $s_{j,k}$ so that for all $x \in [x_{j,0}, x_{j,n_j-1}]$, the function $f_j(x)$ can be written as in (4.40).*

Proof. The function $f_j(x)$, being a spline, takes the form of a polynomial of degree $\widetilde{p}-1$ on each subinterval $[x_{j,k}, x_{j,k+1}]$, while $\widetilde{p}-1$ continuity conditions hold in each knot except for the two end points. This amounts to a dimension of $n_j - 1$ subinterval times $\widetilde{p}$ parameters on each subinterval, minus $n_j - 2$ knots times $\widetilde{p}-1$ restrictions, which is $n_j+\widetilde{p}-2$. This is exactly the number of B-splines available on the interval. Linear independence of the B-splines can be verified by calculating for each knot the $\widetilde{p} - 1$ derivatives of all the basis functions, using Lemma 4.2.8, leading to a non-singular matrix of derivative values. The details are beyond the scope of this text. □

Figure 4.1 depicts the basis functions in a **cubic B-spline basis** on $n_j = 7$ knots. Cubic B-splines are B-splines of degree three, i.e., of order four. The dimension of the vector space of spline functions equals $n_j+\widetilde{p}-2 = 7+4-2 = 9$.

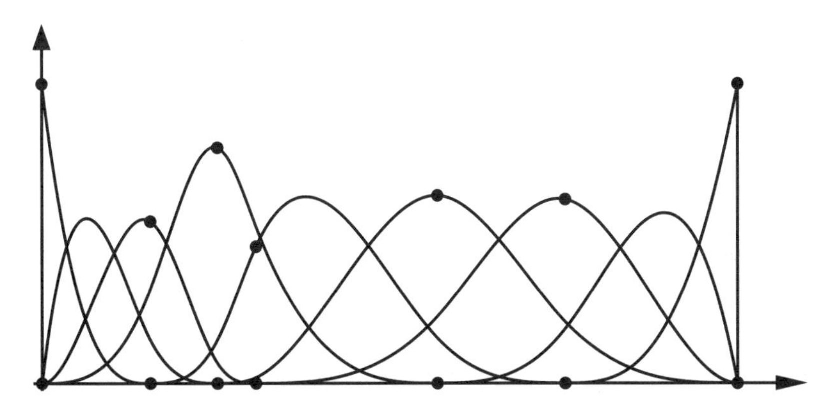

Figure 4.1
Cubic B-spline basis (i.e., $\widetilde{p} = 4$) on $n_j = 7$ knots.

As a consequence of Lemma 4.2.8, a linear combination of the B-splines

$$f_j(x) = \sum_{k\in\mathbb{Z}} s_{j,k}^{[\widetilde{p}]} \varphi_{j,k}^{[\widetilde{p}]}(x), \tag{4.41}$$

has a derivative equal to

$$f_j'(x) = (\widetilde{p}-1) \sum_{k\in\mathbb{Z}} \frac{s_{j,k}^{[\widetilde{p}]} - s_{j,k-1}^{[\widetilde{p}]}}{x_{j,k+\lceil\widetilde{p}/2\rceil-1} - x_{j,k-\lfloor\widetilde{p}/2\rfloor}} \varphi_{j,k-\widetilde{r}'}^{[\widetilde{p}-1]}(x), \tag{4.42}$$

where $\widetilde{r}' = 1 - \mathrm{rem}(\widetilde{p}/2)$.

4.2.4 B-splines and polynomials

For obvious reasons, a polynomial of degree $\widetilde{p}-1$ can be decomposed in a basis of piecewise polynomials of the same degree. The decomposition is, however, nontrivial. This is in contrast to Deslauriers-Dubuc interpolating scaling functions, where, according to (3.25) in Theorem 3.2.1, the function values can be taken as scaling coefficients. In a B-spline basis, Definition 4.2.2 normalises the basis functions so that at least for $q = 0$ the decomposition is trivial,

$$\forall x \in I_j : \sum_{k=0}^{n_j-1} \varphi_{j,k}^{[\widetilde{p}]}(x) = 1, \tag{4.43}$$

for any order $\widetilde{p}$. This property is referred to as the **partition of unity**.

The coefficients of a higher order power function are given by the following theorem, closely related to Marsden's identity (Lee, 1996).

Theorem 4.2.11 *(power coefficients in B-spline bases) For* $q = 0, 1, \ldots, \widetilde{p}-1$, *there exist coefficients* $\widetilde{x}_{j,k}^{[\widetilde{p},q]}$, *so that for* $x \in I_j$,

$$x^q = \sum_{k\in\mathbb{Z}} \widetilde{x}_{j,k}^{[\widetilde{p},q]} \varphi_{j,k}^{[\widetilde{p}]}(x). \tag{4.44}$$

1. For $q = 0$, *these coefficients are one, according to the partition of unity.*

2. For $q = 1, \ldots, \widetilde{p}-1$, *the coefficients can be found by the recursion*

$$\widetilde{x}_{j,k}^{[\widetilde{p},q]} = \widetilde{x}_{j,k-1}^{[\widetilde{p},q]} + \frac{q}{\widetilde{p}-1} \widetilde{x}_{j,k-\widetilde{r}'}^{[\widetilde{p}-1,q-1]} \left(x_{j,k+\lceil \widetilde{p}/2\rceil -1} - x_{j,k-\lfloor \widetilde{p}/2\rfloor}\right), \tag{4.45}$$

where $\widetilde{r}' = 1 - \mathrm{rem}(\widetilde{p}/2)$, *as in (4.42).*

3. In particular, for $q = 1$, *these coefficients satisfy*

$$\widetilde{x}_{j,k}^{[\widetilde{p},1]} = \frac{1}{\widetilde{p}-1} \sum_{i=1-\lfloor \widetilde{p}/2\rfloor}^{\lceil \widetilde{p}/2\rceil -1} x_{j,k+i}. \tag{4.46}$$

4. For $q = \widetilde{p}-1$, *this becomes*

$$\widetilde{x}_{j,k}^{[\widetilde{p},\widetilde{p}-1]} = \prod_{i=1-\lfloor \widetilde{p}/2\rfloor}^{\lceil \widetilde{p}/2\rceil -1} x_{j,k+i}. \tag{4.47}$$

Proof.

1. The result follows from induction, using (4.37).
2. Expression (4.45) follows from (4.42), applied for $f_j(x) = x^q$. The left hand side of (4.42) can then be written as

$$qx^{q-1} = q \sum_{k\in\mathbb{Z}} \widetilde{x}_{j,k-\widetilde{r}'}^{[\widetilde{p}-1,q-1]} \varphi_{j,k-\widetilde{r}'}^{[\widetilde{p}-1]}(x),$$

whereas the right hand side becomes

$$qx^{q-1} = (\widetilde{p}-1) \sum_{k\in\mathbb{Z}} \frac{\widetilde{x}_{j,k}^{[\widetilde{p},q]} - \widetilde{x}_{j,k-1}^{[\widetilde{p},q]}}{x_{j,k+\lceil \widetilde{p}/2\rceil -1} - x_{j,k-\lfloor \widetilde{p}/2\rfloor}} \varphi_{j,k-\widetilde{r}'}^{[\widetilde{p}-1]}(x).$$

Identification of the terms in the decomposition leads to (4.45).

3. With $q = 1$, solving the recursion of (4.45) for $\widetilde{x}_{j,k}^{[\widetilde{p},1]}$ leads to a telescoping sum, which simplifies to

$$\widetilde{x}_{j,k}^{[\widetilde{p},1]} = \frac{1}{\widetilde{p}-1} \sum_{i=1-\lfloor \widetilde{p}/2\rfloor}^{\lceil \widetilde{p}/2\rceil -1} x_{j,k+i} - \frac{1}{\widetilde{p}-1} \sum_{i=1-\lfloor \widetilde{p}/2\rfloor}^{\lceil \widetilde{p}/2\rceil -1} x_{j,i} + x_{j,0}^{[\widetilde{p},1]}.$$

Moreover, $\widetilde{x}_{j,k}^{[\widetilde{p},1]}$ must be independent from the $\widetilde{p}-1$ knots $x_{j,i}$ around $x_{j,0}$ if $k > \widetilde{p}$. So take $x_{j,i}$ symmetric around $x_{j,0} = 0$. Then, obviously,

$$\frac{1}{\widetilde{p}-1} \sum_{i=1-\lfloor \widetilde{p}/2 \rfloor}^{\lceil \widetilde{p}/2 \rceil - 1} x_{j,i} = 0.$$

Finally, the initial value $x_{j,0}^{[\widetilde{p},1]} = 0$, as the corresponding basis function is even. This leads to (4.46).

4. The proof for (4.47) is similar to that for (4.46), following an induction argument on $\widetilde{p}$.

□

The following lemma and corollary combine the power function coefficients above with the bounded support from Theorem 4.2.3. The combination fully characterises the basis, thus motivating the use of the power function coefficients $\widetilde{x}_{j,m}^{[\widetilde{p},q]}$ as a defining property in the design of a refinement scheme for B-splines.

Lemma 4.2.12 *(power coefficients and bounded support) Let $x_{j,k}$ for $k = 0, \ldots, n_j - 1$ be the knots at level j, and let $\varphi_{j,k}^{[\widetilde{p}]}(x)$ be a set of basis functions associated with these knots, satisfying the following two conditions.*

1. The support of $\varphi_{j,k}^{[\widetilde{p}]}(x)$ equals $S_{j,k} = [x_{j,l_k}, x_{j,r_k})$, with l_k and r_k as in Theorem 4.2.3.

2. There exist coefficients $a_{j,k}^{[\widetilde{p},q]}$ so that all power functions up to degree $\widetilde{p}-1$ on $[x_{j,0}, x_{j,n_j})$ can be represented as a linear combination of the $\varphi_{j,k}^{[\widetilde{p}]}(x)$.

Then on each interval $(x_{j,k}, x_{j,k+1})$ the $\widetilde{p}$ basis functions $\varphi_{j,\ell}^{[\widetilde{p}]}(x)$ with $\ell \in \{k + 1 - \lceil \widetilde{p}/2 \rceil, \ldots, k + \lfloor \widetilde{p}/2 \rfloor\}$ take the form of $\widetilde{p}$ linearly independent polynomials.

Proof. On the interknot interval $(x_{j,k}, x_{j,k+1})$ we have $\widetilde{p}$ nonzero basis functions, because

$$\begin{aligned} &(x_{j,k}, x_{j,k+1}) \subset S_{j,\ell} = \left[x_{j,\ell-\lfloor \widetilde{p}/2 \rfloor}, x_{j,\ell+\lceil \widetilde{p}/2 \rceil}\right] \\ &\Leftrightarrow k+1 \leq \ell + \lceil \widetilde{p}/2 \rceil \text{ and } k \geq \ell - \lfloor \widetilde{p}/2 \rfloor \\ &\Leftrightarrow k+1-\lceil \widetilde{p}/2 \rceil \leq \ell \leq k + \lfloor \widetilde{p}/2 \rfloor . \end{aligned}$$

On this interval, the $\widetilde{p}$ power functions are linearly independent and they are linear combinations of the $\widetilde{p}$ nonzero basis functions. As a result, the $\widetilde{p} \times \widetilde{p}$ matrix that maps the basis functions onto the power functions is invertible. Hence, the basis functions on that subinterval are linear combinations of power functions, thus taking the form of polynomials on each subinterval.
□

Lemma 4.2.12 implies that the $\widetilde{p}$ nonzero basis functions are really nonzero on the interval $(x_{j,k}, x_{j,k+1})$. In other words, the support of the basis functions stretches over exactly $\widetilde{p}$ interknot intervals, neither more nor less.

The argument above constructs a basis of piecewise polynomials for a given set of $a_{j,k}^{[\widetilde{p},q]}$, as long as all $\widetilde{p} \times \widetilde{p}$ matrices constructed with the elements from the set are invertible. The continuity conditions fix the choice of $a_{j,k}^{[\widetilde{p},q]}$ to be equal to the elements $\widetilde{x}_{j,k}^{[\widetilde{p},q]}$ in Theorem 4.2.11.

Corollary 4.2.13 *(B-splines defined by power coefficients) Let $x_{j,k}$ for $k = 0, \ldots, n_j - 1$ be the knots $x_{j,k}$ at level j, and let $\varphi_{j,k}^{[\widetilde{p}]}(x)$ be a set of basis functions associated with these knots, satisfying the bounded support condition of Theorem 4.2.3 and the power coefficient condition of Theorem 4.2.11. Then $\varphi_{j,k}^{[\widetilde{p}]}(x)$ must be the B-spline basis.*

Proof. By Lemma 4.2.12 the basis functions must be piecewise polynomials. The continuity conditions at the knots are fulfilled because the power coefficients are fixed as in Theorem 4.2.11, and because the choice of the coefficients uniquely defines the associated basis of piecewise polynomials. □

4.2.5 B-spline refinement

In order to find a two-scale equation for B-splines, we first verify its existence (Qu and Gregory, 1992; Daubechies et al., 2001, Eq.(26), page 497).

Theorem 4.2.14 *(existence of B-spline refinement) On a nested multilevel grid, B-spline basis functions at level j are refinable; i.e., there exists a refinement matrix $H_j^{[\widetilde{p}]}$ so that, for all $x \in I_j$,*

$$\varphi_{j,k}^{[\widetilde{p}]}(x) = \sum_{\ell=0}^{n_{j+1}} H_{j,\ell,k}^{[\widetilde{p}]} \varphi_{j+1,\ell}^{[\widetilde{p}]}(x), \tag{4.48}$$

or, writing $\Phi_j^{[\widetilde{p}]}(x)$ for the row vector of $\widetilde{p}$ order B-splines in $\boldsymbol{x}_j$,

$$\Phi_j^{[\widetilde{p}]}(x) = \Phi_{j+1}^{[\widetilde{p}]}(x)\mathbf{H}_j^{[\widetilde{p}]}. \tag{4.49}$$

Proof. This is an immediate consequence of Theorems 4.2.3 and 4.2.10. A B-spline on a grid at level j is a piecewise polynomial with knots in $x_{j,k}$. Since these knots are also knots at level $j+1$, the B-spline is also a piecewise polynomial at level $j+1$, and so it can be written as a linear combination of the B-spline basis at that resolution level. □

Equation (4.48) is an instance of a two-scale equation, whose generic matrix-vector form is stated in (2.64).

The objective is to find this refinement matrix. With this matrix, we can construct the B-spline basis using subdivision instead of the recursion (4.37). One of the advantages of subdivision is of course its immediate connection to a multiscale decomposition. Another benefit is that it will be easier to deal with the boundaries of a finite interval.

The matrix $\mathbf{H}_j^{[\widetilde{p}]}$ will be band-limited.

Lemma 4.2.15 *(Band-limited refinement matrices)*[5] *Let* $\varphi_{j,k}^{[\widetilde{p}]}(x)$*, for* $k = 0, \ldots, n_j - 1$ *be the order* $\widetilde{p}$ *B-spline basis associated to the knots* $x_{j,k}$*. Let* $S_{j,k}$ *denote the support of* $\varphi_{j,k}^{[\widetilde{p}]}(x)$*, as specified in Theorem 4.2.3. Then an entry* $H_{j,\ell,k}^{[\widetilde{p}]}$ *in the refinement matrix* $\mathbf{H}_j^{[\widetilde{p}]}$ *may be different from zero only if* $S_{j+1,\ell} \subset S_{j,k}$*.*

Proof. Suppose that $H_{j,\ell,k}^{[\widetilde{p}]}$ would be nonzero for some fine scaling function outside the support of the coarse scaling function. Then we would have a nontrivial combination of the zero function using the nonzero segments of the piecewise polynomials, which is impossible by the result of Lemma 4.2.12. □

For the B-spline basis, Lemma 4.2.15 translates as follows.

Corollary 4.2.16 *The columns of the matrix* $H_j^{[\widetilde{p}]}$ *in (4.48) can have at most* $\widetilde{p}+1$ *nonzero elements. In particular*

$$H_{j,\ell,k}^{[\widetilde{p}]} \neq 0 \Rightarrow 2k - \lfloor \widetilde{p}/2 \rfloor \leq \ell \leq 2k + \lceil \widetilde{p}/2 \rceil . \tag{4.50}$$

Proof. The support of a B-spline $\varphi_{j,k}^{[\widetilde{p}]}(x)$ is $S_{j,k} = [x_{j,k-\lfloor \widetilde{p}/2 \rfloor}, x_{j,k+\lceil \widetilde{p}/2 \rceil}]$. We have

$$\begin{aligned}
S_{j+1,\ell} \subset S_{j,k} \quad &\Leftrightarrow \quad \left[x_{j+1,\ell-\lfloor \widetilde{p}/2 \rfloor}, x_{j+1,\ell+\lceil \widetilde{p}/2 \rceil}\right] \subset \left[x_{j,k-\lfloor \widetilde{p}/2 \rfloor}, x_{j,k+\lceil \widetilde{p}/2 \rceil}\right] \\
&\Leftrightarrow \quad \left[x_{j+1,\ell-\lfloor \widetilde{p}/2 \rfloor}, x_{j+1,\ell+\lceil \widetilde{p}/2 \rceil}\right] \\
&\qquad \subset \left[x_{j+1,2k-2\lfloor \widetilde{p}/2 \rfloor}, x_{j+1,2k+2\lceil \widetilde{p}/2 \rceil}\right] \\
&\Leftrightarrow \quad \left\{ \begin{array}{l} \ell - \lfloor \widetilde{p}/2 \rfloor \geq 2k - 2\lfloor \widetilde{p}/2 \rfloor \text{ and} \\ \ell + \lceil \widetilde{p}/2 \rceil \leq 2k + 2\lceil \widetilde{p}/2 \rceil \end{array} \right. \\
&\Leftrightarrow \quad \ell \in \{2k - \lfloor \widetilde{p}/2 \rfloor, \ldots, 2k + \lceil \widetilde{p}/2 \rceil\}.
\end{aligned}$$

As $\#\{2k - \lfloor \widetilde{p}/2 \rfloor, \ldots, 2k + \lceil \widetilde{p}/2 \rceil\} = \lceil \widetilde{p}/2 \rceil + \lfloor \widetilde{p}/2 \rfloor + 1 = \widetilde{p} + 1$, the number of nonzero elements in column k is bounded by $\widetilde{p} + 1$. □

4.2.6 B-spline lifting schemes

Given the band structure of the B-spline refinement matrix $\mathbf{H}_j^{[\widetilde{p}]}$, we can establish the band structures of its even and odd submatrices, for use in Theorem 4.1.3, leading to the factorings (4.29) and (4.33). The bandwidths of $\mathbf{H}_{j,e}^{[\widetilde{p}]}$ and $\mathbf{H}_{j,o}^{[\widetilde{p}]}$ determine the number of bidiagonal lifting operations, leading to the following result.

[5]The band-limited refinement does not hold for general compactly supported refinable bases. As an example, consider transformed spline bases as in Section 3.1.6. making sure that the basis transform matrices are band limited. Then the transformed basis has compact support, however with full refinement matrices.

Lemma 4.2.17 *The factoring of a B-spline refinement matrix* $\mathbf{H}_j^{[\widetilde{p}]}$ *consists of* $u = \lfloor(\widetilde{p}+1)/4\rfloor$ *update steps and* $u+r$ *prediction steps, where* $r = \lceil\widetilde{p}/2\rceil - 2u$ *is a binary value. If* $r = 0$*, then the factoring is given by (4.29); if* $r = 1$*, then the factoring is given by (4.33).*

Filling in the nonzero elements of the transform turns out to be more intuitive in the step by step lifting approach than straight away in the refinement matrix. Indeed, the factoring (4.29) of the refinement matrix in the two-scale equation (4.49) leads to matrix product expression, of which the first factors read

$$
\begin{aligned}
\Phi_j^{[\widetilde{p}]}(x) &= \Phi_{j+1}^{[\widetilde{p}]}(x)\mathbf{H}_j^{[\widetilde{p}]} \\
&= \begin{bmatrix} \Phi_{j+1,e}^{[\widetilde{p}]}(x) & \Phi_{j+1,o}^{[\widetilde{p}]}(x) \end{bmatrix} \begin{bmatrix} \mathbf{I}_{n_j} & -\mathbf{U}_j^{[1]} \\ \mathbf{0} & \mathbf{I}_{n'_j} \end{bmatrix} \begin{bmatrix} \mathbf{I}_{n_j} & \mathbf{0} \\ \mathbf{P}_j^{[1]} & \mathbf{I}_{n'_j} \end{bmatrix} \cdots \\
&= \begin{bmatrix} \Phi_{j+1,e}^{[\widetilde{p}]}(x) - \Phi_{j+1,o}^{[\widetilde{p}]}(x)\mathbf{U}_j^{[1]} & \Phi_{j+1,o}^{[\widetilde{p}]}(x) \end{bmatrix} \begin{bmatrix} \mathbf{I}_{n_j} & \mathbf{0} \\ \mathbf{P}_j^{[1]} & \mathbf{I}_{n'_j} \end{bmatrix} \cdots \\
&= \ldots
\end{aligned}
$$

From this development, it can be seen that the first lifting step, which is an update $\mathbf{U}_j^{[1]}$ in this example, replaces the odd indexed basis functions. The next step, which is a prediction $\mathbf{P}_j^{[1]}$, will affect the even indexed basis functions. The sequence of prediction steps gradually transforms $\Phi_{j+1,e}^{[\widetilde{p}]}(x)$ into $\Phi_j^{[\widetilde{p}]}(x)$.

Because all lifting steps are linear operations, and because the lifting scheme is imposed to factor a B-spline refinement, the intermediate basis functions are known to be spline functions on the knots $\boldsymbol{x}_{j+1}$. As a result, all lifting steps of a B-spline refinement preserve the dual vanishing moments. What changes throughout the lifting scheme is the subset of knots from $\boldsymbol{x}_{j+1}$ on which each basis function rests. The odd knots in which the functions of $\Phi_{j+1,e}^{[\widetilde{p}]}(x)$ have a discontinuous $(\widetilde{p}-1)$st derivative are taken out and replaced by the same number of even knots further away, i.e., on a wide scale, in $\Phi_j^{[\widetilde{p}]}(x)$. The bidiagonal prediction matrices $\mathbf{P}_j^{[s]}$ can be designed such that the $(\widetilde{p}-1)$st derivative jumps in two odd knots are zeroed. The update matrices $\mathbf{U}_j^{[s]}$ annihilate the same derivative jumps for the functions on the odd branch, because otherwise, nonzero derivative jumps at these odd knots would be reintroduced into the even branch functions at the subsequent lifting step $\mathbf{P}_j^{[s+1]}$.

Remark 4.2.18 *The* $(\widetilde{p}-1)$*st derivative conditions at the odd knots can also be stated at once, leading to a set of equations for the elements in the columns of the refinement matrix. As such the refinement matrix can be found without using a lifting factoring. Nevertheless, this approach leads to heavy expressions, in particular when the number of vanishing moments is large.*

The lifting factoring approach keeps the expressions relatively simple, independent from the number of vanishing moments.

As a conclusion, at every lifting step in a B-spline refinement, both input and output scaling functions are B-splines. Each lifting step changes two knots for two other knots, in a gradual process from fine to coarse scale knots.

This principle is further developed for general $\widetilde{p}$ order B-splines in Jansen (2016), using the power function coefficients in Theorem 4.2.11. Section 4.2.7 has more details on a few examples.

4.2.7 Examples of B-spline refinement schemes

For **cubic splines**, i.e., $\widetilde{p} = 4$, we have five nonzeros in each column of $\mathbf{H}_j^{[4]}$. We now develop the refinement matrix for cubic splines, using the general factoring into lifting steps of Sections 4.1.3 and 4.1.4. From (4.18) we know that the even rows of the refinement matrix must factor as

$$\mathbf{H}_{j,e}^{[4]} = \mathbf{D}_j - \mathbf{U}_j \mathbf{H}_{j,o}^{[4]},$$

where $\mathbf{D}_j$ is a diagonal matrix, and $\mathbf{H}_{j,o}^{[4]}$ is bidiagonal. Putting $\mathbf{P}_j^* = \mathbf{H}_{j,o}^{[4]} \mathbf{D}_j^{-1}$, as prescribed in Theorem 4.1.4, the primitive wavelet transform (4.30) becomes

$$\begin{bmatrix} \mathbf{H}_j^{[4]} & \mathbf{G}_j^{[4,0]} \end{bmatrix} = \begin{bmatrix} \mathbf{I}_{n_j} & -\mathbf{U}_j \\ \mathbf{0} & \mathbf{I}_{n'_j} \end{bmatrix} \begin{bmatrix} \mathbf{I}_{n_j} & \mathbf{0} \\ \mathbf{P}_j^* & \mathbf{I}_{n'_j} \end{bmatrix} \begin{bmatrix} \mathbf{D}_j & \mathbf{0} \\ \mathbf{0} & \mathbf{I}_{n'_j} \end{bmatrix}.$$

An equivalent factoring puts $\mathbf{P}_j = \mathbf{H}_{j,o}^{[4]}$ in

$$\begin{bmatrix} \mathbf{H}_j^{[4]} & \mathbf{G}_j^{[4,0]} \end{bmatrix} = \begin{bmatrix} \mathbf{I}_{n_j} & -\mathbf{U}_j \\ \mathbf{0} & \mathbf{I}_{n'_j} \end{bmatrix} \begin{bmatrix} \mathbf{D}_j & \mathbf{0} \\ \mathbf{0} & \mathbf{I}_{n'_j} \end{bmatrix} \begin{bmatrix} \mathbf{I}_{n_j} & \mathbf{0} \\ \mathbf{P}_j^* & \mathbf{I}_{n'_j} \end{bmatrix}. \tag{4.51}$$

The implementation of (4.51) for a cubic spline wavelet transform can be visualised using the flow charts in Figures 4.2 and 4.3. It deviates slightly from a scheme based on (4.30) and Theorem 4.1.4, where the positions of the final prediction $\mathbf{P}_j^*$ and the diagonal matrix $\mathbf{D}_j$ would be switched.

The cubic spline wavelet transform of (4.25) – (4.26) has three matrices $\mathbf{P}_j$, $\mathbf{U}_j$, and $\mathbf{D}_j$. The update elements are given by

$$\begin{aligned} U_{j,k,k-1} &= -\frac{x_{j+1,2k+2} - x_{j+1,2k+1}}{x_{j+1,2k+2} - x_{j+1,2k-2}}, \\ U_{j,k,k} &= -\frac{x_{j+1,2k-1} - x_{j+1,2k-2}}{x_{j+1,2k+2} - x_{j+1,2k-2}}. \end{aligned} \tag{4.52}$$

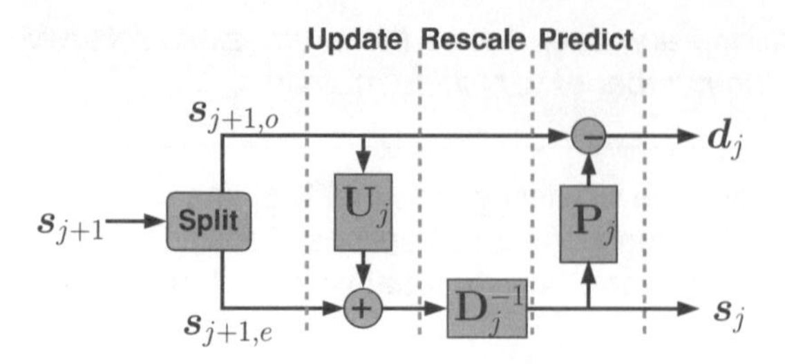

Figure 4.2
Lifting scheme for a wavelet transform with quadratic or cubic B-splines as scaling bases.

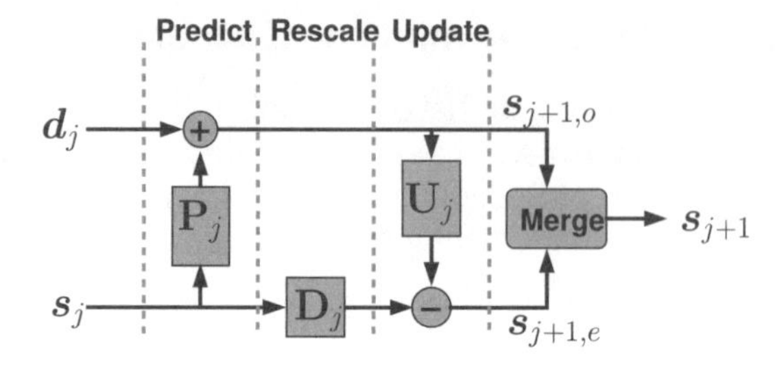

Figure 4.3
Inverse lifting scheme for a wavelet transform with quadratic or cubic B-splines as scaling bases.

The prediction becomes

$$\begin{aligned} P_{j,k,k} &= \frac{x_{j+1,2k+4} - x_{j+1,2k+1}}{x_{j+1,2k+4} - x_{j+1,2k-2}}, \\ P_{j,k,k+1} &= \frac{x_{j+1,2k+1} - x_{j+1,2k-2}}{x_{j+1,2k+4} - x_{j+1,2k-2}}. \end{aligned} \tag{4.53}$$

The **quadratic refinement** matrix $\mathbf{H}_j^{[3]}$ has four nonzeros in each of its columns. Both $\mathbf{H}_{j,e}^{[3]}$ and $\mathbf{H}_{j,o}^{[3]}$ have two nonzeros in each column. In principle, we can propose a factoring

$$\mathbf{H}_{j,o}^{[3]} = \mathbf{D}_j + \mathbf{P}_j \mathbf{H}_{j,e}^{[3]}.$$

As this leaves $\mathbf{H}_{j,e}^{[3]}$ with two nonzeros in each column, further factoring is necessary, according to the analysis in Section 4.1.4. It is better to go by

$$\mathbf{H}_{j,e}^{[3]} = \mathbf{D}_j - \mathbf{U}_j \mathbf{H}_{j,o}^{[3]}.$$

The factoring in Section 4.1.3 assumed that when column l of $\mathbf{H}_{j,e}$ has nonzeros between $k_1(l)$ and $k_2(l)$, then the same column in $\mathbf{H}_{j,o}$ has nonzeros between $k_1(l)$ and $k_2(l)-1$. This can be seen from the summation bounds in Equation (4.21). In our case, we have nonzeros in rows $l-1$ and l of column l in $\mathbf{H}_{j,o}$. In order to apply (4.21), we must take $k_1(l) = l-1$ and $k_2(l) = l+1$. Since $\mathbf{H}_{j,e}$ has only two nonzeros in column l – more precisely on rows l and $l+1$ – we find $\mathbf{H}_{j,e}(k, l_2(k)) = 0$, and thus, using (4.22), $U_{j,k,k} = 0$. The matrix $\mathbf{U}_j$ has only one nonzero in each column, which is $U_{j,k,k-1}$.

A forward wavelet transform based on quadratic spline scaling functions can thus be found by using (4.25), imposing $U_{j,k,k} = 0$. This leads to

$$s_{j,k} = D_{j,k,k}^{-1}(s_{j+1,2k} + U_{j,k,k-1}s_{j+1,2k-1}),$$

with

$$D_{j,k,k} = \frac{x_{j+1,2k+1} - x_{j+1,2k-1}}{x_{j+1,2k+2} - x_{j+1,2k-1}} \tag{4.54}$$

$$U_{j,k,k-1} = -\frac{x_{j+1,2k+2} - x_{j+1,2k+1}}{x_{j+1,2k+2} - x_{j+1,2k-1}}. \tag{4.55}$$

For the prediction we find

$$\begin{aligned} P_{j,k,k} &= \frac{x_{j+1,2k+4} - x_{j+1,2k+1}}{x_{j+1,2k+4} - x_{j+1,2k}}, \\ P_{j,k,k+1} &= \frac{x_{j+1,2k+1} - x_{j+1,2k}}{x_{j+1,2k+4} - x_{j+1,2k}}. \end{aligned} \tag{4.56}$$

The flow charts are the same as for the cubic B-spline wavelets in Figures 4.2 and 4.3.

For the **splines of degree five**, we have seven nonzeros in each column of $\mathbf{H}_j^{[6]}$; this is four nonzeros in $\mathbf{H}_{j,o}^{[6]}$ and three in nonzeros in $\mathbf{H}_{j,e}^{[3]}$. The first step of the factoring is thus

$$\mathbf{H}_{j,o}^{[6]} = \mathbf{H}_{j,o}^{[6,1]} + \mathbf{P}_j^{[1]}\mathbf{H}_{j,e}^{[6]},$$

where $\mathbf{H}_{j,o}^{[6,1]}$ has two nonzeros. The second step is

$$\mathbf{H}_{j,e}^{[6]} = \mathbf{D}_j - \mathbf{U}_j\mathbf{H}_{j,o}^{[6,1]}.$$

As we have a diagonal matrix on the even branch, we can put $\mathbf{H}_{j,o}^{[6,1]} = \mathbf{P}_j^{[2]}$, and propose the following inverse wavelet transform.

$$\begin{aligned} \boldsymbol{s}_{j+1,o}^{[1]} &= \mathbf{P}_j^{[2]}\boldsymbol{s}_j + \boldsymbol{d}_j \\ \boldsymbol{s}_{j+1,e}^{[1]} &= \mathbf{D}_j\boldsymbol{s}_j \\ \boldsymbol{s}_{j+1,e} &= \boldsymbol{s}_{j+1,e}^{[1]} - \mathbf{U}_j\boldsymbol{s}_{j+1,o}^{[1]} \\ \boldsymbol{s}_{j+1,o} &= \boldsymbol{s}_{j+1,o}^{[1]} + \mathbf{P}_j^{[1]}\boldsymbol{s}_{j+1,e} \end{aligned}$$

The forward and inverse transforms are represented in Figures 4.4 and 4.5.

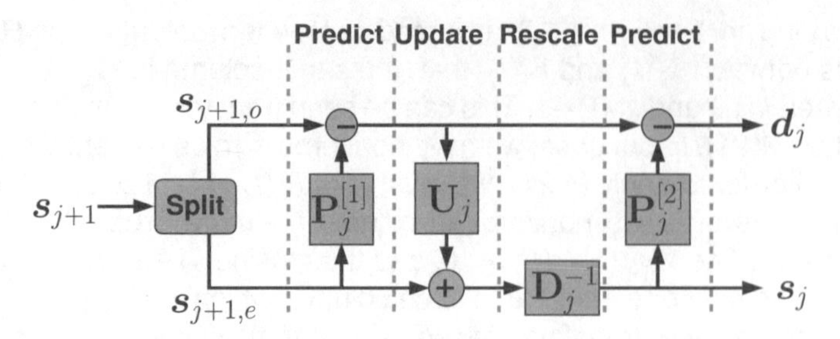

Figure 4.4
Lifting scheme for a wavelet transform with degree 5 B-splines as scaling bases.

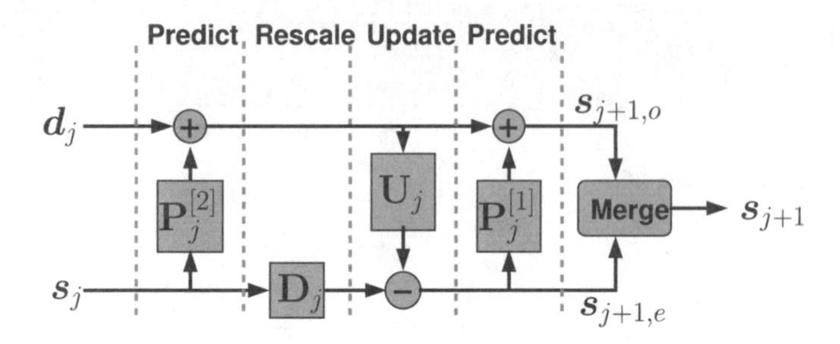

Figure 4.5
Inverse lifting scheme for a wavelet transform with degree 5 B-splines as scaling bases.

4.2.8 Final updates for B-spline wavelet transforms

The previous sections have developed a refinement scheme for B-spline scaling functions. According to Theorem 4.1.7, this refinement scheme can be incorporated into a full wavelet transform by adding one final update step. The design of the final update may focus on adding primal vanishing moments, as proposed in Section 2.1.7 and further discussed in Section 3.1.2. Alternative designs, discussed in Sections 3.1.4 and 3.1.5 control the propagation of the variance throughout a multiscale processing of data with noise. For both options, vanishing moments and variance control, this section provides some technical details.

The design of a B-spline wavelet transform with primal vanishing moments, as in (2.41), requires values for the moments of the B-spline scaling functions. These values can be obtained from partial integration of the definition

$$M_{j,k}^{(\widetilde{p},q)} = \int_{\infty}^{\infty} \varphi_{j,k}^{[\widetilde{p}]}(x)x^q \, dx = -\frac{1}{q+1} \int_{\infty}^{\infty} \frac{d}{dx} \varphi_{j,k}^{[\widetilde{p}]}(x)x^{q+1} \, dx.$$

Application of Lemma 4.2.8 leads to the recursion

$$M_{j,k}^{(\widetilde{p},q)} = \frac{\widetilde{p}-1}{q+1}\left[\frac{M_{j,k+\mathrm{rem}(\widetilde{p}/2)}^{(\widetilde{p}-1,q+1)}}{x_{j,k+\lceil\widetilde{p}/2\rceil} - x_{j,k-\lfloor\widetilde{p}/2\rfloor+1}} - \frac{M_{j,k-1+\mathrm{rem}(\widetilde{p}/2)}^{(\widetilde{p}-1,q+1)}}{x_{j,k+\lceil\widetilde{p}/2\rceil-1} - x_{j,k-\lfloor\widetilde{p}/2\rfloor}}\right]. \tag{4.57}$$

For a B-spline wavelet transform with variance control, as in Section 3.1.5, we need an expression for the Gram matrix in (3.8). Let $\widetilde{\boldsymbol{X}}_J^{[\widetilde{p}]}$ be the $(n+\widetilde{p}-2)\times\widetilde{p}$ matrix with the power coefficients in the B-spline basis of Theorem 4.2.11,

$$\left[\widetilde{\boldsymbol{X}}_J^{[\widetilde{p}]}\right]_{k,q} = \widetilde{x}_{j,k}^{[\widetilde{p},q]},$$

and let $\boldsymbol{x}^{[\widetilde{p}]}$ be a row vector of power functions, i.e., $\left[\boldsymbol{x}^{[\widetilde{p}]}\right]_q = x^q$. Then we have

$$\Phi_J^{[\widetilde{p}]}(x)\widetilde{\boldsymbol{X}}_J^{[\widetilde{p}]} = \boldsymbol{x}^{[\widetilde{p}]}.$$

Defining

$$\mathbf{C}_{J,k} = \int_{x_{J,k-1}}^{x_{J,k}} \Phi_J^{[\widetilde{p}]^\top}(x)\Phi_J^{[\widetilde{p}]}(x)\,dx$$

we have

$$\mathbf{C}_J = \sum_{k=1}^{n_J-1}\mathbf{C}_{J,k},$$

while the $(n+\widetilde{p}-2)\times\widetilde{p}$ matrices $\mathbf{C}_{J,k}$ have each only a $\widetilde{p}\times\widetilde{p}$ nonzero submatrix. The nonzeros can be found from

$$\widetilde{\boldsymbol{X}}_J^{[\widetilde{p}]^\top}\mathbf{C}_{J,k}\widetilde{\boldsymbol{X}}_J^{[\widetilde{p}]} = \int_{x_{J,k-1}}^{x_{J,k}} \boldsymbol{x}^{[\widetilde{p}]^\top}\boldsymbol{x}^{[\widetilde{p}]}\,dx,$$

which, after removal of the zero lines and columns, amounts to $\widetilde{p}$ systems of $\widetilde{p}$ equations in $\widetilde{p}$ unknowns, the elements of the columns of the nonzero submatrix of $\mathbf{C}_{J,k}$.

Figure 4.6 collects a few wavelet bases for the refinement of the cubic B-spline basis in Figure 4.1. The coarse scale basis functions of Figure 4.1 on $n_j = 7$ knots, together with any of the cubic spline wavelet bases of Figure 4.6, generate the space of cubic splines at a finer resolution of $n_{j+1} = 7+6$ knots. The boundary wavelets are left out as they require special treatment.

4.3 Limitations of the construction by lifting

4.3.1 Identification of the lifting steps

Although all wavelet transforms can be factored into lifting steps, the construction of B-spline refinement through the lifting factoring illustrates that

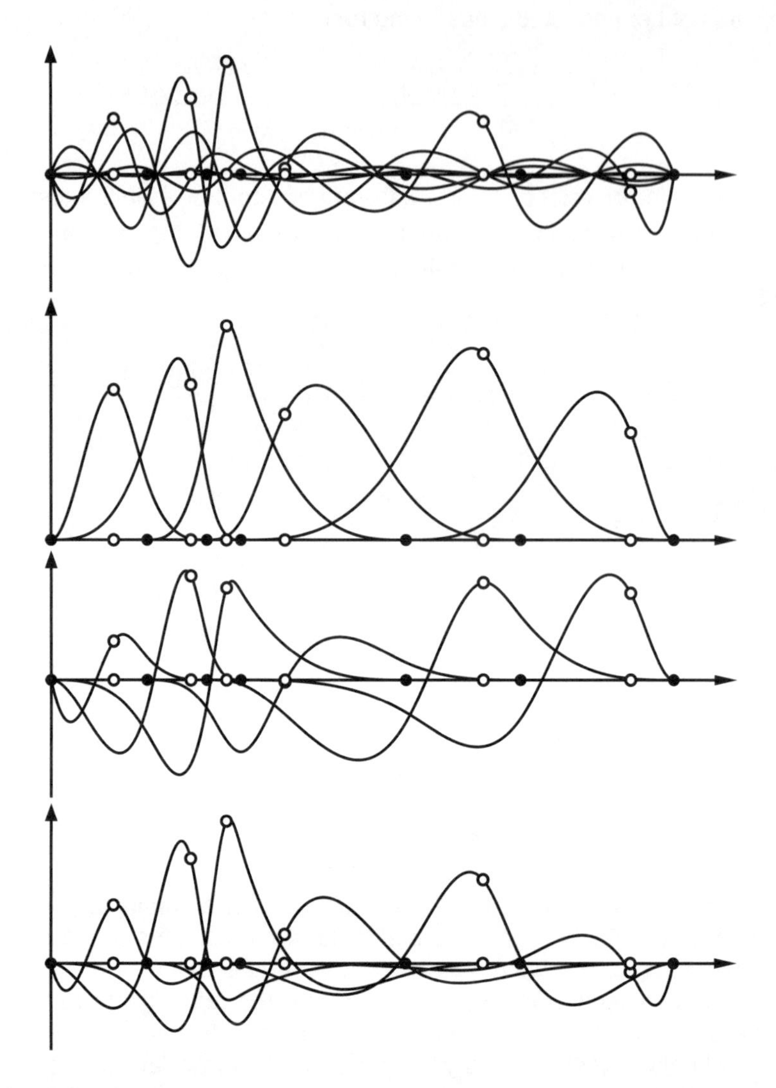

Figure 4.6
Cubic B-spline wavelets (i.e., $\widetilde{p} = 4$) refining the coarse scale grid of $n_j = 7$ knots (black dots) from Figure 4.1 by inserting $n'_j = n_{j+1} - n_j = 6$ knots (white centered dots). Top panel: semi-orthogonal basis. These spline wavelets do not have compact support. Second panel: lifting scheme without final update. The wavelets take the form of scaling functions (using a detail filter arising from the factoring, $\mathbf{G}_j^{[0]}$, just slightly different than the identity matrix). Third panel: using a final update for one vanishing moment. Bottom panel: using a final update for two vanishing moments.

the search for the factoring is nontrivial. Unlike for the Deslauriers-Dubuc refinement, the B-spline lifting does not proceed through vanishing moments. One might wonder whether a simpler lifting scheme for B-splines could exist by adding vanishing moments in every lifting step. This is not possible, as follows from a closer look at (4.14). Indeed, while lifting the refinement matrix $\mathbf{H}_j$ with a prediction $\mathbf{P}_j$, the detail matrix $\mathbf{G}_j$ remains untouched. Also the analysis matrix $\widetilde{\mathbf{H}}_j$ stays the same. The scaling basis matching the lifted refinement then operates with the same scaling coefficients as the old refinement (before lifting, that is). The power functions in the space spanned by the old and lifted scaling bases should have the same coefficients. As this is not the case according to Theorem 4.2.11, a prediction step that simply lifts the order (i.e., the number of dual vanishing moments) of a B-spline refinement cannot exist. This explains why the construction of B-spline refinement is much more complicated than that of Deslauriers-Dubuc interpolating refinement. Careful design of the B-spline lifting steps ensures the preservation of vanishing moments throughout the factored refinement scheme, but lifting cannot be used to "lift higher" B-splines when it comes to vanishing moments. An alternative construction for filterbanks, discussed in Chapter 5, is based on convolution sums. In contrast to lifting and factoring into lifting steps, convolution automatically preserves and easily enhances the number of vanishing moments. In practice, however, the technique of convolution is limited to the case of equidistant knots.

4.3.2 Boundary issues in B-spline refinement

As stated in Lemma 4.2.9, a full B-spline basis defined on n_j knots consists of $n_j+\widetilde{p}-2$ basis functions. Refinement of this basis by inserting a knot between each pair of existing knots almost doubles the knots, as $n_{j+1} = 2n_j - 1$, but the number of basis functions grows at a slightly lower rate. Therefore, going back from fine to coarse scale, less than half of the fine scaling functions are marked as odd in the splitting or subsampling stage. The partitioning into evens and odds cannot be binary or alternating. As a result, the bidiagonal lifting factoring of Theorem 4.1.3, based on a binary split, does not fully apply. This explains the outcome of a lifting based refinement scheme for B-spline basis functions, illustrated in Figure 4.7. The n_j basis functions depicted in a black line are obtained by the refinement scheme with bidiagonal lifting steps. The remaining $\widetilde{p}-2$ basis functions near the boundaries of the interval, depicted in grey, although refinable in accordance to Theorem 4.2.14, cannot be retrieved with a bidiagonal lifting scheme as in Theorem 4.1.3.

While it can be conjectured that more advanced lifting schemes would do the job, the refinement coefficients near the boundaries can be obtained numerically from two small systems of linear equations, following from function

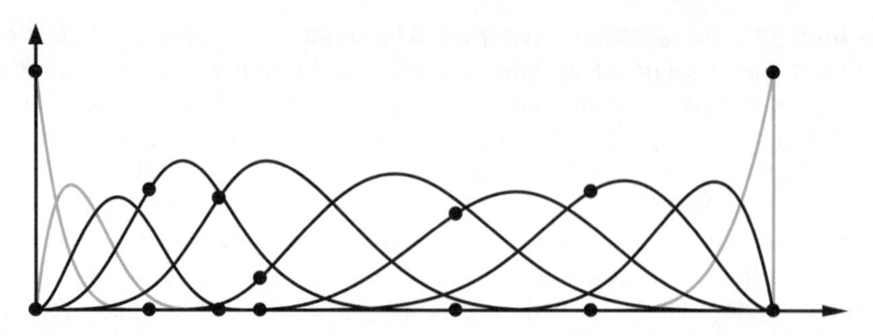

Figure 4.7
B-spline basis of order $\widetilde{p} = 5$ on $n_j = 7$ knots. The basis functions depicted in black are obtained by a refinement scheme based on bidiagonal lifting. The remaining functions in grey, near the boundaries of the interval, cannot be found with bidiagonal lifting.

evaluations of coarse and fine scaling functions $\varphi_{j,k}(x)$ and $\varphi_{j+1,k}(x)$ in $\widetilde{p}$ points near the boundaries[6].

4.4 Concluding summary on primal and dual vanishing moments

Having explored the construction of wavelet transforms using the lifting scheme, it is good to review the properties of transform and basis functions. In particular, it is important to understand the origin and impact of primal and dual vanishing moments.

- **Dual vanishing moments** stand for the order of approximation of the scaling basis. From a statistical point of view, the approximation in a basis is closely related to the **bias** of an estimator in that basis. As an example, cubic B-splines or cubic Deslauriers-Dubuc schemes have four dual vanishing moments, as counting starts at zero. Dual vanishing moments are a property of the scaling or refinable basis, hence they are also a characteristic of the refinement scheme, and thus a

[6]Function evaluations in points all over the interval would theoretically allow for a system defining the complete refinement matrix, but solving such a large system would be computationally complex. Therefore the analytical solution provided by the lifting factoring is used in the interior of the interval.

property of the refinement matrices $\mathbf{H}_j$. From the perfect reconstruction in (2.62), it follows that the dual vanishing moments can be found also in the dual wavelet matrix $\widetilde{\mathbf{G}}_j$ and hence in the dual wavelet basis functions.

The dual vanishing moments may affect the primal wavelet matrix $\mathbf{G}_j$ and the primal basis functions $\Psi_j(x)$ as well, but only if the lifting factoring contains at least one update step other than the final update discussed in 4.1.5.

For practical use, all wavelet transforms should have at least one dual vanishing moment. The refinement scheme is then **affine** (see Section 3.1.7), the scaling basis is then said to satisfy the **partition of unity** (see Section 4.2.4). When the transform aims at being grid dependent, the second dual vanishing moment for perfect reconstruction of linear functions should also be included. Otherwise, the transform cannot reconstruct the identity function defined on the grid, which means that the lack of equidistance in the grid shows up in any reconstruction on that grid.

- **Primal vanishing moments** are not a property of the multiresolution refinement. It appears in the design of the detail matrix $\mathbf{G}_j$ for a given refinement matrix $\mathbf{H}_j$, or, equivalently, in the choice of the detail basis $\Psi_j(x)$ for a given refinable basis $\Phi_j(x)$. The choice of the number of primal vanishing moments, and of $\Psi_j(x)$ in general, is more related to the **variance** propagation properties of the transform. As explained in Section 3.1.1, at least one primal vanishing moment should always be provided.

5

Dyadic wavelets

The construction of wavelet transforms in Chapter 2 hinges on the lifting scheme. Earlier constructions without lifting were limited to the case of equidistant knots. Working on equidistant knots obviously limits the degrees of freedom in the design of a decomposition and the range of applications. On the other hand, it provides a firmly established theoretical framework, thanks to the availability of powerful analysis tools such as the Fourier analysis for data on equidistant knots. Also, the two-scale equation is simplified to a form for which the solution is easier to find and analyse.

Sections 5.1 and 5.2 apply the methods of Chapter 2 to equidistant knots. The following sections, from Section 5.4 forward, resume the construction of wavelet transforms on equidistant knots from scratch.

5.1 Filterbanks on equispaced knots

Writing the forward and inverse filterbanks (2.43), (2.43), and (2.46) as a sum, and denoting $n'_j = n_{j+1} - n_j$ as before, we get

$$s_{j,k} = \sum_{l=0}^{n_{j+1}-1} \widetilde{H}_{j,l,k} s_{j+1,l}, \tag{5.1}$$

$$d_{j,k} = \sum_{l=0}^{n_{j+1}-1} \widetilde{G}_{j,l,k} s_{j+1,l}, \tag{5.2}$$

$$s_{j+1,k} = \sum_{l=0}^{n_j-1} H_{j,k,l} s_{j,l} + \sum_{l=0}^{n'_j-1} G_{j,k,l} d_{j,l}. \tag{5.3}$$

We now consider the case of equidistant knots, which, without loss of generality, are supposed to be given by $x_{j,k} = (k+1)/n_j$ for $k = 0, 1, \ldots, n_j - 1$. Furthermore, all vectors of coefficients are supposed to be extended periodically so that $s_{j,k} = s_{j,k \pm n_j}$ and $d_{j,k} = d_{j,k \pm n'_j}$. The sizes of the matrices $\widetilde{\mathbf{H}}_j$, $\widetilde{\mathbf{G}}_j$, $\mathbf{H}_j$, and $\mathbf{G}_j$, are enlarged accordingly, while the knots are extended by $x_{j,k \pm n_j} = x_{j,k} \pm 1$. The extended index set at level j is denoted as $\mathbb{I}_j$. Its cardinality is denoted by $\overline{n}_j$.

DOI: 10.1201/9781003265375-5

We assume that the filter matrices depend only on the ratios of interknot distances $(x_{j+1,k}-x_{j+1,k'})/(x_{j+1,l}-x_{j+1,l'})$, for $k,k',l,l' \in \mathbb{I}_{j+1}$. On equidistant knots, this implies that $H_{j,k,l} = H_{0,k-2l,0}$ for $k \in \mathbb{I}_{j+1}$ and $l \in \mathbb{I}_j$, with $k-2l \in \mathbb{I}_0$. For this to hold, resolution zero must be chosen so that $\mathbb{I}_0$ is large enough to avoid any boundary effect. More precisely, the number of nonzeros in each column of $\mathbf{H}_j$ must be smaller than the number of knots before the extension, n_0. Because of the periodic extension, we also have a few nonzero values in the upper right and lower left corners of the extended matrices, but these elements are far beyond the original ranges $\{0,1,\ldots,n_j\}$ and $\{0,1,\ldots,n_{j+1}\}$. Obviously, a similar expression as $H_{j,k,l} = H_{0,k-2l,0}$ can be written for the other matrices. As a general conclusion one column is enough to define each of the matrices in the forward and inverse transforms. For $\mathbf{H}_j$, we define the sequence $\boldsymbol{h}$ so that $h_k = H_{0,k,0}$ if $k \in \mathbb{I}_0$ and $h_k = 0$ if $k \in \mathbb{Z}\backslash\mathbb{I}_0$. Similarly, we define the sequences $\boldsymbol{g}$, $\widetilde{\boldsymbol{h}}$, and $\widetilde{\boldsymbol{g}}$ to correspond to the columns of $\mathbf{G}_0$, $\widetilde{\mathbf{H}}_0$ and $\widetilde{\mathbf{G}}_0$.

Then the forward and inverse transforms can be written as

$$s_{j,k} = \sum_{l=-\infty}^{\infty} \widetilde{h}_{l-2k} s_{j+1,l}, \tag{5.4}$$

$$d_{j,k} = \sum_{l=-\infty}^{\infty} \widetilde{g}_{l-2k} s_{j+1,l}, \tag{5.5}$$

$$s_{j+1,k} = \sum_{l=-\infty}^{\infty} h_{k-2l} s_{j,l} + \sum_{l=-\infty}^{\infty} g_{k-2l} d_{j,l}. \tag{5.6}$$

We now introduce the reversed vector $\overline{\boldsymbol{h}}$, defined by $\overline{h}_k = \widetilde{h}_{-k}$. Then, denoting by $x_{e,k} = x_{2k}$ the kth element of the even subvector $\boldsymbol{x}_e$ of vector $\boldsymbol{x}$, and by $x_{o,k} = x_{2k+1}$ the kth element of the odd subvector, we can write

$$\begin{aligned} s_{j,k} &= \sum_{l=-\infty}^{\infty} \overline{h}_{2k-l} s_{j+1,l} \\ &= \sum_{l=-\infty}^{\infty} \overline{h}_{2k-2l} s_{j+1,2l} + \sum_{l=-\infty}^{\infty} \overline{h}_{2k-2l-1} s_{j+1,2l-1} \\ &= \sum_{l=-\infty}^{\infty} \overline{h}_{e,k-l} s_{j+1,e,l} + \sum_{l=-\infty}^{\infty} \overline{h}_{o,k-l-1} s_{j+1,o,l}. \end{aligned}$$

This amounts to two convolution sums,

$$\boldsymbol{s}_j = \overline{\boldsymbol{h}}_e * \boldsymbol{s}_{j+1,e} + \mathbf{S}(\overline{\boldsymbol{h}}_o * \boldsymbol{s}_{j+1,o}), \tag{5.7}$$

where the convolution $\boldsymbol{x} * \boldsymbol{y}$ stands for the sequence with elements

$$(\boldsymbol{x} * \boldsymbol{y})_k = \sum_{l=-\infty}^{\infty} x_l y_{k-l} = \sum_{l=-\infty}^{\infty} x_{k-l} y_l, \tag{5.8}$$

while $\mathbf{S}x$ stands for the sequence or vector with shifted elements $(\mathbf{S}x)_k = x_{k-1}$. Likewise, we can find

$$\begin{aligned} \boldsymbol{d}_j &= \overline{\boldsymbol{g}}_e * \boldsymbol{s}_{j+1,e} + \overline{\boldsymbol{g}}_o * (\mathbf{S}\boldsymbol{s}_{j+1,o}), & (5.9) \\ \boldsymbol{s}_{j+1,e} &= \boldsymbol{h}_e * \boldsymbol{s}_j + \boldsymbol{g}_e * \boldsymbol{d}_j, & (5.10) \\ \boldsymbol{s}_{j+1,o} &= \boldsymbol{h}_o * \boldsymbol{s}_j + \boldsymbol{g}_o * \boldsymbol{d}_j, & (5.11) \end{aligned}$$

which is nothing else but the equidistant version of the **polyphase** representation of the wavelet forward and inverse transform, discussed in Section 4.1.1.

A different scheme (Mallat, 1989; Meyer, 1992; Vetterli and Herley, 1992; Strang and Nguyen, 1996) follows by considering $\boldsymbol{s}_j$ and $\boldsymbol{d}_j$ as binary subsamples of longer versions $\boldsymbol{s}_j = (\downarrow 2)\boldsymbol{s}'_j$ and $\boldsymbol{d}'_j = (\downarrow 2)\boldsymbol{d}'_j$, where the notation $(\downarrow 2)$ stands for binary subsampling, as defined in (2.3). The values taken away by the downsampling can be chosen to be given by

$$s'_{j,2k+1} = \sum_{l=-\infty}^{\infty} \overline{h}_{2k+1-l} s_{j+1,l}, \qquad (5.12)$$

Together with $s'_{j,2k} = s_{j,k}$, the long vector $\boldsymbol{s}'_j$ can then be written as $\boldsymbol{s}'_j = \overline{\boldsymbol{h}} * \boldsymbol{s}_{j+1}$. Similarly we take $\boldsymbol{d}'_j = \overline{\boldsymbol{g}} * \boldsymbol{s}_{j+1}$, leading to the forward filterbank

$$\begin{aligned} \boldsymbol{s}_j &= (\downarrow 2)\overline{\boldsymbol{h}} * \boldsymbol{s}_{j+1}, & (5.13) \\ \boldsymbol{d}_j &= (\downarrow 2)\overline{\boldsymbol{g}} * \boldsymbol{s}_{j+1}. & (5.14) \end{aligned}$$

Obviously, a software package would not really compute the further unused values of $s'_{j,2k+1}$ and $d'_{j,2k+1}$.

For the inverse transform, we insert zeros into $\boldsymbol{s}_j$ and $\boldsymbol{d}_j$ instead of the values in (5.12). The expansion of a vector with zeros, known as **upsampling**, can be represented by a matrix operation $\boldsymbol{s}''_j = (\uparrow 2)\boldsymbol{s}_j \Leftrightarrow s''_{j,2k} = s_{j,k}$ and $s''_{j,2k+1} = 0$. It holds that $(\downarrow 2)(\uparrow 2) = \mathbf{I}_{n_j}$ and $(\uparrow 2) = (\downarrow 2)^\top$, both properties extending to other than binary subsampling, i.e., for $\widetilde{\mathbf{J}}_j^\top \neq (\downarrow 2)$ in (2.2). The inverse transform (5.6) is now written as

$$s_{j+1,k} = \sum_{l=-\infty}^{\infty} h_{k-l} s''_{j,l} + \sum_{l=-\infty}^{\infty} g_{k-l} d''_{j,l}.$$

On the vector level this is

$$\boldsymbol{s}_{j+1} = \boldsymbol{h} * (\uparrow 2)\boldsymbol{s}_j + \boldsymbol{g} * (\uparrow 2)\boldsymbol{d}_j. \qquad (5.15)$$

Substitution of (5.15) into (5.13) and (5.14) leads to the **perfect reconstruction** conditions. Denoting by δ the Kronecker delta, we have

$$(\downarrow 2)(\overline{\boldsymbol{h}} * \boldsymbol{h})(\uparrow 2) = \boldsymbol{\delta} \quad \text{or} \quad \sum_{l \in \mathbb{Z}} \widetilde{h}_l h_{l-2k} = \delta_k, \tag{5.16}$$

$$(\downarrow 2)(\overline{\boldsymbol{h}} * \boldsymbol{g})(\uparrow 2) = \mathbf{0} \quad \text{or} \quad \sum_{l \in \mathbb{Z}} \widetilde{h}_l g_{l-2k} = 0, \tag{5.17}$$

$$(\downarrow 2)(\overline{\boldsymbol{g}} * \boldsymbol{h})(\uparrow 2) = \mathbf{0} \quad \text{or} \quad \sum_{l \in \mathbb{Z}} \widetilde{g}_l h_{l-2k} = 0, \tag{5.18}$$

$$(\downarrow 2)(\overline{\boldsymbol{g}} * \boldsymbol{g})(\uparrow 2) = \boldsymbol{\delta} \quad \text{or} \quad \sum_{l \in \mathbb{Z}} \widetilde{g}_l g_{l-2k} = \delta_k. \tag{5.19}$$

The expressions in terms of the coefficients follow from the fact that $(\downarrow 2)\mathbf{A}(\uparrow 2)$ is the submatrix of $\mathbf{A}$ containing all elements at even rows and even columns. With $\mathbf{A}$ the matrix representing a convolution of two time-invariant, linear filters, all rows of $\mathbf{A}$ are shifted versions of each other[1]. As an example, the condition $(\downarrow 2)(\overline{\boldsymbol{g}} * \boldsymbol{h})(\uparrow 2) = \mathbf{0}$ in (5.18) simplifies to a single row, $\sum_{l \in \mathbb{Z}} \widetilde{g}_l h_{l-k} = 0$, but only for the even columns, i.e., for even values of k.

For a given refinement filter $\boldsymbol{h}$, (5.18) puts a condition on the dual detail filter $\widetilde{g}$. A solution can be proposed of the form

$$\widetilde{g}_l = (-1)^l h_{s-l}. \tag{5.20}$$

Based on the Fourier analysis of this solution, discussed in Section 6.1.2, a pair of filters satisfying (5.20) are named **quadrature mirror filters**. The condition in (5.18) becomes

$$\sum_{l \in \mathbb{Z}} (-1)^l h_{s-l} h_{l-2k} = 0.$$

With $\ell = s - l + 2k$, the two terms

$$(-1)^l h_{s-l} h_{l-2k} + (-1)^\ell h_{s-\ell} h_{\ell-2k} = (-1)^l h_{s-l} h_{l-2k} + (-1)^{s-l+2k} h_{l-2k} h_{s-l}$$

annihilate each other if s is odd. Two pairs of quadrature mirror filters, $(\boldsymbol{h}, \widetilde{\boldsymbol{g}})$ and $(\widetilde{\boldsymbol{h}}, \boldsymbol{g})$, thus satisfy automatically half of the perfect reconstruction conditions, namely (5.17) and (5.18). The most common choice of a quadrature mirror filter is to take $s = 1$, i.e., $\widetilde{g}_l = (-1)^l h_{1-l}$ and $g_l = (-1)^l \widetilde{h}_{1-l}$.

Exercise 5.1.1 *Let $\widetilde{\mathcal{H}}_j$ be a $n_{j+1} \times n_{j+1}$ matrix so that $\widetilde{\mathbf{H}}_j$ is the submatrix of $\widetilde{\mathcal{H}}_j$ consisting of its even columns. The odd columns being free to fill in, the matrix $\widetilde{\mathcal{H}}_j$ can be chosen so that on equidistant knots, the above mentioned property $\widetilde{H}_{j,k,l} = \widetilde{H}_{0,k-2l,0}$ extends to $\widetilde{\mathcal{H}}_{j,k,l} = \widetilde{\mathcal{H}}_{0,k-l,0}$, making $\widetilde{\mathcal{H}}_j$ a **Toeplitz matrix** (See definition page 216). Define $\overline{\mathbf{H}}_j = \widetilde{\mathcal{H}}_j^\top$ and*

[1] A is then a Toeplitz matrix, see definition page 216.

$\overline{\mathcal{H}}_j = \widetilde{\mathcal{H}}_j^\top$. *Then show that the forward wavelet transform in (5.13) is given by* $\overline{\mathbf{H}}_j \boldsymbol{s}_{j+1} = (\downarrow 2)\overline{\mathcal{H}}_j \boldsymbol{s}_{j+1}$. *Show that* $\overline{\mathbf{H}}_j$ *is also a Toeplitz matrix. As a conclusion, a convolution sum can be written as a matrix vector product with a Toeplitz matrix and flipping a sequence in a convolution sum amounts to taking the transpose Toeplitz matrix.*

Exercise 5.1.2 *Expression (4.35), used in Exercise 4.1.8, provides a way to use the factoring of a refinement matrix into lifting steps in the development of a wavelet transform for given primal and dual refinements. Exercise 4.1.10 finds a factoring for a wavelet transform starting from the primal refinement and detail matrices. Now, Expression (5.20) of a quadrature mirror filter suggests to design a wavelet transform from the primal refinement* $\mathbf{H}_j$ *and the dual detail* $\widetilde{\mathbf{G}}_j$. *Explain how the factoring into lifting steps can be used to find the missing matrices* $\widetilde{\mathbf{H}}_j$ *and* $\mathbf{G}_j$.

5.2 The equispaced two-scale and wavelet equations

Perfect reconstruction equispaced filterbanks, discussed in Section 5.1, can be applied in a recursive way, leading to a multiscale analysis. There is, however, no guarantee that a reconstruction from such a recursive filterbank converges to a smooth function. This section introduces the basis functions, termed father and mother functions, that play a crucial role in the reconstruction from a dyadic wavelet scheme. The basis functions are defined by the equispaced version of the two-scale equation introduced in Section 2.2.4. The question of whether the reconstruction filterbank actually converges will be investigated in Section 5.5.5, where we will solve the equispaced two-scale equations.

The general two-scale equation (2.64) can be written as

$$\varphi_{j,l}(x) = \sum_{k=0}^{n_{j+1}-1} H_{j,k,l}\varphi_{j+1,k}(x).$$

As Section 5.1 has argued that on equidistant grids, we have $H_{j,k,l} = h_{k-2l}$, independent from j, the refinement equation becomes

$$\varphi_{j,l}(x) = \sum_{k=-\infty}^{\infty} h_{k-2l}\varphi_{j+1,k}(x). \tag{5.21}$$

The solution of this system is simplified by the following result.

Theorem 5.2.1 *If there exists a constant* c *and a function* $\varphi(x)$ *so that*

$$\varphi(x) = c\sum_{k=-\infty}^{\infty} h_k\varphi(2x-k), \tag{5.22}$$

then a solution of (5.21) is given by

$$\varphi_{j,l}(x) = c^j \varphi(2^j x - l).$$

Proof. We develop the right hand side of (5.21):

$$\begin{aligned}
\sum_{k=-\infty}^{\infty} h_{k-2l} \varphi_{j+1,k}(x) &= c^{j+1} \sum_{k=-\infty}^{\infty} h_{k-2l} \varphi\left(2^{j+1}x - k\right) \\
&= c^j c \sum_{m=-\infty}^{\infty} h_m \varphi\left(2^{j+1}x - 2l - m\right) \\
&= c^j c \sum_{m=-\infty}^{\infty} h_m \varphi\left(2\left(2^j x - l\right) - m\right) \\
&= c^j \varphi\left(2^j x - l\right) = \varphi_{j,l}(x).
\end{aligned}$$

□

Integration in both parts of (5.22) leads immediately to the following result.

Lemma 5.2.2 *If the constant c in (5.22) exists, then it is given by*

$$c = \frac{2}{\sum_{k=-\infty}^{\infty} h_k}. \tag{5.23}$$

If we want c to be one, then the coefficients of the refinement equation should sum up to two. In many applications, we prefer that the scaling functions are normalised across the scales, meaning that

$$\|\varphi_{j,k}\|_q^q = \int_{-\infty}^{\infty} |\varphi_{j,k}(x)|^q \, dx$$

does not depend on j. This is realised if

$$\int_{-\infty}^{\infty} |c\varphi(2x)|^q \, dx = \int_{-\infty}^{\infty} |\varphi(x)|^q \, dx \Leftrightarrow c = 2^{1/q}. \tag{5.24}$$

Remark 5.2.3 *Although (5.24) is formulated in terms of c, the condition should rather be interpreted as a normalisation of $\mathbf{h}$. Indeed, from (5.23), it follows immediately that*

$$\sum_{k=-\infty}^{\infty} h_k = 2^{1-1/q}.$$

On a regular grid of knots, the two-scale equation reduces to the form of (5.22). This expression involves one scaling function, termed the **father function** or **refinable function** from which all other scaling functions are derived by **translation** (shifting) and **dilation** (stretching). The admissible translations and dilations are **dyadic**, meaning that the dilations are an integer power of two, 2^j, for some $j \in \mathbb{Z}$, while the translations are over a distance of the form $k2^{-j}$, an integer multiple of an integer power of two. Although subdivision may still be used as a numerical technique for solving (5.22), the smoothness or continuity of the father function is fully determined by the refinement equation at one resolution level.

In a similar way, the **mother function** of an equispaced wavelet transform is defined by

$$\psi(x) = d \sum_{k=-\infty}^{\infty} g_k \varphi(2x - k). \tag{5.25}$$

All wavelet functions are then of the form

$$\psi_{j,l}(x) = d^j \psi\left(2^j x - l\right).$$

Exercise 5.2.4 *Show that this is indeed the case.*

As on equispaced knots all scaling functions are dyadic dilations and translations of one father function, it holds for $f(x) = \Phi_j(x)\boldsymbol{s}_j$ that a dyadic translation $f(x - 2^{-j}k)$ can also be decomposed in the same scaling basis, while a dyadic dilation $f(2^i x)$ belongs to the space spanned by $\Phi_{j+i}(x)$. Because of the two-scale equation, the function space defined by $\Phi_j(x)$ is a subspace of the function space defined by $\Phi_{j+1}(x)$. The wavelet transform thus operates on layers of nested function spaces of dyadically dilated and translated functions. This structure of nested function spaces is termed a multiresolution analysis. Section 5.4 prepares the formal definition of a multiresolution analysis. It is the starting point for the classical development of dyadic wavelet transform.

5.3 Wavelet transforms on an equispaced setting

We now develop a couple of examples of equispaced filterbanks and corresponding wavelet transforms. In Section 5.3.1, we revisit the lifting scheme with interpolating prediction. Next, in Section 5.3.2, we develop B-spline wavelet transforms on regular grids.

5.3.1 Equispaced Deslauriers-Dubuc schemes

The interpolating scaling basis of the Deslauriers-Dubuc refinement scheme has been developed in Section 3.2 based on the two-step lifting scheme in Section 2.1.7. The filterbank version of this lifting scheme has been established in Section 2.2.2.

In order to simplify the expressions of the filterbank in the equispaced case, we first introduce the $n'_j \times n_j$ complementary update matrix $\mathbf{U}_j^o$, where $n'_j = n_{j+1} - n_j$, and the $n_j \times n'_j$ complementary prediction matrix $\mathbf{P}_j^o$. These matrices are defined by a lifting scheme where the roles of the evens and odds are switched. For the complementary subsampling $\widetilde{\mathbf{J}}_j^{o\top}$, introduced in Section 2.2.2, we know that $\widetilde{\mathbf{J}}_j^{o\top}\widetilde{\mathbf{J}}_j = \mathbf{0}$, but also $\widetilde{\mathbf{J}}_j^{o\top}\widetilde{\mathbf{J}}_j^o = \mathbf{I}_{n'_j}$, while $\widetilde{\mathbf{J}}_j^\top\widetilde{\mathbf{J}}_j = \mathbf{I}_{n_j}$. Then the calculation of the coarse scaling operation in (2.49) can be rewritten as

$$\begin{aligned}
\widetilde{\mathbf{H}}_j^\top &= (\mathbf{I}_{n_j} - \mathbf{U}_j\mathbf{P}_j)\widetilde{\mathbf{J}}_j^\top + \mathbf{U}_j\widetilde{\mathbf{J}}_j^{o\top} = (\widetilde{\mathbf{J}}_j^\top - \widetilde{\mathbf{J}}_j^\top\widetilde{\mathbf{J}}_j\mathbf{U}_j\mathbf{P}_j\widetilde{\mathbf{J}}_j^\top) + \widetilde{\mathbf{J}}_j^\top\widetilde{\mathbf{J}}_j\mathbf{U}_j\widetilde{\mathbf{J}}_j^{o\top} \\
&= \widetilde{\mathbf{J}}_j^\top\left[\mathbf{I}_{n_{j+1}} - (\widetilde{\mathbf{J}}_j\mathbf{U}_j\mathbf{P}_j\widetilde{\mathbf{J}}_j^\top + \widetilde{\mathbf{J}}_j^o\mathbf{U}_j^o\mathbf{P}_j^o\widetilde{\mathbf{J}}_j^{o\top}) + (\widetilde{\mathbf{J}}_j\mathbf{U}_j\widetilde{\mathbf{J}}_j^{o\top} + \widetilde{\mathbf{J}}_j^o\mathbf{U}_j^o\widetilde{\mathbf{J}}_j^\top)\right]
\end{aligned}$$

Defining the $n_j \times n_j$ matrices

$$\begin{aligned}
\overline{\mathbf{U}}_j &= \widetilde{\mathbf{J}}_j\mathbf{U}_j\widetilde{\mathbf{J}}_j^{o\top} + \widetilde{\mathbf{J}}_j^o\mathbf{U}_j^o\widetilde{\mathbf{J}}_j^\top \\
\overline{\mathbf{P}}_j &= \widetilde{\mathbf{J}}_j^o\mathbf{P}_j\widetilde{\mathbf{J}}_j^\top + \widetilde{\mathbf{J}}_j\mathbf{P}_j^o\widetilde{\mathbf{J}}_j^{o\top},
\end{aligned}$$

we have $\overline{\mathbf{U}}_j\overline{\mathbf{P}}_j = \widetilde{\mathbf{J}}_j\mathbf{U}_j\mathbf{P}_j\widetilde{\mathbf{J}}_j^\top + \widetilde{\mathbf{J}}_j^o\mathbf{U}_j^o\mathbf{P}_j^o\widetilde{\mathbf{J}}_j^{o\top}$, and so

$$\widetilde{\mathbf{H}}_j^\top = \widetilde{\mathbf{J}}_j^\top\left[\mathbf{I}_{n_{j+1}} - \overline{\mathbf{U}}_j\overline{\mathbf{P}}_j + \overline{\mathbf{U}}_j\right].$$

The multiplication of $\mathbf{P}_j$ with $\widetilde{\mathbf{J}}_j^\top$ on the right inserts zero columns, so that $\mathbf{P}_j\widetilde{\mathbf{J}}_j^\top$ takes input from the even components of $\boldsymbol{s}_{j+1}$ only. Likewise, the even columns of $\mathbf{P}_j^o\widetilde{\mathbf{J}}_j^{o\top}$ are all zero. As a result, the nonzeros in each row of $\overline{\mathbf{P}}_j$ are separated by a zero, so that each row of $\overline{\mathbf{P}}_j$ takes input either from the even or from the odd components of $\boldsymbol{s}_{j+1}$. The same argument holds for $\overline{\mathbf{U}}_j$. Both $\overline{\mathbf{P}}_j$ and $\overline{\mathbf{U}}_j$ are said to be linear operations *à trous*, with holes (zeros), that is.

In the equidistant setting, all rows of $\overline{\mathbf{P}}_j$ are shifts of each other, the same property holding for $\overline{\mathbf{U}}_j$ as well. Matrix multiplication then amounts to a convolution, leading to

$$\boldsymbol{s}_j = \widetilde{\mathbf{H}}_j^\top\boldsymbol{s}_{j+1} = (\downarrow 2)(\boldsymbol{\delta} - \overline{\boldsymbol{u}} * \overline{\boldsymbol{p}} + \overline{\boldsymbol{u}})\boldsymbol{s}_{j+1},$$

from which identification with (5.13) yields the conclusion

$$\overline{\boldsymbol{h}} = \boldsymbol{\delta} - \overline{\boldsymbol{u}} * \overline{\boldsymbol{p}} + \overline{\boldsymbol{u}} \tag{5.26}$$

In this expression, $\overline{\boldsymbol{p}}$ and $\overline{\boldsymbol{u}}$ are *à trous* filters, containing the elements of one row in the equidistant versions of $\mathbf{P}_j$ and $\mathbf{U}_j$ respectively, alternated with zeros. In a similar way, we find, starting from (2.50),

$$\begin{aligned}\widetilde{\mathbf{G}}_j^\top &= \widetilde{\mathbf{J}}_j^{o\top} - \mathbf{P}_j\widetilde{\mathbf{J}}_j^\top = \widetilde{\mathbf{J}}_j^{o\top}\left[\mathbf{I}_{n_{j+1}} - (\widetilde{\mathbf{J}}_j^o\mathbf{P}_j\widetilde{\mathbf{J}}_j^\top + \widetilde{\mathbf{J}}_j\mathbf{P}_j^o\widetilde{\mathbf{J}}_j^{o\top})\right] \\ &= \widetilde{\mathbf{J}}_j^{o\top}\left[\mathbf{I}_{n_{j+1}} - \overline{\mathbf{P}}_j\right]. \end{aligned} \tag{5.27}$$

In a scheme where boundary issues are resolved through periodic extensions, we can write $\widetilde{\mathbf{J}}_j^{o\top} = \widetilde{\mathbf{J}}_j^\top\widetilde{\mathbf{L}}_j$, where $\widetilde{\mathbf{L}}_j$ is a $n_j \times n_j$ superdiagonal matrix with $\widetilde{L}_{j;i,i+1} = 1$, for all $i = 1, 2, \ldots, n_j$. In the equidistant setting, this becomes $\overline{\boldsymbol{g}} = \boldsymbol{\delta}_1 * (\boldsymbol{\delta} - \overline{\boldsymbol{p}})$, where $\boldsymbol{\delta}_1$ is the Kronecker delta with $\delta_{1,1} = 1$ and $\delta_{1,k} = 0$ for any $k \neq 1$. The resulting $\overline{\boldsymbol{g}}$ is an *à trous* filter, with holes at the odd indices, except for the central value, which is $\overline{g}_{-1} = \widetilde{g}_1 = 1$.

For the inverse transform, rewriting starts from (2.57), now moving the upsampling $\widetilde{\mathbf{J}}_j$ or $\widetilde{\mathbf{J}}_j^o$ to the right.

(5.28)

$$\mathbf{H}_j = \widetilde{\mathbf{J}}_j + \widetilde{\mathbf{J}}_j^o\mathbf{P}_j = \left[\mathbf{I}_{n_{j+1}} + \overline{\mathbf{P}}_j\right]\widetilde{\mathbf{J}}_j, \tag{5.29}$$

$$\mathbf{G}_j = \widetilde{\mathbf{J}}_j^o(\mathbf{I} - \mathbf{P}_j\mathbf{U}_j) - \widetilde{\mathbf{J}}_j\mathbf{U}_j = \left[\mathbf{I}_{n_{j+1}} - \overline{\mathbf{U}}_j\overline{\mathbf{P}}_j - \overline{\mathbf{U}}_j\right]\widetilde{\mathbf{J}}_j^o \tag{5.30}$$

In the equidistant setting, we have $\boldsymbol{h} = \boldsymbol{\delta} + \overline{\boldsymbol{p}}$, which is an *à trous* filter, just like $\overline{\boldsymbol{g}}$. Comparing $\overline{\boldsymbol{g}}$ with $\boldsymbol{h}$, we see that $\overline{g}_{-i} = \widetilde{g}_i = (-1)^i h_{1-i}$, so $\widetilde{\boldsymbol{g}}$ with $\boldsymbol{h}$ is a pair of quadrature mirror filters. Finally, we obtain $\boldsymbol{g} = \boldsymbol{\delta}_1 * (\boldsymbol{\delta} - \overline{\boldsymbol{u}} * \overline{\boldsymbol{p}} - \overline{\boldsymbol{u}})$.

5.3.2 Equispaced B-spline wavelet transforms

The wavelet transforms constructed within the framework of the Haar and triangular hat scaling functions are both Deslauriers-Dubuc and B-spline wavelet transforms. As for the triangular hat wavelets, the prediction is by linear interpolation, so $\overline{p}_{-1} = 1/2 = \overline{p}_1$. The moments are given by

$$\begin{aligned} M_{j,k}^{(q)} &= \int_{-\infty}^{\infty}\varphi_{j,k}(x)x^q dx = \int_{-\infty}^{\infty}\varphi(2^j x - k)x^q dx \\ &= 2^{-j(q+1)}\int_{-\infty}^{\infty}\varphi(u)(u+k)^q du. \end{aligned}$$

This implies that $M_{j,k}^{(0)} = 2^{-j}M_{0,0}^{(0)}$ and $M_{j,k}^{(1)} = 2^{-2j}\left[M_{0,0}^{(1)} + kM_{0,0}^{(0)}\right]$. As $\varphi(u)$ is the triangular hat with knots $-1, 0, 1$, we have $M_{0,0}^{(0)} = 1$ and $M_{0,0}^{(1)} = 0$. Consequently, $M_{j,k}^{(0)} = 2^{-j}$ and $M_{j,k}^{(1)} = 2^{-2j}k$. Following (2.34), an update for two primal vanishing moments is given by the set of two equations (for $q \in \{0, 1\}$),

$$0 = M_{j+1,2k+1}^{(q)} - U_{j;k,k}M_{j,k}^{(q)} - U_{j;k+1,k}M_{j,k+1}^{(q)},$$

which is $\begin{cases} 0 &= 2^{-(j+1)} - U_{j;k,k}2^{-j} - U_{j;k+1,k}2^{-j} \\ 0 &= 2^{-2(j+1)}(2k+1) - U_{j;k,k}2^{-2j}k - U_{j;k+1,k}2^{-2j}(k+1). \end{cases}$

It then follows that $\overline{u}_{-1} = 1/4 = \overline{u}_1$. The filterbank can then be found using the expressions of Section 5.3.1. In particular, we have

$$\boldsymbol{h}_{(-1,\ldots,1)} = \left(\frac{1}{2}, 1, \frac{1}{2}\right) \tag{5.31}$$

$$\widetilde{\boldsymbol{h}}_{(-2,\ldots,2)} = \left(-\frac{1}{8}, \frac{1}{4}, \frac{3}{4}, \frac{1}{4}, -\frac{1}{8}\right). \tag{5.32}$$

The higher order equispaced B-spline refinement filters $\boldsymbol{h}^{[p]}$ can be proven to be given by $\boldsymbol{h}^{[p]} = \boldsymbol{h}^{[p-1]} * \boldsymbol{h}^{[1]}/2$, where $\boldsymbol{h}^{[1]}$ is the Haar refinement filter, i.e., $\boldsymbol{h}^{[1]}_{(0,1)} = (1,1)$. We thus find the quadratic refinement

$$\boldsymbol{h}^{[3]}_{(-1,\ldots,2)} = \left(\frac{1}{4}, \frac{3}{4}, \frac{3}{4}, \frac{1}{4}\right) \tag{5.33}$$

and the cubic refinement

$$\boldsymbol{h}^{[3]}_{(-2,\ldots,2)} = \left(\frac{1}{8}, \frac{1}{2}, \frac{3}{4}, \frac{1}{2}, \frac{1}{8}\right) \tag{5.34}$$

and so on. These equispaced B-spline wavelets are known as the **Cohen-Daubechies-Feauveau** (CDF) spline wavelets (Cohen et al., 1992). In the literature, the equispaced B-spline wavelet with p primal and $\widetilde{p}$ dual vanishing moments is referred to as the CDF(p,$\widetilde{p}$)-wavelet. Section 7.2.1 develops a system of linear equations defining the equispaced B-spline refinement without going through the lifting scheme.

Exercise 5.3.1 *Let $f(x)$ and $g(x)$ be two functions in $L_2(\mathbb{R})$, and let $h(x)$ be defined by the* ***convolution integral****.*

$$h(x) = (f * g)(x) = \int_{-\infty}^{\infty} f(u)g(x-u)du = \int_{-\infty}^{\infty} f(x-u)g(u)du.$$

Take $f(x)$ to the Haar scaling function with knots in 0 and 1, and $g(x)$ to be an equispaced B-spline on the knots $\{0,1,\ldots,\widetilde{p}\}$. Then simplify the expression of the convolution integral for these functions. Find $h'(x)$ and use Lemma 4.2.8 to show that $h(x)$ is a B-spline on the knots $\{0,1,\ldots,\widetilde{p}+1\}$.

Exercise 5.3.2 *Let X_i be a sample of independent uniform random variables on $[0,1]$ and let $Y_{\widetilde{p}} = \sum_{i=1}^{\widetilde{p}} X_i$. Then show that the density function of $Y_{\widetilde{p}}$ is a B-spline on the knots $\{0,1,\ldots,\widetilde{p}\}$. The density of $Y_{\widetilde{p}}/\sqrt{\widetilde{p}}$ is a B-spline on the knots $\{k\sqrt{\widetilde{p}}, k = 0,1,\ldots,\widetilde{p}\}$. For $\widetilde{p} \to \infty$ this B-spline converges to a Gaussian Bell curve.*

5.4 Riesz bases

In Section 5.3, we have developed the equispaced case of the spline and interpolating wavelet transforms discussed in previous chapters. This section prepares Section 5.5, in which the concept of multiresolution analysis is presented as a mathematical framework for the construction of equidistant wavelet transforms. This section thus marks a restart in the discussion of discrete wavelet transforms.

5.4.1 Algebraic bases

Before we can define a multiresolution analysis, we first review notions with respect to bases in function spaces, starting from the very basics. A **basis** of a **vector space** $\mathcal{V}_n$ is a set of linearly independent vectors that spans all elements of the vector space. Formally, the set of vectors $\boldsymbol{\Phi} = \{\varphi_k, k = 1, 2, \ldots, n\}$ is an **algebraic** or **Hamel** basis of $\mathcal{V}_n$ if for every $\boldsymbol{f} \in \mathcal{V}_n$ there *exists* a *unique* vector $\boldsymbol{s} \in \mathbb{R}^n$ of scalars s_k so that

$$\boldsymbol{f} = \sum_{k=1}^{n} s_k \varphi_k = \boldsymbol{\Phi s}. \tag{5.35}$$

The cardinality of the basis $\boldsymbol{\Phi}$, denoted by n, is also termed the dimension of the vector space $\mathcal{V}_n$. There is a slight abuse of notation in (5.35), as we assume that the elements in the set $\boldsymbol{\Phi}$ are ordered, making $\boldsymbol{\Phi}$ a row vector, rather than a set of basis vectors. The vector $\boldsymbol{s}$ owes its existence to the basis being a spanning set, while its uniqueness is due to the linear independence of the basis. Uniqueness is equivalent to the condition that the zero vector $\boldsymbol{0} \in \mathcal{V}_n$ must have no other than the trivial decomposition with all $s_k = 0$.

5.4.2 Bases in infinite dimensions

The scaling basis at a given resolution level is an example of an algebraic basis. In a multiresolution analysis, scaling bases are refined up to an infinitely fine resolution, leading to a vector space $\mathcal{B}$ with infinite dimension. In a vector space of infinite dimension, it is interesting to have the equipment for the analysis of convergence of a sequence of vectors and for the calculation of its limit. When the cardinality of the basis is infinite, the basis decomposition (5.35) becomes an **expansion**, which is an infinite sum

$$\boldsymbol{f} = \sum_{k=1}^{\infty} s_k \varphi_k. \tag{5.36}$$

In order to write the equality sign in (5.36), we have to make sure that both sides of it exist and are unique.

Definition 5.4.1 *(Schauder basis) A* **countable basis** *or* **Schauder basis** *of a vector space $\mathcal{B}$ is a countable set $\Phi \subset \mathcal{B}$ of vectors φ_k for which the expansion (5.36) exists and is unique for every $f \in \mathcal{B}$.*

The notion of a countable basis requires some further development. Indeed, the infinite sum (5.36) should be understood as a limit transition of an n-term approximation, $f = \lim_{n\to\infty} f_n$, where

$$f_n = \sum_{k=1}^{n} \alpha_k \varphi_k.$$

The convergence of the vector f_n to the vector f needs to be defined in terms of a sequence of real numbers, obtained by measuring the distance between f_n and f,

$$\lim_{n\to\infty} \|f - f_n\|_{\mathcal{B}} = 0. \tag{5.37}$$

Although a more general notion of distance is possible, the definition in (5.37) uses a distance or metric induced by a norm. For this we need $\mathcal{B}$ to be a normed vector space.

In a **complete** vector space any converging sequence of vectors has its limit within the vector space. More precisely, let f_n be a Cauchy sequence; this is a sequence so that for any $\varepsilon > 0$, all distances $\|f_n - f_m\|_{\mathcal{B}} < \varepsilon$ for n, m beyond a certain number. As the members of f_n grow closer to each other, they converge to a limit f, which must be in $\mathcal{B}$.

A complete normed vector space is termed a **Banach** space. In a Banach space $\mathcal{B}$, the existence of the expansion on the right hand side of (5.36) ensures that $f \in \mathcal{B}$. The existence of a countable basis provides a countable, dense subspace of the vector space. A space with a countable, dense subspace is a **separable** space. As a summary, a Schauder basis may exist in a complete or in an incomplete but always in a separable space.

Example 5.4.2 *The space $L_2([0,1])$ of square integrable functions is a Banach*[2] *space. As explained in Section 3.1.1, a multiresolution basis with nonzero integrals admits nontrivial expansions of the zero functions. Therefore, such a basis cannot possibly be a Schauder basis, at least not for $L_2([0,1])$.*

5.4.3 Unconditional bases

Even if the space $\mathcal{B}$ admits a countable basis, adding up an infinite number of components may be a delicate operation. For this to be understood, we introduce the notion of unconditional bases.

[2] It is actually a Hilbert space, and thus a Banach space.

Definition 5.4.3 *(Unconditional basis) An* ***unconditional basis*** *of a separable Banach space $\mathcal{B}$ is a Schauder basis for which*

$$\sum_{k=1}^{\infty} s_k \varphi_k \in \mathcal{B} \Leftrightarrow \sum_{k=1}^{\infty} |s_k| \, \varphi_k \in \mathcal{B}. \tag{5.38}$$

Defining the sequence of vectors $\boldsymbol{v}_k = s_k \varphi_k$, the basis is unconditional if the series of vectors $\sum_{k=1}^{\infty} \boldsymbol{v}_k$ converges unconditionally. A series is said to converge unconditionally if $\sum_{k=1}^{\infty} b_k \boldsymbol{v}_k$ converges for any sequence of signs $b_k \in \{-1, 1\}$. Unconditionally convergence is also equivalent to the property that any permuted series $\sum_{k=1}^{\infty} \boldsymbol{v}_{\mathrm{perm}(k)}$ converges to the same sum. A sufficient but not necessary condition for unconditional convergence is absolute convergence, meaning that the series of real numbers $\sum_{k=1}^{\infty} \|\boldsymbol{v}_k\|_{\mathcal{B}}$ converges. For series of real numbers, absolute convergence and unconditional convergence are equivalent.

Example 5.4.4 *The power set $\{x^k | k \in \mathbb{N}\}$ is not an unconditional basis for $L_2([-1, 1])$. Indeed, consider the square integrable function $f(x) = 1/(1+x^2)$ for which the Taylor expansion leads to the n-term approximation*

$$f_n(x) = \sum_{k=0}^{n} (-1)^k x^{2k} = \sum_{k=0}^{n} (-x^2)^k.$$

The error function is given by

$$R_n(x) = f_n(x) - f(x) = \frac{1 - (-x^2)^{n+1}}{1 + x^2} - \frac{1}{1 + x^2} = (-1)^n x^{2n+2} / (1 + x^2).$$

When $n \to \infty$, the L_2-norm of the error function converges to zero,

$$\lim_{n \to \infty} \|R_n\|_2^2 = \lim_{n \to \infty} \int_{-1}^{1} [R_n(x)]^2 dx = 0.$$

Replacing all coefficients in the approximation by their absolute values we get

$$g_n(x) = \sum_{k=0}^{n} x^{2k} = \frac{1 - x^{2n+2}}{1 - x^2},$$

for which the pointwise limit is given by $g(x) = 1/(1 - x^2)$. This function is not square integrable; hence its approximation cannot possibly converge in $L_2([-1, 1])$. A simple switch of signs leads to a totally different conclusion. Procedures, such as wavelet thresholding, that take decisions or make selections based on coefficients' magnitudes would not be able to distinguish between these two cases. The convergence in the case of alternating signs *is due to the mutual compensation of the contributions with positive and negative coefficients. A simple reordering of the series could move the negative coefficients towards the end, for instance imposing that the nth negative coefficient appears only at position n^2. The pointwise convergence is then dominated by the positive coefficients*

$$h_n(x) = \sum_{k=0}^{n} x^{4k} = \frac{1 - x^{4n+4}}{1 - x^4},$$

to the non-square integrable function $1/(1 - x^4)$.

An unconditional basis can be used in the definition of a coefficient norm

$$\|\boldsymbol{s}\| = \| \sum_{k=1}^{\infty} |s_k| \varphi_k \|_{\mathcal{B}}.$$

The coefficient norm is finite if and only if the vector norm is finite. In other words, there exist strictly positive constants γ and Γ such that

$$\frac{1}{\sqrt{\Gamma}} \|\boldsymbol{s}\| \leq \|\boldsymbol{f}\|_{\mathcal{B}} \leq \frac{1}{\sqrt{\gamma}} \|\boldsymbol{s}\|. \tag{5.39}$$

In an unconditional basis, the norm of the coefficients is a good measure for the norm of the object they represent.

5.4.4 Euclidean spaces

The conditions for an unconditional basis can be further simplified in vector spaces equipped with an inner product.

A **Euclidean space** is a normed vector space, where the norm is induced by an inner product, denoted by $\langle \boldsymbol{f}, \boldsymbol{g} \rangle \in \mathbb{R}$. The induced norm is then $\|\boldsymbol{f}\|_{\mathcal{H}} = \sqrt{\langle \boldsymbol{f}, \boldsymbol{f} \rangle}$. The inner product enables the definition of orthogonality, $\boldsymbol{f} \perp \boldsymbol{g} \Leftrightarrow \langle \boldsymbol{f}, \boldsymbol{g} \rangle = 0$. It also provides a way to find the coefficients $\boldsymbol{s}$ for a given vector $\boldsymbol{f}$ by solving the system of n equations

$$\langle \boldsymbol{f}, \varphi_k \rangle = \sum_{l=1}^{n} s_l \langle \varphi_l, \varphi_k \rangle. \tag{5.40}$$

In an orthogonal basis this system simplifies to $s_k = \langle \boldsymbol{f}, \varphi_k \rangle / \|\varphi_k\|_{\mathcal{H}}^2$. A **Hilbert** space is a complete Euclidean space.

For a basis in a separable Hilbert space, a property similar to that of an unconditional basis is defined by taking the ℓ_2 sequence norm instead of the induced norm in (5.39). The ℓ_2 sequence norm is defined by

$$\|\boldsymbol{s}\|_2 = \sqrt{\sum_{k=1}^{\infty} |s_k|^2}, \tag{5.41}$$

leading to the following definition.

Definition 5.4.5 *(Riesz basis) A basis in a separable Hilbert space $\mathcal{H}$ is a* ***Riesz basis*** *if there exist strictly positive and finite constants γ and Γ such that for all $\boldsymbol{f} \in \mathcal{H}$,*

$$\frac{1}{\Gamma}\|\boldsymbol{s}\|_2^2 \leq \|\boldsymbol{f}\|_{\mathcal{H}}^2 \leq \frac{1}{\gamma}\|\boldsymbol{s}\|_2^2. \tag{5.42}$$

Unless explicitly stated otherwise, further use of (5.42) will assume the tightest possible values of γ and Γ (the largest possible value of γ and the smallest possible value of Γ, that is). These tightest values are known as the **Riesz constants**.

In a Hilbert space, replacing the induced norm of (5.39) by the ℓ_2 sequence norm has no fundamental effect. Apart from a normalisation issue, Riesz bases and unconditional bases are the same, as stated in the following theorem (Heil, 2011, Theorem 7.13, Section 7.2).

Theorem 5.4.6 *In a separable Hilbert space $\mathcal{H}$, a basis $\boldsymbol{\Phi}$ is a Riesz basis if and only if it is unconditional and all basis functions are* quasi-normalised, almost normalised, *or* bounded, *meaning that there exist strictly positive and finite constants a and A such that for all $\varphi_k \in \boldsymbol{\Phi}$,*

$$a \leq \|\varphi_k\|_{\mathcal{H}}^2 \leq A. \tag{5.43}$$

An equivalent definition of a Riesz basis uses inner products with the basis vectors, rather than the coefficients.

Theorem 5.4.7 *In a separable Hilbert space $\mathcal{H}$, a basis $\boldsymbol{\Phi}$ is a Riesz basis with Riesz constants γ and Γ if and only if*

$$\gamma \sum_{k=1}^{\infty} \langle \boldsymbol{f}, \varphi_k \rangle^2 \leq \|\boldsymbol{f}\|_{\mathcal{H}}^2 \leq \Gamma \sum_{k=1}^{\infty} \langle \boldsymbol{f}, \varphi_k \rangle^2. \tag{5.44}$$

Theorem 5.4.7 reflects the fact that being a Riesz basis is a property of the basis only; it does not depend on the construction of the coefficients. Therefore, this version is in a sense more appealing from a mathematical point of view.

5.4.5 Duality and orthogonality

The proof of Theorem 5.4.7 (Mallat, 2001) is based on Riesz's representation theorem. As a matter of fact, Theorem 5.4.7 hinges on the existence of a dual basis $\widetilde{\boldsymbol{\Phi}}$ provided by the following result (Dahmen, 1996, Remark 1.2).

Theorem 5.4.8 *In a separable Hilbert space $\mathcal{H}$, a basis $\boldsymbol{\Phi}$ is a Riesz basis with Riesz constants γ and Γ if and only if there exists a dual Riesz basis $\widetilde{\boldsymbol{\Phi}} \subset \mathcal{H}$ with Riesz constants $1/\Gamma$ and $1/\gamma$, such that for any $\boldsymbol{f} \in \mathcal{H}$ the sequence $s \in \ell_2$ in the expansion $\boldsymbol{f} = \boldsymbol{\Phi}s$ (5.36) is found by $s = \langle \widetilde{\boldsymbol{\Phi}}, \boldsymbol{f} \rangle$, i.e.,*

$$s_k = \langle \widetilde{\varphi}_k, \boldsymbol{f} \rangle. \tag{5.45}$$

The dual Riesz basis is for the *analysis* of a vector, which is the computation of the coefficients of the expansion (5.36) as inner products with the dual basis elements. The primal Riesz basis is for the *reconstruction*. Using the inner products with elements of the primal basis in Theorem 5.4.7 thus amounts to switching the roles of dual and primal bases.

An orthonormal basis is a special case of a Riesz basis where the Riesz constants are equal to one. The resulting equality $\|\boldsymbol{f}\|_{\mathcal{H}} = \|s\|_2$ is sometimes named Plancherel's identity or Parseval's identity, referring to results in Fourier analysis, stated in Theorem 6.2.1 in Section 6.1. In fact, a Riesz basis can be transformed into an orthogonal basis using a topological isomorphism, a continuous and continuously invertible bijection, that is. In this sense, a Riesz basis is "almost" an orthogonal basis. Once it has been identified, the dual basis provides a solution of the linear system (5.40), which is equally easy and fast as an orthogonal basis.

5.4.6 Examples of Riesz bases

The above introduced notions can be illustrated with a couple of examples and counter-examples in sequence and function spaces.

In **finite dimensional spaces**, the concepts and results above reduce to more basic facts. Taking $\mathbb{R}^n$ as a representative of an n dimensional Euclidean space, we recycle the notation $\boldsymbol{\Phi}$ for the ordered set or *vector* of basis vectors, replacing the unordered set above. The vector of basis vectors forms an $n \times n$ real, invertible matrix, where $\widetilde{\boldsymbol{\Phi}} = \boldsymbol{\Phi}^{-1}$. The inner product $s = \langle \widetilde{\boldsymbol{\Phi}}, \boldsymbol{f} \rangle$ in Theorem 5.4.8 for finding the coefficients then reduces to a matrix-vector product $s = \widetilde{\boldsymbol{\Phi}}\boldsymbol{f} = \boldsymbol{\Phi}^{-1}\boldsymbol{f}$. By definition, the usual matrix norm satisfies the inequalities

$$\begin{aligned} \|s\|_2^2 &\leq \|\boldsymbol{\Phi}^{-1}\|_2^2 \|\boldsymbol{f}\|_2^2, \\ \|\boldsymbol{f}\|_2^2 &\leq \|\boldsymbol{\Phi}\|_2^2 \|s\|_2^2, \end{aligned}$$

allowing us to identify the Riesz constants in (5.42) as $\gamma = \|\boldsymbol{\Phi}\|_2^2$ and $\Gamma = 1/\|\boldsymbol{\Phi}^{-1}\|_2^2$. As the basis $\boldsymbol{\Phi} \subset \mathbb{R}^n$ has been selected without any special precautions, we conclude that in a finite dimensional Euclidean space, every classical (Hamel) basis is a Schauder basis and a Riesz basis. This conclusion is in line with the fact that every non-singular matrix can be orthogonalised using Gram-Schmidt orthogonalisation, i.e., QR factorisation.

The Riesz constants are directly related to the **condition number** $\kappa(\Phi)$ of the matrix of basis vectors,

$$\kappa(\Phi) = \kappa\left(\Phi^{-1}\right) = \|\Phi\|_2 \cdot \|\Phi^{-1}\|_2 = \sqrt{\frac{\Gamma}{\gamma}}. \tag{5.46}$$

A finite Euclidean space itself is also always a Hilbert space.

The set of all **sequences** of real numbers, endowed with the usual pointwise addition and the pointwise scalar multiplication, forms a vector space. There are many norms measuring the "size" of a sequence, most notably the ℓ_p sequence norms, defined by

$$\|\boldsymbol{f}\|_p = \left(\sum_{k=1}^{\infty} |f_k|^p\right)^{1/p}. \tag{5.47}$$

The ℓ_2 norm, already introduced in (5.41), is induced by the inner product

$$\langle \boldsymbol{f}, \boldsymbol{g} \rangle = \sum_{k=1}^{\infty} f_k g_k.$$

With $\boldsymbol{e}_k$ a canonical vector, consider the basis vectors $\varphi_k = (\boldsymbol{e}_1 + \boldsymbol{e}_k)/\sqrt{2}$ for $k = 2, 3, \ldots$ and $\varphi_1 = \boldsymbol{e}_1$. This basis is not a Riesz basis. Indeed, taking as coefficients the harmonic sequence $s_k = 1/k$, we find for the expansion (5.36) that the first component of $\boldsymbol{f}$ is unbounded, and hence $\boldsymbol{f}$ cannot possibly have a finite ℓ_2 norm. Nevertheless, the coefficient norm is bounded, as

$$\sum_{k=1}^{\infty} \frac{1}{k^2} = \frac{\pi^2}{6}.$$

A simple change of signs leads to the alternating harmonic sequence $s_k = (-1)^{k+1}/k$, for which the first component of $\boldsymbol{f}$ amounts to

$$f_1 = 1 + \left(\sum_{k=2}^{\infty} s_k\right)/\sqrt{2} = (1 - 1/\sqrt{2}) + \frac{1}{\sqrt{2}}\left(\sum_{k=1}^{\infty} \frac{(-1)^{k+1}}{k}\right) = 1 - 1/\sqrt{2} + \log(2),$$

so that $\boldsymbol{f} \in \ell_2$.

From the example we can conclude that unconditional bases are essential in applications that involve operations on the coefficients of an expansion (5.36). Indeed, in an unconditional basis, vectors can be reconstructed by adding the processed coefficients in arbitrary order. From the magnitude of the coefficients it can be assured that the reconstruction will be a member of the Banach space, meaning that it has a finite norm.

In **function spaces**, the elements are real functions defined on a given domain $D \subset \mathbb{R}$. Function spaces can be defined by imposing some degree of smoothness. A classical example is the vector space of infinitely differentiable functions on D, $\mathcal{C}^\infty(D)$. Larger classes are defined by all r times continuously differentiable functions on D, $\mathcal{C}^r(D)$, of which the vector space of all continuous functions, $\mathcal{C}^0(D)$, is the largest. If D is a closed interval or any other compact set, then the membership $f \in \mathcal{C}^r(D)$ implies that f is bounded. Boundedness can be expressed in terms of the uniform norm, $\|f\|_\infty < \infty$, where the uniform norm is defined as $\|f\|_\infty = \sup_{x \in D} |f(x)|$. The vector spaces $\mathcal{C}^r(D)$ are complete under the uniform norm, since a uniformly convergent sequence of continuous functions is also continuous. Alternatives to the uniform norm, useful in $\mathcal{C}^r(D)$, are the L_p norms, defined by the Lebesgue integral expression,

$$\|f\|_p = \left(\int_D |f(x)|^p dx \right)^{1/p}, \tag{5.48}$$

where $1 \leq p < \infty$. Setting $p = 2$ leads to the previously encountered $L_2(D)$ spaces. The limiting case $p = \infty$ leads again to the uniform norm. The space $\mathcal{C}^r(D)$ is not complete under any of these norms, as it is easy to construct sequences of continuous functions converging to a non-continuous function with respect to the L_p norm. The L_p norm can be used to define the Lebesgue or $L_p(D)$ function spaces containing all functions with $\|f\|_p < \infty$. The Lebesgue spaces are complete under the corresponding norm, at least if the integral in (5.48) is the Lebesgue integral, not the Riemann integral.

With $D = [0,1]$, the trigonometric system $\{1, \cos(2\pi kx), \sin(2\pi kx) | k = 1, 2, 3, \ldots\}$, or its complex version $\{\exp(2\pi kix) | k \in \mathbb{N}\}$, provides a Schauder basis for $L_2(D)$. As a counter-example, we have the hierarchical basis of Section 3.1.1, which admits nontrivial expansions of the zero function.

With $D = [0,1]$ (or any other closed interval), the power set $\{x^k | k \in \mathbb{N}\}$ is a countable basis of $\mathcal{C}^0(D)$. This is the Weierstrass (or Stone-Weierstrass) approximation theorem: for every continuous function on a closed interval there exists a uniformly converging series of polynomial approximations[3]. The basis is, however, not unconditional, as can be concluded from the expansion with the harmonic and alternating harmonic sequences. We have the uniformly convergent series on $D = [0,1]$,

$$\log(1+x) = \sum_{k=1}^{\infty} \frac{(-1)^{k-1}}{k} x^k,$$

while

$$-\log(1-x) = \sum_{k=1}^{\infty} \frac{1}{k} x^k,$$

[3]Uniformly converging polynomial approximations can be constructed, for instance, using a sequence of bases of Bernstein polynomials. On $D = [0,1,]$ this is $f_n(x) = \sum_{k=0}^{n} f(k/n) B_{n,k}(x)$ with $B_{n,k}(x) = \binom{n}{k} x^k (1-x)^{n-k}$.

holds in pointwise sense, but not in the uniform norm. As discussed in Example 5.4.4, the set of power functions is not an unconditional basis for $L_2(D)$ either.

Exercise 5.4.9 *Prove that the Taylor series of* $\log(1+x)$ *converges uniformly on* $D = [0, 1]$.

5.5 Multiresolution analysis

The formal definition of a multiresolution analysis provides a framework for the construction of wavelet decompositions on equidistant knots (Mallat, 1989; Meyer, 1992). The next sections present the formal definition and its immediate corollaries.

5.5.1 From multiresolution to the two-scale equation

Definition 5.5.1 *(Multiresolution analysis) A sequence of nested, closed subspaces* $V_j \subset L_2([0,1]), j = L, \ldots, J$ *is called a* ***multiresolution analysis*** *(MRA) if*

$$\forall j \in \mathbb{Z} : V_j \subset V_{j+1}, \tag{5.49}$$

$$\overline{\lim_{j\to\infty} V_j} = \overline{\bigcup_{j\in\mathbb{Z}} V_j} = L_2[0,1], \tag{5.50}$$

$$\lim_{j\to-\infty} V_j = \bigcap_{j\in\mathbb{Z}} V_j = \{0\}, \tag{5.51}$$

$$f(x) \in V_j \Leftrightarrow f(2x) \in V_{j+1}, j \in \mathbb{Z}, \quad \textit{(scale invariance)} \tag{5.52}$$

$$f(x) \in V_0 \Leftrightarrow f(x+k) \in V_0, k \in \mathbb{Z}, \quad \textit{(shift invariance)} \tag{5.53}$$

$$\exists \varphi(x) \in V_0 : \{\varphi(x-k)\}_{k\in\mathbb{Z}} \quad \textit{is a } \text{Riesz } \textit{basis for } V_0 \tag{5.54}$$

From this definition, we find that $f_1(x) \in V_1$ if and only if $f_1(x/2) \in V_0$, meaning that there exist coefficients $s_{1,k}$ so that

$$f_1(x/2) = \sum_{k=-\infty}^{\infty} s_{1,k}\varphi(x-k) \text{ or } f_1(x) = \sum_{k=-\infty}^{\infty} s_{1,k}\varphi(2x-k).$$

Hence, $\{2^{j/2}\varphi(2^j x - k)\}_{k\in\mathbb{Z}}$ is a Riesz basis of V_j, where the basis functions have the same norm at each level j, thanks to the normalisation factor $2^{j/2}$. As the subspaces V_j are nested, the father function $\varphi(x)$ can be decomposed in a linear combination of the basis in V_1, $\{\varphi(2x-k)\}_{k\in\mathbb{Z}}$, leading to the equispaced two-scale equation in (5.22).

5.5.2 The partition of unity

From the equispaced two-scale equation, the next result follows.

Theorem 5.5.2 *(Partition of unity) If $\varphi(x)$ is Riemann integrable and satisfies an equispaced two-scale equation (5.22), then*

$$\sum_{k\in\mathbb{Z}} \varphi(x-k) = \int_{-\infty}^{\infty} \varphi(s)ds. \tag{5.55}$$

Proof. Denoting

$$S_J(x) = \sum_{k\in\mathbb{Z}} \varphi\left(\frac{x-k}{2^J}\right),$$

we find that $S_J(x)$ is periodic, as $S_J(x+1) = S_J(x)$ and furthermore,

$$\begin{aligned}
S_J(x) &= \sum_{k\in\mathbb{Z}} \varphi\left(\frac{x-k}{2^J}\right) = \sum_{k\in\mathbb{Z}} c \sum_{l\in\mathbb{Z}} h_l \varphi\left(\frac{x-k}{2^{J-1}} - l\right) \\
&= c \sum_{l\in\mathbb{Z}} h_l \sum_{k\in\mathbb{Z}} \varphi\left(\frac{x-k-2^{J-1}l}{2^{J-1}}\right) \\
&\quad \text{denoting } k + 2^{J-1}l = n \\
&= c\left(\sum_{l\in\mathbb{Z}} h_l\right) \sum_{n\in\mathbb{Z}} \varphi\left(\frac{x-n}{2^{J-1}}\right) = 2S_{J-1}(x) = 2^J S_0(x).
\end{aligned}$$

So $S_0(x) = 2^{-J}S_J(x) = \lim_{J\to\infty} 2^{-J}S_J(x) = \lim_{J\to\infty} \sum_{l\in\mathbb{Z}} 2^{-J}\varphi\left(\frac{l}{2^J} + \frac{x}{2^J}\right)$.

For $0 \leq x \leq 1$, it holds that $\frac{l}{2^J} + \frac{x}{2^J} \in \left[\frac{l}{2^J}, \frac{l+1}{2^J}\right]$.

We thus have the limit of a Riemann-sum, the integral $S_0(x) = \displaystyle\int_{-\infty}^{\infty} \varphi(s)ds$.

□

The implication of this result is two-fold. First, it states that the scaling basis in a multiresolution analysis must **reproduce constant functions** at all scales.

The second implication of Theorem 5.5.2 holds specifically for dyadic wavelet transforms. It says that the normalisation $\Phi_j(x)\mathbf{1}_j = 1$ is equivalent to the normalisation of the integral of the father function $\int_{-\infty}^{\infty} \varphi(s)ds = 1$. As in the non-equispaced case, there is not a single father function, this result has no extension beyond the equispaced case.

5.5.3 The partition of unity on non-equidistant knots

In a lifting scheme on non-equispaced knots, the constant reproduction does not follow automatically, although in practice the property holds in almost any lifting based wavelet transform. More precisely, the lifting scheme without the final update, which defines the refinement and hence the scaling basis, should produce zero detail coefficients whenever the input scaling coefficients represent a constant function at finest scale. In most schemes, a constant function is represented by a constant vector of scaling coefficients. Constant reproduction then means that the refinement of a constant vector yields a constant vector; i.e., the refinement is affine, as defined in Section 3.1.7. The subsequent discussion develops how this holds in particular for interpolating and B-spline bases.

Let $\Phi_j(x)$ be an interpolating scaling basis, such as the Deslauriers-Dubuc basis; then at scale j we have that $\varphi_{j,k}(x_{j,l}) = \delta_{k,l}$. The basis follows from an interpolating refinement, which, by definition, reproduces constants, $\mathbf{H}_j \mathbf{1}_j = \mathbf{1}_{j+1}$. The constant reproduction can be substituted into the two-scale equation, to find $\Phi_j(x)\mathbf{1}_j = \Phi_{j+1}(x)\mathbf{H}_j = \Phi_{j+1}(x)\mathbf{1}_{j+1}$. By recursion, this becomes $\Phi_j(x)\mathbf{1}_j = \Phi_J(x)\mathbf{1}_J$, or in full

$$\sum_{k=1}^{n_j} \varphi_{j,k}(x) = \sum_{k=1}^{n_J} \varphi_{J,k}(x),$$

for arbitrarily fine resolution J. At the knots at level J, we thus have

$$\sum_{k=1}^{n_j} \varphi_{j,k}(x_{J,l}) = \sum_{k=1}^{n_J} \varphi_{J,k}(x_{J,l}) = \sum_{k=1}^{n_J} \delta_{k,l} = 1,$$

from which constant reproduction follows at all levels by letting $J \to \infty$ (and assuming that the refinement leaves no holes).

If a constant function is represented by a non-constant vector of coefficients $\boldsymbol{s}_j$, then the scaling functions can be renormalised so that the coefficients are constant. Then the constant reproduction holds at coefficient level, i.e., $\mathbf{H}_j\mathbf{1}_j = \mathbf{1}_{j+1}$. The normalisation of B-splines in Definition 4.2.2 holds as an example, leading to the partition of unity as in Section 4.2.4, see (4.43).

5.5.4 The partition of unity in terms of the refinement coefficients

The partition of unity is equivalent to the following statement in terms of the two-scale equation coefficients.

Theorem 5.5.3 *(Partition of unity, coefficient version) If the Riemann integrable function $\varphi(x)$ satisfies an equispaced two-scale equation (5.22), then*

$$\sum_{k\in\mathbb{Z}} h_{2k} = \sum_{k\in\mathbb{Z}} h_{2k+1}. \tag{5.56}$$

Proof. Let $\varphi(x)$ be a refinable function, then it holds that

$$\sum_{l\in\mathbb{Z}} \varphi(x-l) = \int_{-\infty}^{\infty} \varphi(s)ds.$$

By rearranging terms we obtain

$$\begin{aligned}
\sum_{l\in\mathbb{Z}} \varphi(x-l) &= c\sum_{l\in\mathbb{Z}}\sum_{k\in\mathbb{Z}} h_k\varphi(2x-2l-k) \\
&= c\sum_{l\in\mathbb{Z}}\sum_{k\in\mathbb{Z}} h_{2k}\varphi(2x-2l-2k) \\
&\quad +c\sum_{l\in\mathbb{Z}}\sum_{k\in\mathbb{Z}} h_{2k+1}\varphi(2x-2l-2k-1) \\
&= c\sum_{m\in\mathbb{Z}}\sum_{k\in\mathbb{Z}} h_{2k}\varphi(2x-2m) + c\sum_{m\in\mathbb{Z}}\sum_{k\in\mathbb{Z}} h_{2k+1}\varphi(2x-2m-1) \\
&= c\sum_{k\in\mathbb{Z}} h_{2k} \sum_{m\in\mathbb{Z}} \varphi(2x-2m) \\
&\quad +c\sum_{k\in\mathbb{Z}} h_{2k+1} \sum_{m\in\mathbb{Z}} \varphi(2x-2m-1) \qquad (5.57)
\end{aligned}$$

On the other hand, the partition of unity leads straightforwardly to

$$\int_{-\infty}^{\infty} \varphi(s)ds = \sum_{m\in\mathbb{Z}} \varphi(2x-m) = \sum_{m\in\mathbb{Z}} \varphi(2x-2m) + \sum_{m\in\mathbb{Z}} \varphi(2x-2m-1). \qquad (5.58)$$

Considering (5.57) and (5.58) as two expansions of a constant function into the basis $\{\varphi(2x-m); m \in \mathbb{Z}\}$, identification of the even and odd coefficients in both expansions leads to (5.56). □

In combination with (5.23), the partition of unity can be further developed into[4]

$$\sum_{k\in\mathbb{Z}} h_{2k} = \sum_{k\in\mathbb{Z}} h_{2k+1} = \frac{1}{2}\sum_{k\in\mathbb{Z}} h_k = \frac{1}{c}. \qquad (5.59)$$

Example 5.5.4 *As an example, consider the refinement equation (5.22) with* $h_0 = 1 = h_2$*, while all other elements of* $\boldsymbol{h}$ *are zero. This is the upsampled sequence for the Haar refinement. Whereas the Haar scaling function is given by* $\varphi(x) = \chi_{[0,1[}(x)$*, a solution to the refinement equation with zeros inserted is* $\varphi(x) = \chi_{[0,2[}(x)$*. This father function satisfies the partition of unity (5.55), but not the coefficient version (5.56). The paradox is explained by the fact that the proof of (5.56) rests upon the assumption that the shifts of* $\varphi(x)$ *constitute a basis, and here we can write the zero function as a nontrivial expansion:*

$$0 = \sum_{k\in\mathbb{Z}} \varphi(x-2k) - \sum_{k\in\mathbb{Z}} \varphi(x-2k-1).$$

[4]The expression of the partition of unity in terms of refinement coefficient holds specifically for equispaced scaling bases. It is a special case of a specific expression of vanishing moments in terms of refinement coefficients on equispaced knots. The impact of these expressions is further discussed in Section 5.5.7, more precisely in the discussion on Corollary 5.5.16.

5.5.5 Solving the two-scale equation

On non-equidistant grids, there are two ways to define the scaling functions. One option is to use an existing refinable basis, as in Chapter 4. Alternatively, the basis of scaling functions follows from the refinement itself, where an artificial superresolution as in Section 3.2.3 can be introduced in order evaluate the basis functions in an arbitrary point. The downside of this approach is that the scaling functions are probably dependent on the intermediate refinement grids.

None of these issues plays a role when the scaling functions come from a single father function. The basis functions can be defined through the refinement, and yet a single father function fixes the smoothness and other properties at all resolution levels. There are a couple of procedures for the numerical evaluation of that father function from its refinement equation.

5.5.5.1 Iteration of the two-scale equation

A straightforward, iterative procedure constructs a row of functions $\varphi^{[i]}(x)$ by

$$\varphi^{[i]}(x) = c \sum_{k \in \mathbb{Z}} h_k \varphi^{[i-1]}(2x - k), \tag{5.60}$$

starting off from, for instance, $\varphi^{[0]}(x)$ the characteristic function on $[0, 1]$. Convergence, let alone uniform convergence, cannot be taken for granted. This iteration is somehow complementary to the one proposed in (3.29). Indeed, there we interpolate knot values $\overline{\Phi}_j^{[i-1]}(x_{i,l})$ using a continuous extension of the refinement operation, whereas here we work with $\varphi^{[i]}(x)$ in continuous values x, and with the discrete refinement coefficients.

5.5.5.2 Support of the refinable function

From the iteration of the two-scale equation, we find the following result.

Theorem 5.5.5 *Let the Riemann integrable function $\varphi(x)$ satisfy a two-scale equation (5.22), with $h_k = 0$ if $k \notin \{s_l, s_l + 1, \ldots, s_r\}$, then* $\mathrm{supp}(\varphi) \subset [s_l, s_r]$.

Proof. Assuming first that the support of $\varphi(x)$ is bounded, we can denote it by $[a, b]$. Then the support of $\varphi(2x - k)$ is given by $[(a + k)/2, (b + k)/2]$. Hence, at the level of the supports, the two-scale equation becomes

$$[a, b] = \bigcup_{k=s_l}^{s_r} \left[\frac{a+k}{2}, \frac{b+k}{2}\right] = \left[\frac{a+s_l}{2}, \frac{b+s_r}{2}\right],$$

from which it follows that $a = (a + s_l)/2 \Leftrightarrow a = s_l$ and $b = s_r$. Secondly, assuming that the support of $\varphi(x)$ is unbounded and using the integrability, we can find for any $\varepsilon > 0$ a pair (a, b) so that $\varphi(x) = \varphi_{ab}(x) + \varphi_\infty(x)$, where $\mathrm{supp}(\varphi_{ab}) \subset [a, b]$, while $\left|\int_{-\infty}^{\infty} \varphi_\infty(x) dx\right| < \varepsilon$. As $\varphi(x)$ solves the two-scale equation (5.22), we have

$$\varphi(x) = c\sum_{k=s_l}^{s_r} h_k\varphi_{ab}(2x-k) + c\sum_{k=s_l}^{s_r} h_k\varphi_{\infty}(2x-k).$$

Defining $\varphi_{ab}^{[0]}(x) = \varphi_{ab}(x)$, and

$$\varphi_{ab}^{[i]}(x) = c\sum_{k=s_l}^{s_r} h_k\varphi_{ab}^{[i-1]}(2x-k),$$

and with a similar definition for $\varphi_{\infty}^{[i]}(x)$, we see that

$$\varphi(x) = c\sum_{k=s_l}^{s_r} h_k\varphi_{ab}^{[i]}(2x-k) + c\sum_{k=s_l}^{s_r} h_k\varphi_{\infty}^{[i]}(2x-k).$$

Also, with $[a_i, b_i]$ denoting the support of $\varphi_{ab}^{[i]}(x)$, the iterative definition implies that

$$[a_i, b_i] = \bigcup_{k=s_l} s_r \left[\frac{a_{i-1}+k}{2}, \frac{b_{i-1}+k}{2}\right] = \left[\frac{a_{i-1}+s_l}{2}, \frac{b_{i-1}+s_r}{2}\right],$$

from which it follows that $a_i = (a_{i-1}+s_l)/2$ and $b_i = (b_{i-1}+s_r)/2$, or, by induction,

$$a_i = 2^{-i}a + \left(\sum_{j=1}^{i}\frac{1}{2^j}\right)s_l.$$

From there, we find $\lim_{i\to\infty} a_i = s_l$ and $\lim_{i\to\infty} b_i = s_r$. On the other hand, as the iteration preserves the integrals, we have for all i that

$$\left|\int_{-\infty}^{\infty}\varphi_{\infty}^{[i]}(x)dx\right| < \varepsilon.$$

As a conclusion, we can write $\varphi(x)$ as a sum of a function whose support is bounded by $[s_l, s_r]$ and a function with arbitrarily small integral. □

Note that this result can be seen as complementary to the result in Lemma 4.2.15, applied to the case where knots coincide with integer values.

5.5.5.3 Scaling function values in integer abscissae

Using the bounded support, we can find the exact values of $\varphi(k)$ for $k \in \mathbb{Z}$, in particular the nonzero values for $k \in \{s_l, s_l+1, \ldots, s_r\}$.

Corollary 5.5.6 *Let $\varphi(x)$ satisfy a two-scale equation (5.22), with $h_k = 0$ if $k \notin \{s_l, s_l+1, \ldots, s_r\}$, and define the vector $\boldsymbol{\varphi}_I$ with components $\varphi_{I,m} = \varphi(m+s_l-1)$. Then $\boldsymbol{\varphi}_I$ is the nontrivial solution to the homogeneous linear system*

$$\boldsymbol{\varphi}_I = c\mathbf{H}_I\boldsymbol{\varphi}_I. \tag{5.61}$$

In this system, the band matrix $\mathbf{H}_I$ is given by $H_{I,l,m} = h_{2l-m+s_l-1}$.

Proof. This follows immediately from substituting $x = l$ in the two-scale equation (5.22), $\varphi(l) = c\sum_{k=s_l}^{s_r} h_k\varphi(2l-k) = c\sum_{m=2l-s_r}^{2l-s_l} h_{2l-m}\varphi(m)$, followed by shifting the indices so that the first component coincides with $\varphi(s_l)$. □

As an example, with $s_l = 0$ and $s_r = 5$, the system becomes

$$\begin{bmatrix} \varphi(0) \\ \varphi(1) \\ \varphi(2) \\ \varphi(3) \\ \varphi(4) \\ \varphi(5) \end{bmatrix} = c \begin{bmatrix} h_0 & & & & & \\ h_2 & h_1 & h_0 & & & \\ h_4 & h_3 & h_2 & h_1 & h_0 & \\ & h_5 & h_4 & h_3 & h_2 & h_1 \\ & & & h_5 & h_4 & h_3 \\ & & & & & h_5 \end{bmatrix} \cdot \begin{bmatrix} \varphi(0) \\ \varphi(1) \\ \varphi(2) \\ \varphi(3) \\ \varphi(4) \\ \varphi(5) \end{bmatrix}$$

For the system (5.61) to have a nontrivial solution, we need that the square matrix $\mathbf{H}_I$ has an eigenvalue $1/c$. This condition is fulfilled if the sequence $\boldsymbol{h}$ satisfies the coefficient version (5.59) of the partition of unity. Indeed, each column of $\mathbf{H}_I$ contains either all even or all odd nonzeros of $\boldsymbol{h}$. As a result, $\mathbf{1}^\top/c = \mathbf{1}^\top\mathbf{H}_I$, so $1/c$ is an eigenvalue with left eigenvector $\mathbf{1}^\top$.

Example 5.5.7 *For the Haar filter, we have $h_0 = 1 = h_1$, while all other components are zero. The matrix $\mathbf{H}_I = \mathbf{I}_2$ has a double eigenvalue in 1, with the canonical vectors $(1,0)^\top$ and $(0,1)^\top$ as eigenvectors. These vectors are the evaluations $\varphi(0)$ and $\varphi(1)$ of two independent solutions to the two-scale equation, $\varphi(x) = \chi_{[0,1[}(x)$ and $\varphi(x) = \chi_{]0,1]}(x)$.*

Example 5.5.8 *Reconsider the sequence $\mathbf{h}$ of example 5.5.4. This sequence does not satisfy the coefficient version of the partition of unity, but the corresponding matrix*

$$\mathbf{cH}_I = \begin{bmatrix} 1 & 0 & 0 \\ 1 & 0 & 1 \\ 0 & 0 & 1 \end{bmatrix}$$

has a double eigenvalue 1 (with left eigenvectors different from $\mathbf{1}^\top$). The system (5.61) has two independent solutions, the right eigenvalues of $c\mathbf{H}_I$ for eigenvalue 1. They are given by $(1,1,0)^\top$, corresponding to the solution $\varphi(x) = \chi_{[0,2[}(x)$ and by $(0,1,1)^\top$, corresponding to the solution $\varphi(x) = \chi_{]0,2]}(x)$. Both functions are solutions of the same two-scale equation, but for none of them the shifts $\{\varphi(x-k), k \in \mathbb{Z}\}$ form a linearly independent set.

Example 5.5.9 *Let all elements of $\mathbf{h}$ be zero, except for $h_0 = 1 = h_3$. This sequence does satisfy the coefficient version of the partition of unity, hence the corresponding matrix*

$$c\mathbf{H}_I = \begin{bmatrix} 1 & 0 & 0 & 0 \\ 0 & 0 & 1 & 0 \\ 0 & 1 & 0 & 0 \\ 0 & 0 & 0 & 1 \end{bmatrix}$$

must have 1 among its eigenvalues. As a matter of fact, it has a triple eigenvalue 1, with eigenvectors $(1,0,0,0)^\top$, $(0,1,1,0)^\top$ and $(0,0,0,1)^\top$.

Example 5.5.10 *For the dual filter in the CDF(2,2) wavelet transform, i.e., the equispaced B-spline wavelet transform with two primal and two dual vanishing moments, given in (5.32), the normalisation constant is $c = 2$; hence the matrix $c\mathbf{H}_I$ is given by*

$$c\mathbf{H}_I = \frac{1}{4}\begin{bmatrix} -1 & 0 & 0 & 0 & 0 \\ 6 & 2 & -1 & 0 & 0 \\ -1 & 2 & 6 & 2 & -1 \\ 0 & 0 & -1 & 2 & 6 \\ 0 & 0 & 0 & 0 & -1 \end{bmatrix}$$

The matrix has a double eigenvalue 1, with just one eigenvector, $(0,-1,2,-1,0)^\top$. The sum of the components of the eigenvector is zero. From the partition of unity (5.55), it follows that the integral of $\phi(x)$ is then zero. This is not what we expect from a scaling function. In Section 5.5.6, it will be explained that indeed this dual scaling function cannot be used as a proper, i.e., primal scaling function.

Example 5.5.11 *The equispaced cubic B-spline refinement, given by (5.34), leads to a quite similar matrix,*

$$c\mathbf{H}_I = \frac{1}{8}\begin{bmatrix} 1 & 0 & 0 & 0 & 0 \\ 6 & 4 & 1 & 0 & 0 \\ 1 & 4 & 6 & 4 & 1 \\ 0 & 0 & 1 & 4 & 6 \\ 0 & 0 & 0 & 0 & 1 \end{bmatrix}$$

with a simple eigenvalue 1, for which the eigenvector is $(0,1,4,1,0)^\top/6$. These are the values of the equispaced cubic B-spline at the integer knots.

5.5.5.4 Solving the two-scale equation by dyadic recursion

Once the solution of (5.61) is found, it serves as a starting point in a dyadic recursion scheme for $j = 1, 2, \ldots$

$$\varphi(2^{-j}l) = c\sum_{k=s_l}^{s_r} h_k\varphi(2^{-j+1}l - k). \tag{5.62}$$

This recursion scheme identifies all values of $\varphi(x)$ in a dense subset of the real line.

Example 5.5.12 *Consider again the example 5.5.9 where all elements of $\mathbf{h}$ are zero, except for $h_0 = 1 = h_3$. The outcome of the dyadic (5.62) depends on the starting values in $\varphi(k)$ for $k \in \{0,1,2,3\}$. Starting from the vector $(0,0,0,1)$, we obtain the function for which $\varphi(3k/2^j) = 1$ for any $j = 1,2,\ldots$ and $k \in \{1,2,\ldots,2^j\}$ while $\varphi((3k-1)/2^j) = 0$ and $\varphi((3k-2)/2^j) = 0$. So, the function is discontinuous at all real numbers of $[0,3]$, regardless of how it is* *extended beyond the dyadic points $k2^{-j}$. The same conclusion holds for the outcome from the two other eigenvectors, $(1,0,0,0)$ and $(0,1,1,0)$. Taking linear combinations of the eigenvectors, however, results in an alternative set of eigenvectors. In particular, for the eigenvector $(1,1,1,0)$, the outcome is $\varphi(x) = \chi_{[0,3[}(x)$, while for $(0,1,1,1)$, the dyadic recursion can be extended continuously to $\varphi(x) = \chi_{]0,3]}(x)$. For none of the proposed solutions, however, the integer shifts $\{\varphi(x-k), k \in \mathbb{Z}\}$ form a linearly independent set.*

5.5.5.5 The cascade or subdivision algorithm

The subdivision algorithm of (2.69) and in Section 2.2.5 may also be applied to the equispaced two-scale equation (5.22).
Let $f_j(x) = \sum_{k\in\mathbb{Z}} s_{j,k}\varphi(2^jx-k)$; then refinement leads to the coefficients $s_{j+1,l}$ in $f_j(x) = \sum_{l\in\mathbb{Z}} s_{j+1,l}\varphi(2^{j+1}x-l)$. From the refinement equation (5.22), it follows that

$$\varphi(2^jx-k) = c\sum_{l\in\mathbb{Z}} h_{l-2k}\varphi(2^{j+1}x-l), \tag{5.63}$$

And so $f_j(x) = \sum_{k\in\mathbb{Z}} s_{j,k}\varphi(2^jx-k) = c\sum_{l\in\mathbb{Z}}\sum_{k\in\mathbb{Z}} s_{j,k}h_{l-2k}\varphi(2^{j+1}x-l)$. Hence

$$s_{j+1,l} = c\sum_{k\in\mathbb{Z}} h_{l-2k}s_{j,k}. \tag{5.64}$$

This iterative refinement of the scaling coefficients can be applied in particular for $f_0(x) = \varphi(x)$, hence with s_0 a canonical vector (i.e., a Kronecker delta).

In practice, when the finest resolution level J is finite, subdivision also includes the numerical approximation of the scaling functions $\varphi(2^Jx-k)$. It is common practice to take $\varphi(2^Jx-k) \approx \chi_{[k2^{-J},(k+1)2^{-J}[}(x)$, which amounts to saying that $s_{J,k} \approx f_J(x)$. As explained in Section 8.4.1, this practice ignores the higher order approximation power of scaling bases with more than one dual vanishing moment. It is also unsatisfactory in the refinement used in Example 5.5.12, where $h_0 = 1 = h_3$. In order to obtain $\varphi(x) = \chi_{[0,3[}(x)$, we need to take $\varphi(2^Jx-k) = \chi_{[k2^{-J},(k+3)2^{-J}[}(x)$. With $\varphi(2^Jx-k) = \chi_{[k2^{-J},(k+1)2^{-J}[}(x)$, the father function would be

$$\varphi(x) = \sum_{k=0}^{2^J} \chi_{[3k2^{-J},(3k+1)2^{-J}[}(x),$$

which is zero on two-thirds of each interval $[3k2^{-J}, 3(k+1)2^{-J}]$ and one on the remaining third.

Note that the sum runs over k, whereas in the actual two-scale equation (5.63) it runs over l. The expression (5.64) corresponds to one step in the inverse transform (5.6) with all $d_{j,l} = 0$ and with constant c, given by (5.23) and compensating for the fact that we work with the non-normalised basis $\varphi(2^jx-$

k) instead of $\varphi_{j,k}(x) = c\varphi(2^j x - k)$. Repeated application of (5.64) corresponds to evaluating subdivision product $\Phi_j(x)\boldsymbol{s}_j = \Phi_J(x)\mathbf{H}_J\mathbf{H}_{J-1}\dots\mathbf{H}_j\boldsymbol{s}_j$ in (2.69) from the right to the left. In the equidistant case, the elements of $\Phi_J(x)$ are translations of a single dilated father function. As a result, the evaluation from the left to the right corresponds to the iteration in (5.60), taking as initial value $\varphi^{[0]}(x)$ the first component of the vector of functions in $\Phi_J(2^{-J}x)$. Convergence of the cascade algorithm, for $J \to \infty$, is thus equivalent to convergence of the iteration procedure for $i \to \infty$ in $\varphi^{[i]}(x)$.

5.5.6 Convergence of the refinement procedures

Convergence of the cascade algorithm is important in practice, because the reconstruction of a function from its wavelet coefficients consists in a cascaded refinement from the coarse scaling coefficients plus corrections at each scale from the detail coefficients. In order to assess the convergence of subdivision, we investigate the corresponding iteration scheme (5.60). The iteration scheme can be evaluated in a dense subset of the domain of $\varphi(x)$ by taking dyadic knots $x_{j,k} = k2^{-j}$, for $j = 0, 1, 2, \dots$ and $k = 0, 1, \dots, n_j - 1$, with $n_j = 2^j(s-1)+1$. Restricted to a finite resolution J the iteration can be represented by the matrix operation $\boldsymbol{\varphi}_J^{[i]} = \boldsymbol{H}_J\boldsymbol{\varphi}_J^{[i-1]}$, where $\boldsymbol{\varphi}_J$ is the vector with n_J elements $\varphi(k2^{-J})$ and $\boldsymbol{H}_J$ a squared matrix containing the refinement coefficients. It holds that $\boldsymbol{H}_j$ is a submatrix of $\boldsymbol{H}_{j+1}$, defined by its even rows and even columns $\boldsymbol{H}_j = \boldsymbol{H}_{j+1;e,e}$. Since the odd elements in $\boldsymbol{\varphi}_{j+1,o}$ do not appear on the right hand side of the iteration scheme (5.60), we also have $\boldsymbol{H}_{j+1;e,o} = \mathbf{0}_{j+1;e,o}$, and $\boldsymbol{H}_{j+1;o,o} = \mathbf{0}_{j+1;o,o}$. Using the expression for determinants of block matrices,

$$\det\begin{bmatrix} \mathbf{A}_{1,1} & \mathbf{0}_{1,2} \\ \mathbf{A}_{2,1} & \mathbf{A}_{2,2} \end{bmatrix} = \det(\mathbf{A}_{1,1})\det(\mathbf{A}_{2,2}),$$

we find that $\det(\boldsymbol{H}_J - \lambda\mathbf{I}_{n_J}) = \det(\boldsymbol{H}_0 - \lambda\mathbf{I}_s)\lambda^{n_J - s}$. Furthermore, $\boldsymbol{H}_0 = c\mathbf{H}_I$, the matrix used in the system (5.61) for the values $\varphi(k)$ in $\boldsymbol{\varphi}_I$. The eigenvalues of the iteration matrix $\boldsymbol{H}_J$ are those of $c\mathbf{H}_I$. If $\boldsymbol{H}_J$ has n_J linearly independent eigenvectors, then these eigenvectors constitute the columns of the matrix $\mathbf{E}_J$ in the factorisation

$$\boldsymbol{H}_J = \mathbf{E}_J\boldsymbol{\Lambda}_J\mathbf{E}_J^{-1}, \tag{5.65}$$

where $\boldsymbol{\Lambda}_J$ is a diagonal matrix with the eigenvalues on the diagonal. The convergence of the iteration depends on the eigenvalues, as

$$\boldsymbol{\varphi}_J^{[i]} = \boldsymbol{H}_J\boldsymbol{\varphi}_J^{[i-1]} = \boldsymbol{H}_J^i\boldsymbol{\varphi}_J^{[0]} = \mathbf{E}_J\boldsymbol{\Lambda}_J^i\mathbf{E}_J^{-1}\boldsymbol{\varphi}_J^{[0]}.$$

If no eigenvalue is larger than one in absolute value, then the iteration converges to

$$\boldsymbol{\varphi}_J^{[\infty]} = \mathbf{E}_J\boldsymbol{\Lambda}_J^{\infty}\mathbf{E}_J^{-1}\boldsymbol{\varphi}_J^{[0]}.$$

The diagonal elements of $\mathbf{\Lambda}_J^\infty$ are either one or zero. All eigenvalues strictly smaller than one in absolute value vanish in $\mathbf{\Lambda}_J^\infty$. If $\mathbf{\Lambda}_J^\infty$ contains eigenvalues larger than one, then the iteration does not converge.

An eigenvalue that is a multiple root of the characteristic equation $\det(\boldsymbol{H}_J - \lambda \mathbf{I}_{n_J}) = 0$ is said to have algebraic multiplicity larger than one. It may happen that the dimension of the kernel of $\boldsymbol{H}_J - \lambda \mathbf{I}_{n_J}$, termed the geometric multiplicity, is lower than the algebraic multiplicity. The number of linearly independent eigenvectors is then lower than n_J. As a result, the matrix $\boldsymbol{H}_J$ cannot be diagonalised by the factorisation in (5.65). Instead, the eigenvectors can be completed by so-called generalised eigenvectors to constitute an invertible matrix $\mathbf{E}_J$. With this invertible matrix, the factorisation in (5.65) still holds, but now with a slightly non-diagonal matrix $\mathbf{\Lambda}_J$. For each column c in $\mathbf{E}_J$ that is not an eigenvector, the superdiagonal element is given by $\Lambda_{J;c-1,c} = 1$. As a result, both $\mathbf{\Lambda}_J$ and $\mathbf{\Lambda}_J^i$ take a block diagonal form, where in the case of a double eigenvalue with a simple eigenvector the block looks like

$$\mathbf{\Lambda}_{J;b} = \begin{bmatrix} \lambda_b & 1 \\ 0 & \lambda_b \end{bmatrix} \Rightarrow \mathbf{\Lambda}_{J;b}^i = \begin{bmatrix} \lambda_b & 1 \\ 0 & \lambda_b \end{bmatrix}^i = \begin{bmatrix} \lambda_b^i & i\lambda_b^{i-1} \\ 0 & \lambda_b^i \end{bmatrix}.$$

If the double eigenvalue satisfies $|\lambda_b| < 1$, then this converges to a zero 2×2-matrix. If $|\lambda_b| = 1$, then this submatrix is a source of divergence. Similar conclusions hold for triple or multiple eigenvalues[5]. The discussion is concluded by the following theorem.

Theorem 5.5.13 *The subdivision and iteration procedures converge (pointwise) to a refinable and Riemann integrable function $\varphi(x)$ if the following conditions are met.*

1. *The eigenvalues λ_i of the matrix $c\mathbf{H}_I$, defined in (5.61), are bounded by $|\lambda_i| \leq 1$.*
2. *At least one eigenvalue λ_i equals 1.*
3. *The geometric multiplicity of eigenvalue 1 equals its algebraic multiplicity.*
4. *The component sum of an eigenvector for eigenvalue 1 is different from zero.*

Example 5.5.14 *For the dual filter of the CDF(2,2) transform, discussed in Example 5.5.10, the matrix $c\mathbf{H}_I$ has a double eigenvalue in 1, with just one eigenvector. As a result, the iteration and subdivision do not converge. Figure 2.12 has a version of a CDF(2,2) dual scaling function on a non-equispaced set of knots. As even on an equispaced set of knots there is no convergence, a fortiori we cannot expect convergence on arbitrary knots. The plot in Figure 2.12 must depend on the number of refinement steps.*

[5] the sum of all eigenvalues of $\boldsymbol{H}_J$ is the sum of all eigenvalues of $c\mathbf{H}_I$, which is $c\sum_{k\in\mathbb{Z}} h_k = 2$.

Since the iteration (5.60) preserves the integral, it holds that

$$\int_{-\infty}^{\infty} \varphi^{[i]}(x)dx = \int_{-\infty}^{\infty} \varphi^{[0]}(x)dx.$$

Starting the iteration with integral 1 (for instance, taking $\varphi^{[0]}(x) = \chi_{[0,1[}(x)$*), the values of* $\varphi^{[i]}(x)$ *in a fixed point* x *will grow unbounded, so that* $\varphi^{[i]}(x)$ *normalised by the values, or by the integral of* $|\varphi^{[i]}(x)|$ *tends to the* $\varphi(x)$ *with integral zero that results from the solution by dyadic recursion found in Example 5.5.10.*

A scaling function with zero integral and non-convergent subdivision is definitely not what we want for the reconstruction from a wavelet analysis. This comes on top of the highly unsmooth, fractal-like shape displayed in Figure 2.12. For use in the analysis, as a dual basis function that is, these unpretty characteristics are of far less importance.

5.5.7 Vanishing moments

As discussed in Sections 1.6.4, 2.3, and 3.1.7, it is a common practice to impose a certain number, say $\widetilde{p}$, of dual vanishing moments. This means that polynomials up to degree $\widetilde{p} - 1$ are spanned by the primal scaling basis. On an equispaced set of knots, there exist coefficients $\widetilde{x}_k^{[q]}$ so that

$$x^q = \sum_{k\in\mathbb{Z}} \widetilde{x}_k^{[q]} \varphi(x-k). \tag{5.66}$$

Thanks to dual vanishing moments, the smooth components of a function, which are close to a polynomial, are well approximated by a linear combination of scaling functions too. As a result, the detail coefficients associated to these smooth components are small, allowing the wavelet decomposition to concentrate on singularities.

For $q = 1$, the dual vanishing moment condition amounts to the partition of unity. Hence, (5.66) holds automatically once a Riemann integrable function $\varphi(x)$ has been found for a given refinement equation. The values of $\widetilde{x}_k^{[0]}$ are constant, namely $\left[\int_{-\infty}^{\infty} \varphi(s)ds\right]^{-1}$.

For $q \geq 1$, the proof of the subsequent theorem finds that the values of the coefficients $\widetilde{x}_k^{[q]}$ in the expansion of a power function can be seen as evaluations of a polynomial. Conversely, if the function values of any polynomial used as coefficients lead to a polynomial, then all polynomials can be reproduced within this basis.

Theorem 5.5.15 *(Strang-Fix) Let* $\varphi(x)$ *be a father scaling function, so that* $\{\varphi(x-k), k \in \mathbb{Z}\}$ *constitutes a Riesz basis for a subspace* $V_0 \subset L_2(\mathbb{R})$*. Then for all* $q \in \{0, 1, \ldots, \widetilde{p} - 1\}$*, the function*

$$v^{[q]}(x) = \sum_{k\in\mathbb{Z}} k^q \varphi(x-k),$$

is a polynomial of degree q if and only if for all $q \in \{0, 1, \ldots, \widetilde{p}-1\}$ there exist a sequence of coefficients $\widetilde{x}_k^{[q]}$ so that

$$x^q = \sum_{k \in \mathbb{Z}} \widetilde{x}_k^{[q]} \varphi(x-k).$$

Proof. Suppose that x^q lies in the space spanned by $\{\varphi(x-k), k \in \mathbb{Z}\}$. As $\{\varphi(x-k), k \in \mathbb{Z}\}$ constitutes a Riesz basis, there exists a dual basis to define the coefficients $\widetilde{x}_k^{[q]}$ in the expansion of x^q. Furthermore, since the knots are equidistant, the dual basis consists of translations of a dual father function, thus leading to

$$\widetilde{x}_k^{[q]} = \int_{-\infty}^{\infty} \widetilde{\varphi}(u-k) u^q du = \int_{-\infty}^{\infty} \widetilde{\varphi}(u)(u+k)^q.$$

Consider the function $\widetilde{x}^{[q]}(z) = \int_{-\infty}^{\infty} \widetilde{\varphi}(u)(u+z)^q$; then it is straightforward to check that its $(q+1)$st derivative is zero; hence $\widetilde{x}^{[q]}(z)$ is a polynomial of degree q. The polynomials $\widetilde{x}^{[r]}(z)$ for $r = 0, 1, \ldots, q$ form a linearly independent set, because otherwise the matrix $\widetilde{x}_k^{[r]}$ would be singular, thus leading to the conclusion that the power functions $\{1, x, x^2, \ldots, x^q\}$ would be linearly dependent. Hence, there exist coefficients a_r so that $z^q = \sum_{r=0}^{q} a_r \widetilde{x}^{[r]}(z)$. By linearity and the definition of $\widetilde{x}^{[r]}(z)$, it then holds that $k^q = \sum_{r=0}^{q} a_r \widetilde{x}_k^{[r]}$, and so

$$\sum_{k \in \mathbb{Z}} k^q \varphi(x-k) = \sum_{k \in \mathbb{Z}} \sum_{r=0}^{q} a_r \widetilde{x}_k^{[r]} \varphi(x-k) = \sum_{r=0}^{q} a_r x^r.$$

Conversely, let $v^{[r]}(x)$ be polynomials of degree r. They are linearly independent because the matrix with elements k^r, a Vandermonde matrix, is non-singular and $\{\varphi(x-k), k \in \mathbb{Z}\}$ is a basis. Hence, x^q can be written as a linear combination of $v^{[r]}(x)$, and from there as a linear combination of $\varphi(x-k)$. □

Corollary 5.5.16 *Let $\varphi(x)$ be a father scaling function, so that $\{\varphi(x-k), k \in \mathbb{Z}\}$ constitutes a Riesz basis for a subspace $V_0 \subset L_2(\mathbb{R})$. Then $\varphi(x)$ has $\widetilde{p}$ dual vanishing moments if and only if for all $q = 0, 1, \ldots, \widetilde{p}-1$,*

$$\sum_{k \in \mathbb{Z}} (-1)^k k^q h_k = 0. \tag{5.67}$$

Proof. The function $w^{[q]}(x) = \sum_{k \in \mathbb{Z}} (1-k)^q \varphi(x-k)$ is a linear combination of the polynomials $v^{[r]}(x)$, defined in Theorem 5.5.15, as indeed,

$$w^{[q]}(x) = \sum_{r=0}^{q} \binom{q}{r} (-1)^r v^{[r]}(x)$$

By one step of a forward wavelet transform, we also have

$$w^{[q]}(x) = \sum_{k\in\mathbb{Z}} s_{-1,k}\varphi(2x-k) + \sum_{k\in\mathbb{Z}} d_{-1,k}\psi(2x-k).$$

Independently from the exact choice of the mother function $\psi(x)$, we know that all detail coefficients $d_{-1,k}$ will be zero, since $w^{[q]}(x)$ is polynomial of degree lower than $\widetilde{p}$, and so it belongs to the space spanned by $\{\varphi(x-k)\}$, but also to the space spanned by $\{\varphi(2x-k\}$. As a result, $d_{-1,0} = \sum_{l\in\mathbb{Z}} \widetilde{g}_l(1-l)^q = 0$. By (5.20), a possible choice for the dual filter $\widetilde{g}$ is to take a quadrature mirror filter, so that

$$\sum_{l\in\mathbb{Z}}(-1)^l(1-l)^q h_{1-l} = 0,$$

from which (5.67) follows by taking $k = 1-l$. □

The result in (5.67) generalises the coefficient form of the partition of unity (5.56). Expression (5.67) can be used in the design of refinement equations on equispaced knots with a given number of vanishing moments. The design of refinement equations with vanishing moments directly imposed onto the refinement coefficients is not possible on non-equispaced knots, because in the general, non-equispaced case, there exists no expression of the vanishing moments condition in terms of the refinement coefficients. For instance, in the development of B-spline refinement through its lifting factorisation, the vanishing moments follow automatically from the fact that all intermediate basis functions in the subsequent lifting steps are linear combinations of B-splines, and so they are splines, thus preserving the exact representation of polynomials.

From (5.67) it follows, by linear combination, that

$$\sum_{k\in\mathbb{Z}}(-1)^k v_{\widetilde{p}-1}(k)h_k = 0,$$

for any polynomial $v_{\widetilde{p}-1}(x)$ of degree $\widetilde{p}-1$. In particular, it holds, for any $l\in\mathbb{Z}$, that

$$(-1)^l\sum_{k\in\mathbb{Z}}(-1)^{-k}(l-k)^q h_k = 0,$$

which amounts to $\boldsymbol{h} * \boldsymbol{a}^{[q]} = \boldsymbol{0}$, where $a_k^{[q]} = (-1)^k k^q$. By the associativity of the convolution, we then have the following result.

Theorem 5.5.17 *If $\boldsymbol{h}_0$ has $\widetilde{p}$ vanishing moments, i.e., if $\boldsymbol{h}_0 * \boldsymbol{a}^{[q]} = \boldsymbol{0}$ for $q = 0, 1, \ldots, \widetilde{p}-1$, then for any $\boldsymbol{q}$, the filter $\boldsymbol{h} = \boldsymbol{q} * \boldsymbol{h}_0$ has at least $\widetilde{p}$ vanishing moments. If $\boldsymbol{q}$ is the Haar filter, i.e., if all its elements are zero, except for $q_0 = 1/2 = q_1$, then the filter $\boldsymbol{h} = \boldsymbol{q} * \boldsymbol{h}_0$ has $\widetilde{p}+1$ vanishing moments.*

Proof. For the Haar filter, straightforward calculations show that

$$\sum_{k\in\mathbb{Z}}(-1)^k k^{\widetilde{p}} h_k = \frac{1}{2}\sum_{k\in\mathbb{Z}}(-1)^k\left[k^{\widetilde{p}} - (k+1)^{\widetilde{p}}\right] h_{0,k},$$

where $k^{\widetilde{p}} - (k+1)^{\widetilde{p}}$ is a polynomial of degree $\widetilde{p} - 1$, evaluated in k. By the $\widetilde{p}$ vanishing moments of $\boldsymbol{h}_0$, the outcome is zero. □

The importance of Theorem 5.5.17 lies in the fact that convolution preserves vanishing moments, making it an ideal tool in the design of wavelet transforms with a desired number of vanishing moments. First apply repeated convolution of the Haar filter with itself; then apply convolution with an additional sequence $\boldsymbol{q}$ in order to incorporate other features. The repeated convolution of the Haar filter with itself yields the shortest filters for a given number of vanishing moments. These filters are the CDF B-spline filters. The classical definition of convolution sums hinges on the equidistance of the knots. This is the reason why Chapter 4 has used lifting instead of convolution for the construction of B-spline wavelet transforms. As discussed in Section 4.3 however, lifting does not automatically preserve vanishing moments.

6

Dyadic wavelet design in the frequency domain

This chapter approaches the construction and the performance of a wavelet transform from a frequency, or Fourier, perspective. Frequency or Fourier analysis, introduced in Section 1.2.1, offers a complementary view on the operations and equations. In Section 6.1, the refinement and filterbank operations are first translated into the frequency domain. Next, in Section 6.2, it is explained how a frequency analysis can be used in the design of refinement filters and wavelet transform. For this, several properties and results of the frequency domain are developed. At the end of the chapter, in Section 6.3, a few orthogonal wavelets are constructed in the Fourier domain and then transformed back into the time domain.

As further reading does not rely on results in this section, readers with little affinity for frequency spectra can easily skip the chapter.

6.1 Multiresolution in the frequency domain

As refinement involves both functions of continuous independent variables and sequences or filters, the subsequent analysis will use the continuous as well as the discrete time Fourier transforms.

6.1.1 Refinement in the Fourier domain

The two-scale (5.22) and wavelet (5.25) equations can be transformed into the Fourier domain. Following the definition of a continuous Fourier transform in (1.10), we write

$$\Phi(\omega) = \frac{1}{2\pi} \int_{-\infty}^{\infty} \varphi(x) \exp(-i\omega x) dx;$$

DOI: 10.1201/9781003265375-6

then the two-scale equation (5.22) is transformed into

$$\begin{aligned}
\Phi(\omega) &= \frac{1}{2\pi}\int_{-\infty}^{\infty} c \sum_{k=-\infty}^{\infty} h_k \varphi(2x-k)\exp(-i\omega x)dx \\
&= c \sum_{k=-\infty}^{\infty} h_k \frac{1}{2\pi}\int_{-\infty}^{\infty} \varphi(2x-k)\exp(-i\omega x)dx \\
&= c \sum_{k=-\infty}^{\infty} h_k \frac{1}{2\pi}\frac{1}{2}\int_{-\infty}^{\infty} \varphi(u)\exp(-i\omega(u+k)/2)du \\
&= \frac{c}{2} \sum_{k=-\infty}^{\infty} h_k \exp(-i\omega k/2)\frac{1}{2\pi}\int_{-\infty}^{\infty} \varphi(u)\exp(-i\omega u/2)du \\
&= \frac{c}{2} \sum_{k=-\infty}^{\infty} h_k \exp(-i\omega k/2)\,\Phi\left(\frac{\omega}{2}\right).
\end{aligned}$$

With

$$H(\omega) = \sum_{k=-\infty}^{\infty} h_k \exp(-i\omega k),$$

the Discrete Time Fourier Transform (DTFT) of $\boldsymbol{h}$, as defined in (1.8), we find

$$\Phi(\omega) = \frac{c}{2} H\left(\frac{\omega}{2}\right)\Phi\left(\frac{\omega}{2}\right). \tag{6.1}$$

The value of c, found in (5.23), can be written as $c = 2/H(0)$. Defining $H_c(\omega) = (c/2)H(\omega) = H(\omega)/H(0)$, the repeated refinement in the Fourier domain then reads

$$\Phi(\omega) = H_c\left(\frac{\omega}{2}\right)\Phi\left(\frac{\omega}{2}\right) = \Phi\left(\frac{\omega}{2^J}\right)\prod_{j=1}^{J} H_c\left(\frac{\omega}{2^j}\right) = \Phi(0)\prod_{j=1}^{\infty} H_c\left(\frac{\omega}{2^j}\right). \tag{6.2}$$

The value $\Phi(0) = \frac{1}{2\pi}\int_{-\infty}^{\infty} \varphi(x)dx$ is a normalising constant.

Example 6.1.1 *(Shannon or sinc refinement) The function $H_c(\omega)$ is 2π-periodic. As an example, we consider the characteristic function on the interval $[-\pi/2, \pi/2]$ with its periodic extensions outside $[-\pi, \pi]$. Then by (6.2), the Fourier transformed scaling function is found to be the non-periodic characteristic function on $[-\pi, -\pi]$. By the inverse Fourier transform (1.11), and fixing the normalising constant $\Phi(0) = 1/(2\pi)$, this is equivalent to*

$$\begin{aligned}
\varphi(x) &= \int_{-\infty}^{\infty} \Phi(\omega)\exp(i\omega x)d\omega = \frac{1}{2\pi}\int_{-\pi}^{\pi} \exp(i\omega x)d\omega \\
&= \frac{\exp(i\pi x)-\exp(-i\pi x)}{2\pi i x} = \frac{\sin(\pi x)}{\pi x}.
\end{aligned} \tag{6.3}$$

The coefficients of the two-scale equation can be found from the inverse DTFT (1.9)

$$h_k = \frac{2}{c}\frac{1}{2\pi}\int_{-\pi/2}^{\pi/2} \exp(i\omega k)d\omega = \frac{2}{c\pi}\frac{\sin(\pi k/2)}{k}. \tag{6.4}$$

For $k = 0$, this has to be amended, $h_0 = 1/c$. The other even indexed coefficients are zero, while

$$h_{2k+1} = c\,\frac{(-1)^k 2}{(2k+1)\pi}.$$

Exercise 6.1.2 *Use the elements of Example 6.1.1 to prove that*

$$\int_{-\infty}^{\infty} \frac{\sin(\pi x)}{\pi x}dx = 1.$$

Example 6.1.3 *(Haar refinement) The DTFT of the moving average filter of a Haar transform, $h_0 = 1 = h_1$ is given by*

$$\begin{aligned} H(\omega) &= \frac{1}{2} + \frac{1}{2}\exp(-i\omega) = \frac{\exp(i\omega/2) + \exp(-i\omega/2)}{2}\exp(-i\omega/2) \\ &= \cos(\omega/2)\exp(-i\omega/2). \end{aligned}$$

With $c = 2$ in this case we have $H_c(\omega) = H(\omega)$ in (6.2), hence

$$\Phi(\omega) = \Phi(0)\prod_{j=1}^{\infty}\cos\left(\frac{\omega}{2^{j+1}}\right)\exp\left(-i\omega/2^{j+1}\right).$$

The identity

$$\prod_{k=0}^{n-1}\cos(2^k\alpha) = \frac{\sin(2^n\alpha)}{2^n\sin(\alpha)},$$

which follows from recursive application of the identity

$$\sin(2\alpha^{n-1}) = 2\sin(\alpha^{n-1})\cos(\alpha^{n-1}),$$

leads to the limit

$$\prod_{j=1}^{\infty}\cos\left(\frac{\omega}{2^{j+1}}\right) = \lim_{n\to\infty}\frac{\sin(\omega/2)}{2^n\sin(\omega/2\cdot 2^{-n})} = \frac{\sin(\omega/2)}{\omega/2}.$$

Together with

$$\prod_{j=1}^{\infty}\exp\left(-i\omega/2^{j+1}\right) = \exp\left[-i\omega\sum_{j=1}^{\infty}2^{-(j+1)}\right] = \exp(-i\omega/2),$$

which results in $\Phi(\omega) = \Phi(0)\,[1 - \exp(-i\omega)]\,/(i\omega)$, *while on the other hand, direct calculation of the Fourier transform of* $\varphi(x) = \chi_{[0,1[}(x)$ *reveals that*

$$\Phi(\omega) = \frac{1}{2\pi}\int_0^1 \exp(-i\omega x)dx = \frac{1}{2\pi}\,[1 - \exp(-i\omega)]\,/(i\omega).$$

Note that the Fourier transform cannot distinguish between the two solutions $\varphi(x) = \chi_{[0,1[}(x)$ *and* $\varphi(x) = \chi_{]0,1]}(x)$, *as the set where these two solutions are different has measure zero. For the same reason, the Fourier approach would find only one solution to the two-scale equation of Example 5.5.12.*

Example 6.1.4 *As discussed in Example 5.5.10, the dual filter in the CDF(2,2) wavelet transform has* $c = 2$; *hence the function* $H_c(\omega)$ *in (6.2) is given by*

$$H_c(\omega) = \frac{3}{4} + \frac{1}{2}\cos(\omega) - \frac{1}{4}\cos(2\omega) = [2 + \cos(\omega) - \cos^2(\omega)]/2.$$

As $H_c(\omega) > 1$ *for* $|\omega| \in]0, \pi/2[$, *the subdivision product diverges. The cubic spline refinement of Example 5.5.11 has the function*

$$H_c(\omega) = \frac{3}{8} + \frac{1}{2}\cos(\omega) + \frac{1}{8}\cos(2\omega) = [\cos(\omega) + 1]^2/4,$$

for which (6.2) converges.

6.1.2 Filterbanks in the Fourier domain

Filterbanks can be designed for their properties in the frequency domain. This is why it is interesting to write the operations (5.13), (5.14), and (5.15) in the frequency domain. For this, we list a few properties of the DTFT.

Lemma 6.1.5 *(DTFT of subsampling) Let* $X(\omega)$ *be the DTFT of* $\boldsymbol{x}$; *then the DTFT of* $\boldsymbol{x}_m = (\downarrow m)\boldsymbol{x}$ *is given by*

$$X_m(\omega) = \frac{1}{m}\sum_{d=0}^{m-1} X\left(\frac{\omega + 2\pi d}{m}\right) \tag{6.5}$$

Proof. Using an argument similar to footnote 6 on page 6, it holds for l not a multiple of m that

$$\frac{1}{m}\sum_{d=0}^{m-1}\exp\left[-i(2\pi l/m)d\right] = 0.$$

Indeed, the terms are powers of a complex dth root of unity z, for which $1 + z + \ldots + z^{d-1} = (1 - z^d)/(1 - z) = 0$. For $l = km$, the terms are all equal to one. As a result,

$$\begin{aligned} X_m(\omega) &= \sum_{k\in\mathbb{Z}} x_{mk} \exp\left[-i(\omega/m)mk\right] \\ &= \sum_{l\in\mathbb{Z}} x_l \exp\left[-i(\omega/m)l\right] \cdot \left(\frac{1}{m}\sum_{d=0}^{m-1} \exp\left[-i(2\pi l/m)d\right]\right) \\ &= \frac{1}{m}\sum_{d=0}^{m-1}\sum_{l\in\mathbb{Z}} x_l \exp\left[-i\left(\omega + 2\pi d/m\right)l\right]. \end{aligned}$$

□

Lemma 6.1.6 *(DTFT of flip) Let $X(\omega)$ be the DTFT of $\boldsymbol{x}$; then the DTFT of $\overline{\boldsymbol{x}}$, defined by $\overline{x}_k = x_{-k}$, is given by*

$$\overline{X}(\omega) = \overline{X(\omega)} = X(-\omega). \tag{6.6}$$

Lemma 6.1.7 *(DTFT of alternating sign) Let $X(\omega)$ be the DTFT of $\boldsymbol{x}$; then the DTFT of $\widetilde{\boldsymbol{x}}$, defined by $\widetilde{x}_k = (-1)^k x_k$, is given by*

$$\widetilde{X}(\omega) = X(\omega + \pi). \tag{6.7}$$

Lemma 6.1.8 *(DTFT of shift) Let $X(\omega)$ be the DTFT of $\boldsymbol{x}$; then the DTFT of $\mathrm{S}^m\boldsymbol{x}$, defined by $(\mathrm{S}^m\boldsymbol{x})_k = x_{k-m}$, is given by $\exp(-i\omega m)X(\omega)$.*

Lemma 6.1.9 *(DTFT of upsampling) Let $X(\omega)$ be the DTFT of $\boldsymbol{x}$; then the DTFT of $(\uparrow m)\boldsymbol{x}$ is given by $X(m\omega)$.*

Proof. Let $y_{mk} = x_k$ and $y_l = 0$ for $l \neq mk$. Then

$$Y(\omega) = \sum_{k=-\infty}^{\infty} x_k \exp(-i\omega mk) = X(m\omega).$$

□

Lemma 6.1.10 *(DTFT of convolution sum) Let $X(\omega)$ be the DTFT of $\boldsymbol{x}$, $H(\omega)$ the DTFT of $\boldsymbol{h}$, and $\boldsymbol{y} = \boldsymbol{x} * \boldsymbol{h}$ (defined in (5.8)); then the DTFT of $\boldsymbol{y}$ is given by $Y(\omega) = H(\omega)\cdot X(\omega)$.*

Proof.

$$\begin{aligned} Y(\omega) &= \sum_{m=-\infty}^{\infty}\sum_{k=-\infty}^{\infty} h_{m-k}x_k \exp(-i\omega n) \\ &= \sum_{k=-\infty}^{\infty} x_k \exp(-i\omega k) \sum_{m=-\infty}^{\infty} h_{m-k} \exp(-i\omega(m-k)). \end{aligned}$$

□

Based on these results, and using the 2π periodicity of a DTFT, the filter-bank operations (5.13), (5.14), and (5.15) are transformed into

$$S_j(\omega) = \frac{1}{2}\left[\widetilde{H}\left(-\frac{\omega}{2}\right)S_{j+1}\left(\frac{\omega}{2}\right)+\widetilde{H}\left(\pi-\frac{\omega}{2}\right)S_{j+1}\left(\frac{\omega}{2}+\pi\right)\right], \quad (6.8)$$

$$D_j(\omega) = \frac{1}{2}\left[\widetilde{G}\left(-\frac{\omega}{2}\right)S_{j+1}\left(\frac{\omega}{2}\right)+\widetilde{G}\left(\pi-\frac{\omega}{2}\right)S_{j+1}\left(\frac{\omega}{2}+\pi\right)\right], \quad (6.9)$$

$$S_{j+1}(\omega) = H(\omega)S_j(2\omega)+G(\omega)D_j(2\omega). \quad (6.10)$$

From here, we can formulate the **perfect reconstruction** condition in the Fourier domain, starting from

$$\begin{aligned} S_{j+1}(\omega) &= H(\omega)S_j(2\omega)+G(\omega)D_j(2\omega) \\ &= \frac{1}{2}H(\omega)\left[\widetilde{H}(-\omega)S_{j+1}(\omega)+\widetilde{H}(\pi-\omega)S_{j+1}(\omega+\pi)\right] \\ &+ \frac{1}{2}G(\omega)\left[\widetilde{G}(-\omega)S_{j+1}(\omega)+\widetilde{G}(\pi-\omega)S_{j+1}(\omega+\pi)\right] \\ &= \frac{1}{2}S_{j+1}(\omega)\left[H(\omega)\widetilde{H}(-\omega)+G(\omega)\widetilde{G}(-\omega)\right] \\ &+ \frac{1}{2}S_{j+1}(\omega+\pi)\left[H(\omega)\widetilde{H}(\pi-\omega)+G(\omega)\widetilde{G}(\pi-\omega)\right]. \end{aligned}$$

The left and right hand side are identical if

$$H(\omega)\widetilde{H}(-\omega)+G(\omega)\widetilde{G}(-\omega) = 2, \quad (6.11)$$

$$H(\omega)\widetilde{H}(\pi-\omega)+G(\omega)\widetilde{G}(\pi-\omega) = 0. \quad (6.12)$$

Expressions (6.11) and (6.12) are the Fourier equivalent to the perfect reconstruction condition in (2.58). The single condition in (2.58) translates into a double one here because the matrices in (2.58) include the subsampling and upsampling into the matrix structures.

Taking two pairs of quadrature mirror filters $(\boldsymbol{h},\widetilde{\boldsymbol{g}})$ and $(\widetilde{\boldsymbol{h}},\boldsymbol{g})$, as proposed in (5.20), translates into $\widetilde{G}(\omega) = (-1)\exp(-i\omega)H(\pi-\omega)$ and $G(\omega) = (-1)\exp(-i\omega)\widetilde{H}(\pi-\omega)$, for which half of the perfect reconstruction condition (6.12) is automatically satisfied. The other half, (6.11) reduces to

$$H(\omega)\widetilde{H}(-\omega)+H(\pi+\omega)\widetilde{H}(\pi-\omega) = 2. \quad (6.13)$$

Exercise 6.1.11 *Plug in the definition of a DTFT (1.8) into (6.13) to find that this is actually the Fourier version of (5.16),* $\sum_{l\in\mathbb{Z}}\widetilde{h}_l h_{l-2k} = \delta_k$.

The (dual) **vanishing moments** condition (5.67) can be stated in the Fourier domain as

$$H^{(q)}(\pi) = 0, \quad (6.14)$$

for $q = 0, 1, \ldots, \widetilde{p}-1$. Indeed, $H^{(q)}(\omega) = \sum_{k\in\mathbb{Z}} h_k(-k)^q\exp(-i\omega k)$, hence

$H^{(q)}(\pi) = (-1)^q = \sum_{k\in\mathbb{Z}} h_k k^q (-1)^k$. As a result, $H(\omega)$ must be of the form $H(\omega) = [1 + \exp(-i\omega)]^{\widetilde{p}} Q(\omega)$. If $\boldsymbol{h}$ has a finite number of nonzero elements, then $Q(\omega)$ is a polynomial in $\exp(-i\omega)$.

Using the repeated refinement in the Fourier domain (6.2), the vanishing moment condition can be reformulated for the Fourier scaling basis as

$$\Phi^{(q)}(\pi) = 0, \tag{6.15}$$

again for $q = 0, 1, \ldots, \widetilde{p} - 1$.

6.2 Frequency content of filters and functions

Section 6.1 has investigated the effect of the refinement and filterbanks in the frequency domain. We now focus on the frequency characteristics of the filters in the filterbank and the basis functions in a multiresolution analysis.

6.2.1 Frequency content of filters

The frequency analysis of a filter $\boldsymbol{h}$ in the convolution $\boldsymbol{y} = \boldsymbol{x} * \boldsymbol{h}$ can be interpreted as the response to an oscillation $x_k = \exp(i\omega k)$. Indeed, when the input is an oscillation, then the output is given by

$$\begin{aligned} y_k &= \sum_{l\in\mathbb{Z}} \exp(i\omega l) h_{k-l} = \sum_{m\in\mathbb{Z}} \exp(i\omega(k-m)) h_m \\ &= \exp(i\omega k) \sum_{m\in\mathbb{Z}} \exp(-i\omega m) h_m = \exp(i\omega k) H(\omega) = x_k H(\omega). \end{aligned}$$

In other words, oscillations at the input give rise to oscillations with the same frequency at the output; they are eigenfunctions of the linear, time-invariant system represented by the convolution. The value of the Fourier transform in ω acts as an eigenvalue: it determines whether and how much an oscillation with the corresponding frequency is reduced or amplified. For this reason, the Fourier transform of $\boldsymbol{h}$ is also termed its **frequency response**.

The frequency analysis of a wavelet transform reveals how the refinement and detail filters play a specific role in the analysis and synthesis. The Haar transform with moving average and moving difference filters is illustrative. For the moving average filter with $h_0 = 1/2 = h_1$, the frequency response was found in Example 6.1.3 to be $H(\omega) = \cos(\omega/2)\exp(-i\omega/2)$. The magnitude, $|H(\omega)|$, reaches a maximum at $\omega = 0$. This means that an input with zero pulsation, a constant function, is maximally preserved, while at $\omega = \pi$, the frequency response is zero. The moving average applied to $x_k = \exp(-i\pi k) = (-1)^k$ results in a zero output. From the **partition of**

unity, it follows that $H(\pi) = 0$ for any refinement filter. This is a special case, $q = 0$ of (6.14). Alternatively, it follows from rewriting (5.56) as

$$\sum_{k\in\mathbb{Z}}(-1)^k h_k = \sum_{k\in\mathbb{Z}} h_k \exp(-i\pi k) = H(\pi) = 0. \tag{6.16}$$

All refinement filters have the tendency to suppress high frequencies. They are therefore termed **lowpass filters**. The ideal lowpass filter is that of the Shannon refinement (6.4): its DTFT is the characteristic function on the frequencies around zero.

For the moving difference filter $g_0 = 1 = -g_1$, the frequency response becomes

$$G(\omega) = 1 - \exp(-i\omega) = \sin(\omega/2)\,2i\exp(-i\omega/2),$$

which is zero for $\omega = 0$. This means, not surprisingly, that a constant input is reduced to zero by the moving difference filter. The moving difference filter is an example of a **highpass filter**.

6.2.2 Frequency properties of functions and sequences

Section 6.2.1 has discussed the DTFT as a spectral analysis, where the basic oscillations $\exp(-i\omega x)$ behave as **eigenfunctions** of linear, time-invariant filters. For signals, i.e., functions or sequences, on the other hand, the continuous or discrete Fourier transforms are interpreted as decompositions into frequency spectra; i.e., they write the signal as a **superposition** of elementary oscillations. The combination of the frequency spectrum of the input signal and the DTFT of the linear filter operation are analysed through input-output results, such as the one provided by Lemma 6.1.10. The study of functions or sequences through their frequency contents is known as **harmonic analysis**. At any moment in a signal processing routine, there exists a tight connection between the spectral representation and the result in the original domain, stated by Plancherel's or Parseval's theorem.

Theorem 6.2.1 *(Plancherel) Let $F(\omega)$ and $G(\omega)$ be the Fourier transforms of the (possibly complex valued) functions $f(x)$ and $g(x)$, as defined in (1.10); then*

$$\int_{-\infty}^{\infty} f(x)\overline{g(x)}dx = 2\pi\int_{-\infty}^{\infty} F(\omega)\overline{G(\omega)}d\omega. \tag{6.17}$$

$$\int_{-\infty}^{\infty} |f(x)|^2\,dx = 2\pi\int_{-\infty}^{\infty} |F(\omega)|^2\,d\omega. \tag{6.18}$$

(Parseval) Let $F(\omega)$ and $G(\omega)$ be the DTFT of f and g, as defined in (1.8); then

$$\sum_{k\in\mathbb{Z}} f_k\overline{g}_k = \frac{1}{2\pi}\int_{-\pi}^{\pi} F(\omega)\overline{G(\omega)}d\omega. \tag{6.19}$$

Conversely, let a be the Fourier series of $f(x) \in L_2([0,1])$ as in (1.2); then

$$\int_0^1 |f(x)|^2 dx = \sum_{k=-\infty}^{\infty} |a_k|^2. \tag{6.20}$$

Proof. (Continuous case)

$$\begin{aligned} 2\pi \int_{-\infty}^{\infty} F(\omega)\overline{G(\omega)}d\omega &= 2\pi \int_{-\infty}^{\infty} \frac{1}{2\pi} \int_{-\infty}^{\infty} f(x)\exp(-i\omega x)dx\overline{G(\omega)}d\omega \\ &= \int_{-\infty}^{\infty} f(x) \overline{\int_{-\infty}^{\infty} G(\omega)\exp(i\omega x)d\omega} dx \\ &= \int_{-\infty}^{\infty} f(x)\overline{g(x)}dx. \end{aligned}$$

□

Proof. (Discrete case)

$$\begin{aligned} \int_{-\pi}^{\pi} F(\omega)\overline{G(\omega)}d\omega &= \int_{-\pi}^{\pi} \sum_{s=-\infty}^{\infty} \sum_{t=-\infty}^{\infty} f_s \overline{g}_t \exp[-i\omega(s-t)]d\omega \\ &= \sum_{s=-\infty}^{\infty} \sum_{t=-\infty}^{\infty} f_s \overline{g}_t \int_{-\pi}^{\pi} \exp[-i\omega(s-t)]d\omega. \end{aligned}$$

The latter integral is zero, except when $s = t$, for which it takes the value 2π. The case of the Fourier series expansion is of course completely similar, except for the integration constant 2π, which is replaced by 1. □

Plancherel's theorem essentially states that the Fourier transform preserves angles and norms; it is an orthogonal operation.

6.2.3 Heisenberg's uncertainty principle

Whereas Plancherel's theorem states that the Fourier transform preserves all information on angles, at the same time the focus of the frequency spectrum is quite complementary to that of the original domain. More precisely, Heisenberg's uncertainty principle states the following.

Theorem 6.2.2 *(Heisenberg's uncertainty principle) Let $F(\omega)$ be the Fourier transform of $f(x)$ as defined in (1.10), and define*

$$\begin{aligned} \overline{x}_f &= \frac{\int_{-\infty}^{\infty} |f(x)|^2 x dx}{\int_{-\infty}^{\infty} |f(x)|^2 dx} \\ (\Delta_f x)^2 &= \frac{\int_{-\infty}^{\infty} |f(x)|^2 (x-\overline{x}_f)^2 dx}{\int_{-\infty}^{\infty} |f(x)|^2 dx} \end{aligned}$$

(and similar definitions in the Fourier domain); then:

$$(\Delta_f x)(\Delta_F \omega) \geq \frac{1}{2}. \tag{6.21}$$

Proof. Define the Fourier pair[1] $g(x) = \exp(i\overline{\omega}_F x) f(x - \overline{x}_f)$ and $G(\omega) = \exp(i\overline{x}_f \overline{\omega}_F)\exp(-i\overline{x}_f \omega)F(\omega - \overline{\omega}_F)$; then $\overline{x}_g = 0 = \overline{\omega}_G$ and $(\Delta_f x) = (\Delta_g x)$ and $(\Delta_F \omega) = (\Delta_G \omega)$.

We thus further assume that $\overline{x}_f = 0 = \overline{\omega}_F$. Let $\mathcal{X}$ denote the operator that maps $f(x)$ onto $xf(x)$, and $F(\omega)$ onto $\omega F(\omega)$. Then $(\Delta_f x)^2 = \|\mathcal{X} f\|_2^2/\|f\|_2^2$, where the L_2-norm is given by $\|f\|_2^2 = \int_{-\infty}^{\infty} |f(x)|^2 dx$. This norm is induced by the inner product $\langle f, g\rangle = \int_{-\infty}^{\infty} \overline{f(x)} g(x) dx$.

Furthermore, let $\mathcal{F}$ be the operator that maps $f(x)$ onto its Fourier transform $F(\omega)$. Finally, let $\mathcal{D}$ be the operator that maps a function onto its derivative. Then, by partial integration, using the fact that any function in $L_2(\mathbb{R})$ converges to zero for $x \to \pm\infty$, it holds that[2] $\mathcal{F}(\mathcal{D}f) = i\mathcal{X}F$, so, by Plancherel's theorem, we have $\|\mathcal{D}f\|_2^2/\|f\|_2^2 = \|\mathcal{X}F\|_2^2/\|F\|_2^2$.

It holds also that $(\mathcal{D}\mathcal{X} - \mathcal{X}\mathcal{D})f(x) = [xf(x)]' - xf'(x) = f(x)$. Using partial integration, one finds

$$\begin{aligned} \|f\|_2^2 &= \langle f, (\mathcal{D}\mathcal{X} - \mathcal{X}\mathcal{D})f\rangle = \langle f, \mathcal{D}\mathcal{X}f\rangle - \langle f, \mathcal{X}\mathcal{D}f\rangle \\ &= -\langle \mathcal{D}f, \mathcal{X}f\rangle - \langle \mathcal{X}f, \mathcal{D}f\rangle = -2\mathrm{Re}\langle \mathcal{D}f, \mathcal{X}f\rangle. \end{aligned}$$

Applying the Cauchy-Schwartz inequality leads to $\|f\|_2^2 \leq 2\|\mathcal{D}f\|_2 \cdot \|\mathcal{X}f\|_2$ or

$$\frac{\|\mathcal{D}f\|_2}{\|f\|_2} \cdot \frac{\|\mathcal{X}f\|_2}{\|f\|_2} = \frac{\|\mathcal{X}F\|_2}{\|F\|_2} \cdot \frac{\|\mathcal{X}f\|_2}{\|f\|_2} \geq \frac{1}{2}$$

□

Because of Heisenberg's uncertainty principle, a decomposition of a function into a basis cannot provide at the same time infinitely precise information on the original observations and frequency. In particular, a Fourier decomposition is infinitely precise on the frequency contents of a function. For instance, a slowly decaying frequency spectrum reveals the presence of singularities in the original domain, but no information is given on where exactly the singularities occur. Conversely, the individual observations of a function carry no information on frequencies and singularities.

A wavelet decomposition provides a compromise between precision in the original domain of the observations and the scale at which different features in the function occur. Scale should be interpreted as inverse frequencies here. Most data are sparse in time and frequency, meaning that the data contain a couple of features that occur during a limited period in time and at a limited number of scales. The balanced localisation of a wavelet basis function in the original domain and frequency thus lies at the basis of sparsity and nonlinear processing, as summarised in Section 1.4.

[1] See Exercise 6.2.3.

[2] See Exercise 6.2.5.

The **optimal balance** between precision in frequency and precision in the original domain is obtained if the Heisenberg uncertainty becomes an equality. This is the case if the function $f(x)$ has a Gaussian bell component. Indeed, let $w(x) = \exp(-x^2/2)$ be a Gaussian bell curve; then its Fourier transform is given by[3] $W(\omega) = \exp(-\omega^2/2)/\sqrt{2\pi}$. We have $(\Delta_w x)^2 = 1/2 = (\Delta_W \omega)^2$, leading to an equality in (6.21). Obviously, the bell curve itself is situated at one single place in time and frequency. More specifically, we have $\overline{x}_w = 0$ and $\overline{\omega}_W = 0$. In order to use this function in a time frequency analysis, it is used as a window in a windowed or short time Fourier analysis, which defines a family of functions $f_{\omega,b}(x) = w(x-b)\exp(-i\omega x)$, with dilation parameter b and frequency parameter ω. All these functions share the same time and frequency uncertainties $(\Delta_f x)^2 = 1/2 = (\Delta_F \omega)^2$, while being situated at a unique location in the time-frequency plane.

Exercise 6.2.3 *Let $F(\omega)$ be the Fourier transform of $f(x)$. Then find the Fourier transforms of $f(x-a)$, $\exp(ibx)f(x)$, $\exp(ibx)f(x-a)$, $f(sx)$, and $f(sx-a)$.*

Exercise 6.2.4 *Let $g(x) = f(sx-a)$ and let $F(\omega)$ and $G(\omega)$ denote the Fourier transforms of $f(x)$ and $g(x)$. Use the results of Exercise 6.2.3 to show that dilation and translation have no impact on the area of the Heisenberg uncertainty window, i.e., $(\Delta_f x)(\Delta_F \omega) = (\Delta_g x)(\Delta_G \omega)$.*

Exercise 6.2.5 *Let $F(\omega)$ be the Fourier transform of $f(x)$. Then show that $i\omega F(\omega)$ is the Fourier transform of $f'(x)$. Using Plancherel's theorem, we can express Heisenberg's uncertainty without Fourier transforms, as indeed*

$$(\Delta_F \omega)^2 = \frac{\displaystyle\int_{-\infty}^{\infty} |f'(x)|^2 dx}{\displaystyle\int_{-\infty}^{\infty} |f(x)|^2 dx}.$$

Exercise 6.2.6 *Find the Fourier transform of the function $f(x) = \exp(-x^2/2\sigma^2)$.*

Exercise 6.2.7 *Let $f(x)$ and $g(x)$ be two functions in $L_1(\mathbb{R})$, and let $h(x) = (f * g)(x)$ be the convolution integral, defined in Exercise 5.3.1. Then show that the Fourier transforms of these functions satisfy*

$$H(\omega) = (2\pi)F(\omega)G(\omega).$$

Use this result and Exercise 5.3.1 to show that the Fourier transform of a B-spline on the knots $\{0, 1, \ldots, \widetilde{p}\}$ is given by

$$\Phi(\omega) = \frac{\exp(-i\omega\widetilde{p}/2)}{2\pi}\left(\frac{\sin(\omega/2)}{\omega/2}\right)^{\widetilde{p}}.$$

[3]See Exercise 6.2.6.

Exercise 6.2.8 *Let $f(x)$ be the triangular hat function with knots $-1, 0, 1$. Find $(\Delta_f x)^2$.*

Exercise 6.2.9 *Let $f(x)$ be the triangular hat function with knots $-1, 0, 1$. Find $(\Delta_F \omega)^2$, using the result in Exercise 6.2.5. Compare the results of Exercises 6.2.8 and 6.2.9 to conclude that the triangular hat carries more precise information on time or space than on frequency. Compare the value of $(\Delta_f x)(\Delta_F \omega)$ with the Heisenberg lower bound.*

Exercise 6.2.10 *Let $f(x)$ be the Haar scaling function with knots $0, 1$. Show that $(\Delta_F \omega)^2 = \infty$, using the result in Exercise 6.2.7. As $f(x)$ is not differentiable, the result of Exercise 6.2.5 does not apply. Because of its discontinuities, the Haar basis carries no precise information on frequency.*

Exercise 6.2.11 *Consider a wavelet transform combining a triangular hat refinement matrix $\mathbf{H}_j$ with a Haar detail matrix $\mathbf{G}_j$ on an equidistant set of knots $\{0, 1, \ldots, n_{j+1} - 1\}$. The wavelet functions are given by $\psi_{j,k}(x) = \varphi_{j+1,2k+1}(x) - \varphi_{j+1,2k}(x)$, where $\varphi_{j+1,k}(x)$ are the triangular hat functions. Let $g(x) = \psi(x)$ be the mother function of these wavelets, centered around zero. It then holds that $g(x) = x$ on $[-1, 1]$, $g(x) = (3 - x)/2$ on $[1, 3]$ and $g(x) = -(x+3)/2$ on $[-3, -1]$. Calculate $(\Delta_g x)^2$ and $(\Delta_G \omega)^2$, using the techniques of Exercises 6.2.8 and 6.2.9.*
Note that this detail matrix does not follow from a straightforward lifting design with the triangular hat prediction, followed by a simple update for adding a single primal vanishing moment. That transform can be checked to have a tridiagonal $\mathbf{G}_j$. In order to find the lifting scheme for the combination of a triangular hat refinement and a Haar detail matrix, the procedure of Exercise 4.1.10 can be followed.

6.2.4 Fourier decay and global smoothness

The Fourier transform carries global information on the smoothness of a function in the original domain. This is learned from a few fundamental results.

Theorem 6.2.12 *The Fourier transform of an absolutely integrable function is a uniformly continuous function. Conversely, if the Fourier transform of a function is absolutely integrable, the function is uniformly continuous.*

Proof. Since the forward and inverse transforms basically have the same form, it suffices to prove one of the two implications. Let $f(x) \in L_1(\mathbb{R})$ be an absolutely integrable function with Fourier transform $F(\omega)$. Then

$$\begin{aligned}
|F(u) - F(\omega)| &= \frac{1}{2\pi} \left| \int_{-\infty}^{\infty} f(x) \left(e^{-iux} - e^{-i\omega x} \right) dx \right| \\
&\leq \frac{1}{2\pi} \int_{-\infty}^{\infty} |f(x)| \left| e^{-iux} - e^{-i\omega x} \right| dx \\
&\leq \frac{1}{2\pi} \int_{-M}^{M} |f(x)| \cdot |(u - \omega)x| dx + \frac{1}{2\pi} \int_{|x|>M} 2|f(x)| dx.
\end{aligned}$$

The last step follows from choosing a value M and a double inequality, namely $\left|e^{-iux} - e^{-i\omega x}\right| \leq 2$ and

$$\begin{aligned}\left|e^{-iux} - e^{-i\omega x}\right| &= \sqrt{[\cos(ux) - \cos(\omega x)]^2 + [\sin(ux) - \sin(\omega x)]^2} \\ &= 2|\sin[(u-\omega)x/2]| \leq |(u-\omega)x|.\end{aligned}$$

As $f(x)$ is absolutely integrable, there exists a value M so that for given ε it holds that

$$\frac{1}{2\pi}\int_{|x|>M} 2|f(x)|dx < \varepsilon.$$

On the other hand, denoting by I the absolute integral of $f(x)$, we have

$$\frac{1}{2\pi}\int_{-M}^{M} |f(x)| \cdot |(u-\omega)x|dx \leq \frac{|u-\omega|}{2\pi}\int_{-M}^{M} |f(x)|M dx \leq \frac{|u-\omega|MI}{2\pi}.$$

Taking $\delta = 2\pi\varepsilon/(MI)$, we find for all $|u-\omega| < \delta$ that $|F(u) - F(\omega)| < 2\varepsilon$. □

The absolute integrability in Theorem 6.2.12 can be understood as a condition of decay: for a function to be absolutely integrable, its absolute value has to tend to zero sufficiently fast. On the other side, the uniform continuity stands for a global, not local, form of smoothness. This is a typical result for Fourier analysis. A counterexample is given by the function with discontinuity in Section 1.3.3. The Fourier transform of a function with discontinuity has a slow decay[4]. The Fourier transform does not reveal where exactly the discontinuity is situated, nor does it tell us how many discontinuities there are.

A more general result on global smoothness is based on the notion of Lipschitz continuity, for which we first give a local definition.

Definition 6.2.13 *(Lipschitz continuity) A function $f(x)$ is Lipschitz (or Hölder) continuous with Lipschitz exponent $0 \leq \alpha \leq 1$ at x_0 if there exists a Lipschitz constant L so that for x sufficiently close to x_0,*

$$|f(x) - f(x_0)| \leq L|x - x_0|^\alpha.$$

A function $f(x)$ is Lipschitz (or Hölder) continuous with exponent $\alpha > 1$ if it is differentiable and its derivative is Lipschitz continuous with exponent $\alpha - 1$. The Lipschitz regularity (or smoothness) of a function at a value x_0 is the supremum of all exponents for which the function is Lipschitz continuous at that point.

The notion of Lipschitz regularity is a slight variation and extension of the idea of continuous differentiability. The case $\alpha = 1$ sheds some light on the subtle differences. If a function has Lipschitz regularity one in x_0, then it is necessarily continuous. On the other hand, if a function is continuously differentiable

[4]As long as the function is absolutely integrable, the Fourier transform always has asymptotic decay. This result is known as the Riemann-Lebesgue Lemma. See also Section 9.2.4.

at that point, it has a Lipschitz regularity of at least one. A Lipschitz regularity of one lies between continuity and differentiability. The function $f(x) = |x|$ is Lipschitz one at the origin. It is continuous, but not continuously differentiable. As an example of a continuous function with Lipschitz exponent smaller than one, it is straightforward to consider the function $f(x) = \mathrm{sign}(x)|x|^\alpha$ with $0 < \alpha < 1$ at the origin.

Definition 6.2.14 *(uniform Lipschitz continuity) A function is said to be uniformly Lipschitz continuous with exponent α on a closed interval I if there exists a single Lipschitz constant so that at all points of I the function is Lipschitz α continuous with this exponent.*

The uniform Lipschitz regularity of a function can be identified from the asymptotic decay of its Fourier transform (Mallat, 2001, Theorem 6.1).

Theorem 6.2.15 *If $F(\omega)$, the Fourier transform of $f(x)$, has sufficiently fast decay so that*

$$\int_{-\infty}^{\infty} F(\omega)(1 + |\omega|^\alpha)d\omega < \infty,$$

then $f(x)$ is bounded and uniformly Lipschitz α.

The proof of this theorem (Mallat, 2001) is quite similar to that of Theorem 6.2.12. Once again, the smoothness information provided by the Fourier transform is global or uniform.

Note that the asymptotic decay of a Fourier transform also plays a role in the Heisenberg uncertainty window. The case of the Haar scaling function is illustrative. This basis function has a bounded support, so it is local in the original domain. By the discussion in Example 6.1.3, its Fourier transform has slow decay, so the uncertainty in the frequency domain, $(\Delta_F \omega)$ is infinite[5]. From a pure Fourier point of view, the Haar transform is not a good compromise between time and frequency precision.

6.2.5 Sampling and aliasing

Harmonic analysis also describes the effect of discretisation from continuous signals or subsampling discrete signals. For an equidistant sample from a function on a continuous line we have the following result, known as the **Poisson summation formula**.

Theorem 6.2.16 *Let $F(\omega)$ be the Fourier transform of $f(x)$ as defined in (1.10), then*

$$\sum_{\ell \in \mathbb{Z}} f(x + \ell\Delta) = \frac{2\pi}{\Delta} \sum_{k \in \mathbb{Z}} F\left(\frac{2\pi}{\Delta}k\right) \exp\left(i\frac{2\pi}{\Delta}kx\right). \tag{6.22}$$

[5]See also Exercise 6.2.10.

In particular, for $x = 0$ and $\Delta = 1$, we have

$$\sum_{\ell \in \mathbb{Z}} f(\ell) = 2\pi \sum_{k \in \mathbb{Z}} F(2\pi k). \tag{6.23}$$

Proof. The left hand side in (6.22) is a function $s(x)$ with period Δ, i.e., $s(x + \Delta) = s(x)$. This function can be decomposed into a Fourier series,

$$s(x) = \sum_{k \in \mathbb{Z}} a_k \exp\left[i\frac{2\pi}{\Delta}kx\right],$$

which is a dilated version of the series in (1.2). The coefficients are found by

$$\begin{aligned}
a_k &= \frac{1}{\Delta}\int_0^{\Delta} s(x)\exp\left[-i\frac{2\pi}{\Delta}kx\right] = \frac{1}{\Delta}\sum_{\ell \in \mathbb{Z}}\int_0^{\Delta} f(x+\ell\Delta)\exp\left[-i\frac{2\pi}{\Delta}kx\right]dx \\
&= \frac{1}{\Delta}\sum_{\ell \in \mathbb{Z}}\int_{\ell\Delta}^{(\ell+1)\Delta} f(u)\exp\left[-i\frac{2\pi}{\Delta}ku\right]\exp\left[i\frac{2\pi}{\Delta}k\ell\Delta\right]du \\
&= \frac{1}{\Delta}\int_{\infty}^{\infty} f(u)\exp\left[-i\frac{2\pi}{\Delta}ku\right]\cdot 1\,du = \frac{2\pi}{\Delta}F\left(\frac{2\pi}{\Delta}k\right).
\end{aligned}$$

□

A variant of the Poisson summation formula puts the Fourier series on the other side of the equality sign,

$$\sum_{s \in \mathbb{Z}} f(s\Delta)\exp(-i\omega s) = \frac{2\pi}{\Delta}\sum_{k \in \mathbb{Z}} F\left(\frac{\omega + 2\pi k}{\Delta}\right), \tag{6.24}$$

the proof of which is completely similar to that of Theorem 6.2.16.

Exercise 6.2.17 *Prove (6.24).*

Expression (6.24) is interpreted as the DTFT of a **discretised function of a continuous variable**, $f(s\Delta)$. If $F(\omega)$ has no high frequencies, meaning that $F(\omega) = 0$ for for any $|\omega| > \omega_{\max}$ with $\omega_{\max}$ sufficiently small, then the terms in

$$S(\omega) = \sum_{k \in \mathbb{Z}} F\left(\frac{\omega + 2\pi k}{\Delta}\right),$$

on the right hand side in (6.24) do not overlap. If the terms do not overlap, then $F(\omega)$ can be recovered from the sum by multiplying the sum $S(\Delta\omega)$ with the characteristic function on $[-\pi, \pi]$. In the signal processing literature, $\omega_{\max}/2\pi$ is referred to as the **bandwidth** of a **bandlimited** function. This terminology will not be adopted here, as bandwidth is already in use in statistics in the context of kernel smoothing and in linear algebra when talking about banded matrices.

More precisely, the terms in $k = 1$ and in $k = 0$ do not overlap if there exists no value of ω for which $F((\omega+2\pi)/\Delta) > 0$ and $F(\omega/\Delta) > 0$. This occurs if for any value ω it holds that $\omega/\Delta \geq -\omega_{\max}$ implies that $(\omega+2\pi)/\Delta > \omega_{\max}$. For the value on the border, $\omega = -\omega_{\max}\Delta$ that is, the condition becomes $(-\omega_{\max}\Delta + 2\pi)/\Delta > \omega_{\max}$, leading to the following result

Theorem 6.2.18 *(Nyquist-Shannon) If $f(x)$ has a bounded frequency spectrum, meaning that $F(\omega) = 0$ if $|\omega| > \omega_{\max}$, then $f(x)$ can be reconstructed from the equidistant observations $f(s\Delta)$ if*

$$\omega_{\max} < \frac{1}{2}\frac{2\pi}{\Delta}. \tag{6.25}$$

In (6.25), the value $2\pi/\Delta$ can be interpreted as the pulsation corresponding to sampling rate $1/\Delta$. The Nyquist-Shannon theorem thus states that any signal with frequencies bounded by half of the sampling rate can be reconstructed from the function values. The reconstruction follows from the inverse Fourier transform

$$\begin{aligned}
f(x) &= \int_{-\infty}^{\infty} S(\Delta\omega) I(-\pi \leq \Delta\omega \leq \pi) \exp(i\omega x) d\omega \\
&= \int_{-\pi/\Delta}^{\pi/\Delta} S(\Delta\omega) \exp(i\omega x) d\omega \\
&= \int_{-\pi/\Delta}^{\pi/\Delta} \frac{\Delta}{2\pi} \sum_{s\in\mathbb{Z}} f(s\Delta) \exp(-i\Delta\omega s) \exp(i\omega x) d\omega \\
&= \sum_{s\in\mathbb{Z}} f(s\Delta) \frac{1}{2\pi/\Delta} \int_{-\pi/\Delta}^{\pi/\Delta} \exp\left[i(x-\Delta s)\omega\right] d\omega \\
&= \sum_{s\in\mathbb{Z}} f(s\Delta) \frac{\sin\left[(x-\Delta s)\pi/\Delta\right]}{(x-\Delta s)\pi/\Delta}.
\end{aligned}$$

The observations $f(s\Delta)$ act as coefficients in an expansion using the basis functions $\varphi_{\Delta,s}(x) = \frac{\sin[(x-\Delta s)\pi/\Delta]}{(x-\Delta s)\pi/\Delta}$. These basis functions are interpolating, i.e., $\varphi_{\Delta,k}(\Delta m) = 0$ unless $k = m$.

The exact reconstruction of any function with a bounded frequency spectrum can be seen as a kind of sparsity. A bounded frequency spectrum is, however, a strong condition, and at the same time, it offers not really sparse representations, as it fixes in advance which frequencies can be nonzero. In other words, an approximation or estimation based on bounded frequency spectra belongs to a class of linear methods, very much in a similar way as approaches that assume the unknown function to be a polynomial. Real sparsity occurs when nonlinear or data-adaptive procedures find out where the significant contributions are. In such cases, the sampling rate may be much lower than the Nyquist rate, i.e., twice the highest frequency in the spectrum of $f(x)$.

If there is overlap between the terms in $S(\omega)$, then $F(\omega)$ cannot be identified from the sum $S(\omega)$. This is because high frequencies are mixed with low frequencies, which is aliasing. Aliasing has been discussed in Section 3.1.2, in the context of subsampling or splitting a sequence of fine scale coefficients. The similarity between aliasing in sampling from functions and subsampling from sequences is illustrated by comparing the expressions (6.5) and (6.24). The forward filterbank operations in (6.8) and (6.9) consist of two terms, the second of which depends on the shifted frequency $(\omega + 2\pi)/2$. This is an example of aliasing. The second Fourier transformed perfect reconstruction condition (6.12) annihilates the contamination by the shifted frequencies. It is therefore known as the anti-aliasing condition.

6.3 Orthogonal wavelet transforms in the frequency domain

6.3.1 The sinc wavelet transform

Working in the frequency domain, the most straightforward choice for $H(\omega)$ and $G(\omega)$ is to take two complementary indicator functions. For $\omega \in (-\pi, \pi)$ this is

$$\begin{aligned} H(\omega) = \widetilde{H}(\omega) &= \sqrt{2} \cdot I(-\pi/2 \leq \omega \leq \pi/2), \\ G(\omega) = \widetilde{G}(\omega) &= (-1)\exp(-i\omega)H(\pi - \omega) \\ &= -\sqrt{2}\exp(-i\omega) \cdot I(\pi/2 < |\omega| \leq \pi). \end{aligned}$$

The notation $I(\omega \in A)$ represents the indicator or characteristic function on the set A. This is the Shannon or sinc refinement from Example 6.1.1, taking $c = \sqrt{2}$. The scaling function has been found in (6.3). The proposed dual and detail filters satisfy the perfect reconstruction conditions in (6.11) and (6.12). Since dual and primal filters are the same, the wavelet transform is orthogonal, as has been discussed in Section 3.1.6. The detail basis function in the Fourier domain is given by

$$\begin{aligned} \Psi(\omega) &= G\left(\frac{\omega}{2}\right) \Phi\left(\frac{\omega}{2}\right) = -\sqrt{2}\exp(-i\omega/2)I(\pi < |\omega| \leq 2\pi) \cdot \frac{1}{2\pi} I(|\omega| \leq 2\pi) \\ &= -\frac{\sqrt{2}}{2\pi}\exp(-i\omega/2)I(\pi < |\omega| \leq 2\pi). \end{aligned}$$

By the inverse transform in (1.11), this leads to

$$\begin{aligned}
\psi(x) &= \int_{-\infty}^{\infty} \Psi(\omega)\exp(i\omega x)d\omega \\
&= -\frac{\sqrt{2}}{2\pi}\left(\int_{-2\pi}^{-\pi} \exp\left[i\omega(x-1/2)\right]d\omega + \int_{\pi}^{2\pi} \exp\left[i\omega(x-1/2)\right]d\omega\right) \\
&= \sqrt{2}\left(\frac{\sin\left[\pi(x-1/2)\right]}{\pi(x-1/2)} - 2\frac{\sin\left[2\pi(x-1/2)\right]}{2\pi(x-1/2)}\right).
\end{aligned}$$

Figure 6.1 depicts the sinc (or Shannon) scaling and wavelet functions. The sinc wavelet transform is orthogonal, but the functions have infinite support and very slow decay. The slow decay is due to the discontinuity in the Fourier

Figure 6.1
Father function $\varphi(x)$ (left) and mother function $\psi(x)$ (right) of the sinc wavelet transform.

transform. The sinc wavelet transform is the frequency counterpart of the Haar transform. Whereas a slow decay in the frequency domain may still be acceptable in many applications, the slow decay of a sinc scaling or detail basis function makes it of little practical use for problems where the Fourier domain is not the main focus of the analysis.

6.3.2 The Meyer wavelet transform

Since the slow decay of the sinc wavelet basis functions is explained by the discontinuity in the Fourier spectrum of its father and mother functions, basis functions with faster decay can be designed from Fourier spectra with smoother transitions. The objective is to have fast decay, while keeping the orthogonality of the transform, meaning that dual and primal filters should be the same. With $H(\omega) = \widetilde{H}(\omega)$, the perfect reconstruction condition in (6.13) becomes

$$|H(\omega)|^2 + |H(\pi+\omega)|^2 = 2. \tag{6.26}$$

We used the fact that $H(-\omega)$ and $H(\pi - \omega)$ are the complex conjugates of $H(\omega)$ and $H(\pi + \omega)$ respectively. Condition (6.26) has to be maintained when deviating from the Shannon indicator lowpass $H(\omega)$ to a smoother version. Meyer wavelets take $H(\omega) = 1$ for $-\pi/3 \leq \omega \leq \pi/3$ and $H(\omega) = 0$ for $|\omega| \geq 2\pi/3$. The transition between the low and high frequency responses needs to be chosen. From (6.2) it follows that $\Phi(\omega) = \Phi(0)$, a constant value, for $|\omega| \leq 2\pi/3$. For $|\omega| \geq 4\pi/3$ it is obvious to find $\Phi(\omega) = 0$. The values in between, for $2\pi/3 \leq |\omega| \leq 4\pi/3$ that is, follow from the two-scale equation (6.1). Plugging in $\Phi(\omega/2) = \Phi(0)$ for these values of ω, we find $\Phi(\omega) = H_c(\omega/2)$.

The mother function, satisfying $\Psi(\omega) = G(\omega/2)\Phi(\omega/2)$ will be zero for both $|\omega| \leq 2\pi/3$ and for $|\omega| \geq 8\pi/3$, localising the frequency spectrum of the function in the band $[2\pi/3, 8\pi/3]$, which is wider than the ideal band $[\pi, 2\pi]$ realised in the Shannon wavelet.

7

Design of dyadic wavelets

This chapter discusses a couple of well established examples of wavelet families on equispaced knots. Two properties play a key role: orthogonality and symmetry. Unfortunately, Lemma 7.1.2 reveals that the Haar wavelet basis is the only compactly supported multiscale basis that combines both properties.

7.1 Orthogonal wavelets with compact support

Section 3.1.4 has introduced semi-orthogonal wavelets as a benchmark in the assessment of the variance control in a multiscale transform. The construction of semi-orthogonal wavelets involves orthogonal projections of a fine scale onto a coarser scale, thus leading to detail basis functions with infinite support. Section 3.1.6 has explored the orthogonalisation of a given refinement scheme, leading to fully orthogonal wavelet transforms. In general, these transforms operate within bases of both scaling and detail functions with infinite support.

On equidistant grids, it is possible to develop orthogonal wavelets with compact support. The orthogonality condition is, however, quite restrictive.

7.1.1 Orthogonal refinement

The analysis filterbank, represented by the matrix $\widetilde{\mathbf{W}}_j$ in (2.44) is orthogonal if its transpose equals inverse, $\mathbf{W}_j = \widetilde{\mathbf{W}}_j^{-1} = \widetilde{\mathbf{W}}_j^\top$. By (2.44) and (2.48), this means that the primal and dual filters are the same, $\widetilde{\mathbf{H}}_j = \mathbf{H}_j$ and $\widetilde{\mathbf{G}}_j = \mathbf{G}_j$. The perfect reconstruction conditions in (5.16) to (5.19) can be formulated without the tildes. In particular, (5.16) becomes the **double shift orthogonality** condition

$$\sum_{l\in\mathbb{Z}} h_l h_{l-2k} = \delta_k. \tag{7.1}$$

Double shift orthogonality is a necessary but not a sufficient condition. For instance, the filter $\boldsymbol{h}_{(0,1,2,3)} = (1,0,0,1)$ from Example 5.5.12 is double shift orthogonal, but none of the possible solutions to the refinement equation are

DOI: 10.1201/9781003265375-7

orthogonal to their translations. For instance, the function $\varphi(x) = \chi_{[0,3[}(x)$ is not orthogonal to $\varphi(x-1)$.

Lemma 7.1.1 *Let* $\boldsymbol{h}$ *be a sequence satisfying the double shift orthogonality property. Let* s_l *and* s_k *be the minimum and the maximum of the set* $\{k \in \mathbb{Z} | s_k \neq 0\}$. *Then if* $s = s_r - s_l + 1$ *is finite, it must be even.*

Proof. Suppose $s = 2r+1$ is odd, then with $k = r$, (7.1) becomes $h_{s_r} h_{s_r - 2r} = h_{s_r} h_{s_l} = 0$, from which it follows that at least one element of the pair $\{h_{s_r}, h_{s_l}\}$ must be zero. This is in contrast to the definition of s_l and s_r. □

Lemma 7.1.2 *Let* $\boldsymbol{h}$ *be a sequence satisfying the double shift orthogonality property, with a finite number of nonzero elements* s *as defined in Lemma 7.1.1. If* $\boldsymbol{h}$ *is symmetric, then it has only two nonzero components.*

In other words, symmetric filters cannot possibly be orthogonal, except for the Haar filter and *à trous* versions of the Haar filter.
Proof. Knowing that $s = s_r - s_l + 1$ is even, we write $s = 2r$. For $r = 1$, the statement is trivial, so let $r > 1$. Then with $k = r - 1$, the double shift orthogonality (7.1) becomes $h_{s_r} h_{s_r - 2r + 2} + h_{s_r - 1} h_{s_r - 2r + 1} = h_{s_r} h_{s_l + 1} + h_{s_r - 1} h_{s_l} = 0$. By symmetry, we have $h_{s_r} = h_{s_l}$ and $h_{s_r - 1} = h_{s_l + 1}$, so the double shift orthogonality implies that $2 h_{s_r} h_{s_r - 1} = 0$. Moreover, by definition $h_{s_r} \neq 0$, from which it follows that $h_{s_r - 1} = 0$. Repeating the argument for $k = r-2, r-3, \ldots, 1$ leads to the conclusion that $h_l = 0$ for all $l \in \{s_l + 1, s_l + 2, \ldots, s_r - 1\}$. □

Lemma 7.1.3 *Let* $\boldsymbol{h}$ *be a sequence satisfying the double shift orthogonality property, so that the dyadic two-scale equation (5.22) admits a Riemann integrable function* $\varphi(x)$. *Then*

$$\sum_{k \in \mathbb{Z}} h_{2k} = \sum_{k \in \mathbb{Z}} h_{2k+1} = \frac{\sqrt{2}}{2}. \tag{7.2}$$

Proof. Summing up the double shift orthogonality conditions yields

$$\begin{aligned} 1 &= \sum_{k \in \mathbb{Z}} \sum_{l \in \mathbb{Z}} h_l h_{l-2k} = \sum_{k \in \mathbb{Z}} \sum_{m \in \mathbb{Z}} h_{2m} h_{2m-2k} + \sum_{k \in \mathbb{Z}} \sum_{m \in \mathbb{Z}} h_{2m+1} h_{2m+1-2k} \\ &= \sum_{r \in \mathbb{Z}} \sum_{m \in \mathbb{Z}} h_{2m} h_{2r} + \sum_{r \in \mathbb{Z}} \sum_{m \in \mathbb{Z}} h_{2m+1} h_{2r+1} = \left(\sum_{k \in \mathbb{Z}} h_{2k} \right)^2 + \left(\sum_{k \in \mathbb{Z}} h_{2k+1} \right)^2 \end{aligned}$$

Both terms in the sum are equal because of the partition of unity[1] in (5.56). □

Lemma 7.1.3 provides an alternative to the ℓ_2-normalisation of the refinement coefficients in an orthogonal wavelet transform. The other equations defining these coefficients are all homogeneous.

[1] The partition of unity itself follows automatically from refinement on regular grids. On irregular grids, it has to be imposed explicitly. See Sections 5.5.2 and 5.5.4.

7.1.2 Maxflats and symmlets: orthogonality and vanishing moments

The simplest orthogonal wavelet transform is the Haar transform. By adding vanishing moments, the number of nonzeros in the refinement filters $\boldsymbol{h}$ must increase, since every vanishing moment adds a linear, homogeneous condition of the form (5.67). On the other hand, $2\widetilde{p}$ nonzeros in $\boldsymbol{h}$ generate $\widetilde{p}-1$ nonlinear, double shift orthogonality equations of the form (7.1). With $2\widetilde{p}$ degrees of freedom and $\widetilde{p}-1$ nonlinear homogeneous conditions along with one normalisation, it is reasonable to hope for the existence of a solution if the system is completed by $\widetilde{p}$ linear, homogeneous vanishing moment conditions[2,3]. Let $\mathbf{S}$ be the matrix that shifts the elements of a vector; then each of the double shift orthogonality equations can be written as $\boldsymbol{h}^\top(\mathbf{S}^{2k}\boldsymbol{h}) = 0$, for $k = 1, 2, \ldots, \widetilde{p}-1$. Using Gaussian elimination, the $\widetilde{p}$ vanishing moment conditions (5.67) can be written in the explicit form $\boldsymbol{h}_L = \mathbf{A}_{\widetilde{p}}\boldsymbol{h}_R$, where $\boldsymbol{h}_L$ is the vector containing the first half of the nonzeros in $\boldsymbol{h}$, i.e., $\boldsymbol{h}_L = (h_0, h_1, \ldots, h_{\widetilde{p}-1})$, while $\boldsymbol{h}_R$ contains the second half, $\boldsymbol{h}_R = (h_{\widetilde{p}}, h_{\widetilde{p}+1}, \ldots, h_{2\widetilde{p}})$. The matrix $\mathbf{A}_{\widetilde{p}}$ has elements

$$A_{\widetilde{p};ij} = (-1)^j \binom{\widetilde{p}+j-i-1}{j}\binom{\widetilde{p}+j}{i}.$$

Substitution into the double shift orthogonality conditions leads to the quadratic equations, for $k = 1, 2, \ldots, \widetilde{p}-1$,

$$\boldsymbol{h}_R^\top \left[\begin{array}{cc} \mathbf{A}_{\widetilde{p}}^\top & \mathbf{I}_{\widetilde{p}} \end{array}\right] \mathbf{S}^{2k} \left[\begin{array}{c} \mathbf{A}_{\widetilde{p}} \\ \mathbf{I}_{\widetilde{p}} \end{array}\right] \boldsymbol{h}_R = 0.$$

Example 7.1.4 *With two vanishing moments, we have four nonzeros in $\boldsymbol{h}$. The second half of these coefficients, $\boldsymbol{h}_R = (h_2, h_3)$ satisfies the single quadratic equation $h_2^2 - 3h_3^2 = 0$. This equation follows from elimination of the linear equations from the set of three equations*

$$\left\{\begin{array}{rcl} h_0h_2 + h_1h_3 & = & 0 \\ h_0 - h_1 + h_2 - h_3 & = & 0 \\ -h_1 + 2h_2 - 3h_3 & = & 0. \end{array}\right.$$

After normalisation, the system has four solutions, one of which is $\boldsymbol{h}_{(0,\ldots,3)} = (h_0, h_1, h_2, h_3) = (1+\sqrt{3}, 3+\sqrt{3}, 3-\sqrt{3}, 1-\sqrt{3})/4\sqrt{2}$. The second solution is the flipped sequence (h_3, h_2, h_1, h_0). The other solutions follow from switching signs.

The matrix $c\mathbf{H}_I$, defined in (5.61) has one eigenvalue with one eigenvector, so the subdivision converges. The limiting function, depicted in Figure 7.1,

[2] Existence and uniqueness of a solution to the partly quadratic system cannot be taken for granted.

[3] Note that this approach makes use of the availability of an explicit expression of vanishing moments in the refinement coefficients. Such an expression is not available on irregular point sets, thus making the approach specific for equispaced wavelets.

is the representative with two vanishing moments of the famous ***Daubechies maxflat*** *family. In some part of the literature this representative is referred to as the Daubechies-4, for its number of nonzero elements in* $\boldsymbol{h}$*. In line with the nomenclature of the CDF B-spline wavelets, this text opts for a parametrisation by the number of vanishing moments, thus referring to the father function of Figure 7.1 as the Daubechies-2 or the maxflat-2. The term maxflat comes from the characteristics in the Fourier transform.*

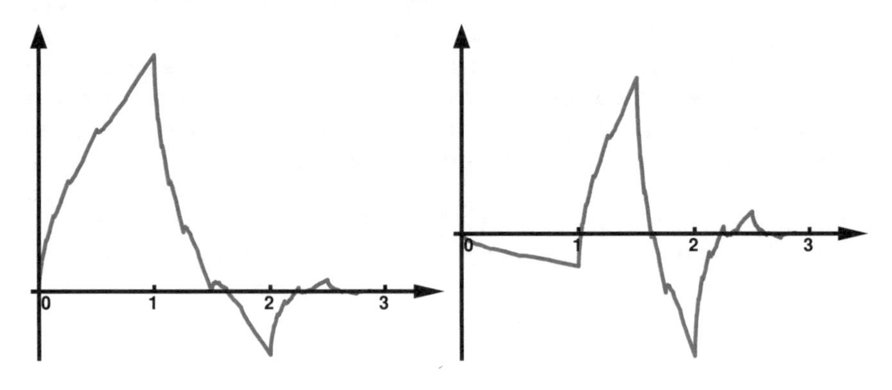

Figure 7.1
Father function $\varphi(x)$ (left) and mother function $\psi(x)$ (right) of the Daubechies wavelet with two vanishing moments.

The example 7.1.4 illustrates the fact that for a given orthogonal wavelet transform, the flipped version of its refinement sequence is also a valid solution. This is because the double shift orthogonality conditions are completely symmetric with respect to flipping and because the vanishing moment conditions in (5.67) are equivalent to the flipped form

$$\sum_{k\in\mathbb{Z}}(-1)^k(2\widetilde{p}-k)^q h_k = 0,$$

for any c. With $\varphi(x)$ the father function for a given refinement, the father function corresponding to the flipped sequence is $\varphi(2\widetilde{p}-x)$.

Exercise 7.1.5 *Use an argument similar to the one in the proof of Corollary 5.5.16 to prove the flipped vanishing moment conditions are equally valid in orthogonal wavelet transforms.*

Example 7.1.6 *With three vanishing moments, we have six nonzeros in* $\boldsymbol{h}$*. The second half of these coefficients,* $\boldsymbol{h}_R = (h_3, h_4, h_5)$ *satisfies the set of two quadratic equations*

$$\left\{\begin{array}{rcl} 6h_3^2+12h_4^2+60h_5^2-20h_3h_4-56h_4h_5+44h_3h_5 & = & 0 \\ -3h_4^2+15h_5^2+h_3h_4-2h_4h_5+3h_3h_5 & = & 0. \end{array}\right.$$

While the normalisation can take place once both sides, $\boldsymbol{h}_L$ and $\boldsymbol{h}_R$, are known, we start off with a preliminary normalisation $h_5 = 1$, leading to a set of two non-homogeneous quadratic equations in h_4 and h_5. From the second equation, h_3 can be isolated, yielding $h_3 = (3h_4^2 + 2h_4 - 15)/(h_4 + 3)$. Substitution into the first equation then leads to $3h_4^4 - 10h_4^2 - 45 = 0$. This equation has two real solutions, which, after renormalisation, are elements in two refinement vectors $\boldsymbol{h}$ that are flipped versions of each other. Two more outcomes follow by replacing the preliminary normalisation $h_5 = 1$ by $h_5 = -1$. These solutions simply switch signs in the earlier vectors. As a result, we end up, once again, with four variations of a single nontrivial filter. The corresponding father and mother functions are depicted in Figure 7.2

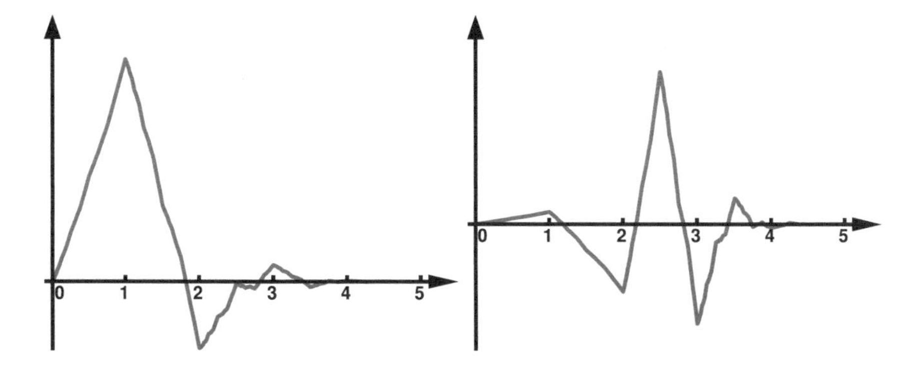

Figure 7.2
Father function $\varphi(x)$ (left) and mother function $\psi(x)$ (right) of the Daubechies wavelet with three vanishing moments.

When the number of vanishing moments is larger than three, then the numerical solution reveals more than just one possible outcome. It is then possible to add further specifications. For instance, the **symmlet** transforms are orthogonal wavelet transforms whose refinement coefficients are as close as possible to being symmetric (recall that full symmetry is unreachable once beyond the Haar stage). The representative with four vanishing moments is depicted in Figure 7.3. Alternatively, the above mentioned **maxflat** transforms focus on the Fourier spectrum of the filters. A full development of the orthogonal wavelet transforms in the Fourier domain is found in Daubechies (1992, Chapter 6), see also Vidakovic (1999, Section 3.4.5). The maxflat with four vanishing moments is depicted in Figure 7.4. It is clearly more asymmetric than its symmlet counterpart in Figure 7.3. The symmlet and maxflat families share their representatives with one, two, and three vanishing moments.

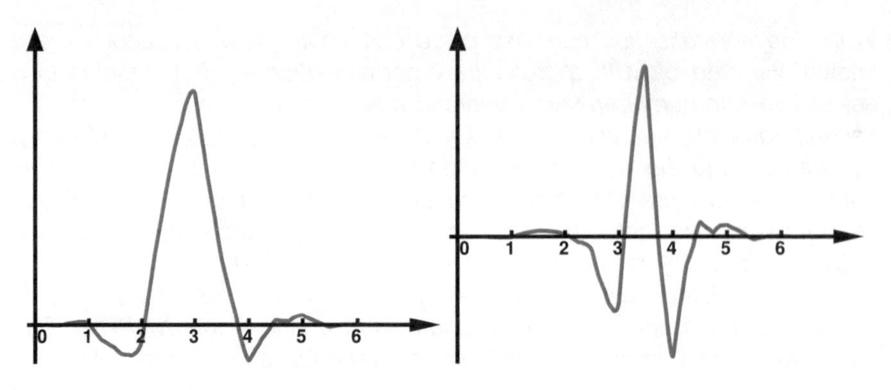

Figure 7.3
Father function $\varphi(x)$ (left) and mother function $\psi(x)$ (right) of the Daubechies symmlet wavelet with four vanishing moments.

7.1.3 Coiflets

The orthogonal wavelet transforms in the previous chapter devoted all degrees of freedom to vanishing moments. Suppose now that, besides $\widetilde{p}$ vanishing moments, we wish to equip a wavelet transform with $\widetilde{r}$ other features, for which the refinement filter has to fulfil $\widetilde{r}$ linear conditions; then the double shift orthogonality can be fulfilled with $2\widetilde{p}+\widetilde{r}$ nonzeros in $\boldsymbol{h}$.

As an example, Daubechies (1993) tackled the problem, raised by Ronald Coifman, to find orthogonal, compactly supported wavelets with a given number of vanishing moments and father functions satisfying

$$\int_{-\infty}^{\infty} \varphi(x)x^q dx = 0, \tag{7.3}$$

for[4] $q = 1, 2, \ldots, \widetilde{r}$. The property in (7.3), which is nothing but a vanishing moment for scaling functions, will turn out to be interesting in linear approximation schemes, as explained in Section 8.4.2. In particular, it will be interesting to take $\widetilde{r} = \widetilde{p} - 1$. The resulting wavelet family has been termed **coiflets**.

By the refinement equation (5.22), the coiflet condition in (7.3) becomes

$$c \sum_{k=-\infty}^{\infty} \int_{-\infty}^{\infty} h_k x^q \varphi(2x-k)dx = 0.$$

This is reformulated as

$$\sum_{k=-\infty}^{\infty} \int_{-\infty}^{\infty} h_k \cdot \left(\frac{u+k}{2}\right)^q \varphi(u)\frac{du}{2}.$$

[4] For $q = 0$, it still holds that $\int_{-\infty}^{\infty} \varphi(x)dx = 1$.

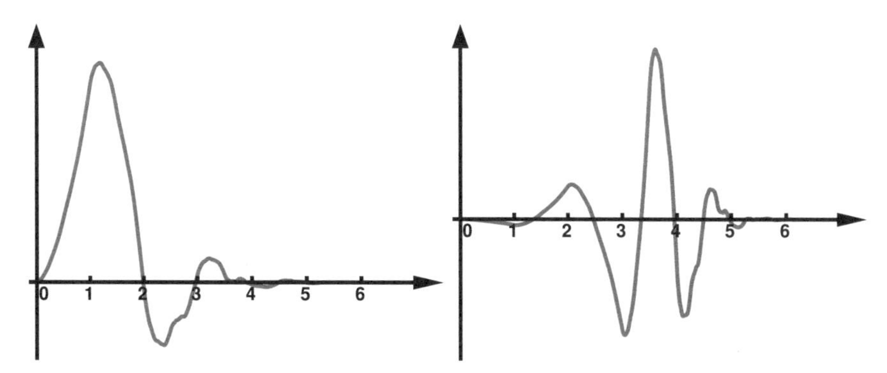

Figure 7.4
Father function $\varphi(x)$ (left) and mother function $\psi(x)$ (right) of the Daubechies maxflat wavelet with four vanishing moments.

From (7.3) it follows that for any polynomial of degree $\widetilde{r}$ or lower,

$$\int_{-\infty}^{\infty} \varphi(u)p(u)du = p(0).$$

Substitution of $p(u) = (u+k)^q$ leads to a set of $\widetilde{r}$ linear conditions,

$$\sum_{k=-\infty}^{\infty} h_k k^q = 0. \tag{7.4}$$

As the coiflet condition (7.3) does not hold for $q = 0$, it is not translation invariant (Daubechies, 1993, page 511), in the sense that for $k \neq 0$,

$$\int_{-\infty}^{\infty} \varphi(x-k)x^q dx \neq 0.$$

This is in contrast to the vanishing moment conditions of (5.67) and to the double shift orthogonality conditions of (7.1), on which the maxflat and symmlet families are based. As a result, unlike for maxflats and symmlets, translating the father function makes a difference here, and so the design of the coiflet father function comprises a careful consideration of the position of its support. In other words, we have the freedom to decide on which interval exactly the father function is nonzero. According to Theorem 5.5.5, this amounts to fixing the locations of the nonzero elements in the refinement filter $\boldsymbol{h}$. With $\widetilde{r} = \widetilde{p} - 1$, the classical definition of coiflets puts the first nonzero element in $h_{-\widetilde{p}}$. It turns out (Daubechies, 1993) that with this choice of the first nonzero and for an even $\widetilde{p}$, a total number of $3\widetilde{p}$ nonzeros is enough to satisfy the $2\widetilde{p}-1$ linear conditions as well as the $3\widetilde{p}/2-1$ double shift orthogonality conditions. For $\widetilde{p} = 2$, the system of 3 linear and 2 quadratic equations, along with the normalisation, has two different solutions. The most symmetric of these two solutions is depicted in Figure 7.5.

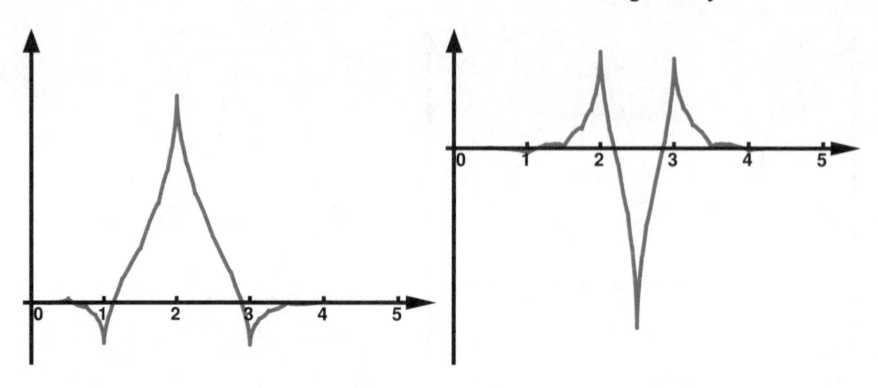

Figure 7.5
Father function $\varphi(x)$ (left) and mother function $\psi(x)$ (right) of the coiflet wavelet with two vanishing moments. These functions are remarkably close to symmetric.

7.2 Symmetric wavelets

In most applications, a multiscale analysis or processing of the flipped observations $Y_{n-i}, i = 0, 1, \ldots, n$ is expected to lead to the same conclusions as the analysis on the original data Y_i. For this to be ensured, it is convenient to work with symmetric basis functions. Flipping the data would then lead to exactly the same coefficients in the expansion, in the reversed order of course.

7.2.1 B-spline filters on regular knots

The simplest case of a nontrivial symmetric refinement equation is to propose $\varphi(x) = c[h_0\varphi(2x) + h_0\varphi(2x-1)]$, where the value of ch_0 follows from the normalisation in Lemma 5.2.2, $ch_0 = 1$. This is the Haar transform. According to Lemma 7.1.2, it is the only symmetric refinement with orthogonal refinement functions. It also has one vanishing moment, as it satisfies the partition of unity, which is in line with Theorem 5.5.2.

Let us now make the refinement filter $\boldsymbol{h}$ one tap longer, keeping the symmetry, i.e., considering (h_0, h_1, h_0). Then the partition of unity must hold for this to be a refinement filter. So, according to Theorem 5.5.2, this means that $h_1 = 2h_0$, leaving us with $(1, 2, 1)$, up to normalisation. A second vanishing moment comes in for free, as indeed, by symmetry, (5.67) in Corollary 5.5.16 is satisfied for $q = 1$. This is the triangular hat scaling function.

Adding one more nonzero in the refinement filter does not introduce more degrees of freedom, since we now have (h_0, h_1, h_1, h_0). With an even number of nonzeros, the symmetric form automatically satisfies the partition of unity. Following (5.67) we can impose a second vanishing moment, $-h_1+2h_1-3h_0$, to find out again that, thanks to the symmetry, we get one more vanishing moment for free. Indeed, The second vanishing moment condition is satisfied for $h_1 = 3h_0$, so we have, up to normalisation, the vector of nonzeros $(1,3,3,1)$. This vector also satisfies the third vanishing moment condition $-h_1 + 4h_1 - 9h_0 = 0$.

In general, the $\widetilde{p}+1$ **binomial coefficients** $h_k = \binom{\widetilde{p}}{k}$, satisfy $\widetilde{p}$ vanishing moment conditions. Since $h_k = h_{\widetilde{p}-k}$ these $\widetilde{p}$ conditions are fulfilled with just $\lfloor \widetilde{p}/2 \rfloor$ degrees of freedom. According to Theorem 5.5.5, the support of the father function $\varphi(x)$ is then $[0, \widetilde{p}]$ (or a subset of it). From Lemma 4.2.12, it then follows that $\varphi(x)$ and its translates are piecewise polynomials. Not surprisingly, the binomial coefficients can be proven to be the refinement coefficients that make the piecewise polynomials meet the knot continuity conditions of **B-splines**. Equispaced B-splines wavelet transforms, known as **Cohen-Daubechies-Feauveau** (CDF) spline wavelets (Cohen et al., 1992), have been constructed from the lifting scheme in Section 5.3.2.

On an equidistant grid, B-splines are symmetric scaling functions, and their refinement coefficients form symmetric synthesis sequences. The design of a CDF spline wavelet transform is a fully linear problem. First, the primal refinement coefficients are found by imposing a given number of dual vanishing moments for a minimal number of nonzeros. The outcome is given by the binomial numbers. Once the primal refinement is known, the perfect reconstruction (PR) conditions are linear expressions in the dual refinement coefficients. The PR conditions are completed again by linear primal moment conditions.

7.2.2 Symmetric primal and dual transforms with less dissimilar lengths

In the construction of the CDF spline wavelets, the primal refinement filter $\boldsymbol{h}$ is designed to minimise the number of nonzeros. The dual refinement filter $\widetilde{\boldsymbol{h}}$ follows afterwards. It has to be long enough so that it fulfills the perfect reconstruction and all further conditions that one wishes to impose. As a result, the dual filters are generally much longer than the primal filters. Therefore, the range of observations from which a given scaling coefficient gets input is much larger than the range of reconstructions to which the same scaling coefficient contributes. In order to reduce the difference in scales between forward and inverse transforms, a pair of primal and dual filters can be designed from a lattice where the numbers of nonzeros in both filters are as close as possible to each other[5].

The most famous representative of this class of symmetric wavelets has four dual and four primal vanishing moments. The CDF B-spline wavelet with

[5]Note that in orthogonal transforms, there is no issue about dissimilar lengths.

these numbers of vanishing moments has a primal refinement filter of length 5 and a dual filter of length 11, thus adding up to 16 degrees of freedom. Dividing 16 parameters into two equisized groups of 8 dual and 8 primal refinement coefficients leads to a contradiction, as can be verified by following the subsequent procedure. A combination of 7 primal and 9 dual refinement coefficients, however, is successful.

These coefficients are found as follows. Let $(h_1, h_2, \ldots, h_7)$ be the nonzeros of the primal refinement filter and $(\widetilde{h}_0, \widetilde{h}_1, \ldots, \widetilde{h}_8)$ the nonzeros of the dual refinement. The first nonzero primal coefficient is h_1, while the the primal filter starts at $\widetilde{h}_0$. If both filters would start at index 0, then the double perfect reconstruction (5.16) with $k = 3$ would read $\widetilde{h}_6 h_0 = 0 \Rightarrow h_0 = 0$ or $\widetilde{h}_6 = 0$, which would contradict the proposed number of zeros. The symmetry is reflected by writing $\boldsymbol{h}_{(1,\ldots,7)} = \mathbf{R}\boldsymbol{h}_{(1,\ldots,4)}$, where the matrix $\mathbf{R}$ has zeros, except for $R_{i,4-|i-4|} = 1$, with $i = 1, 2, \ldots, 7$. Similarly, we have $\widetilde{\boldsymbol{h}}_{(0,\ldots,8)} = \widetilde{\mathbf{R}}\widetilde{\boldsymbol{h}}_{(0,\ldots,4)}$. For symmetry reasons, the four vanishing moment conditions (5.67) lead to a set of only two linearly independent equations. The set can be used to eliminate two elements from $\boldsymbol{h}_{(1,\ldots,4)}$ and two elements from $\widetilde{\boldsymbol{h}}_{(0,\ldots,4)}$, thus writing $\boldsymbol{h}_{(3,4)} = \mathbf{A}\boldsymbol{h}_{(1,2)}$ and $\widetilde{\boldsymbol{h}}_{(3,4)} = \widetilde{\mathbf{A}}\widetilde{\boldsymbol{h}}_{(0,1,2)}$. For the expression of the nonlinear perfect reconstruction conditions, let $\mathbf{S}_{2k}$ be the 9×7 matrix with zeros, except for the elements $S_{2k;i,i+2k} = 1$. Then the perfect reconstruction conditions read

$$\widetilde{\boldsymbol{h}}_{(0,1,2)}^{\top} \begin{bmatrix} \mathbf{I}_2 & \widetilde{\mathbf{A}}^{\top} \end{bmatrix} \widetilde{\mathbf{R}}^{\top} \mathbf{S}_{2k} \mathbf{R} \begin{bmatrix} \mathbf{A} \\ \mathbf{I}_2 \end{bmatrix} \boldsymbol{h}_{(1,2)} = 0,$$

for $k = 1, 2, 3$. The three perfect reconstruction conditions can be written out as a set of three equations that are linear in the six products $\widetilde{h}_i h_j$ with $i = 1, 2, 3$ and $j = 1, 2$. In particular, we have

$$\left[\begin{array}{rrr|rrr} -128 & 63 & -48 & 95 & -48 & 28 \\ 0 & -18 & 4 & 6 & 4 & 1 \\ 0 & 1 & 0 & 1 & 0 & 0 \end{array}\right] \begin{bmatrix} \widetilde{\boldsymbol{h}}_{(0,1,2)} h_1 \\ ---- \\ \widetilde{\boldsymbol{h}}_{(0,1,2)} h_2 \end{bmatrix} = 0.$$

By splitting the 3×6 matrix above into two 3×3 blocks $\mathbf{B}_1$ and $\mathbf{B}_2$, the equation takes the form of an eigenvalue problem,

$$\mathbf{B}_1 \widetilde{\boldsymbol{h}}_{(0,1,2)} h_1 + \mathbf{B}_2 \widetilde{\boldsymbol{h}}_{(0,1,2)} h_2 = 0 \Leftrightarrow -\mathbf{B}_2^{-1}\mathbf{B}_1 \widetilde{\boldsymbol{h}}_{(0,1,2)} = (h_2/h_1)\widetilde{\boldsymbol{h}}_{(0,1,2)},$$

with $\widetilde{\boldsymbol{h}}_{(0,1,2)}$ in the role of the eigenvector of $-\mathbf{B}_2^{-1}\mathbf{B}_1$, corresponding to the eigenvalue h_2/h_1. The matrix $-\mathbf{B}_2^{-1}\mathbf{B}_1$ has one real and two complex eigenvalues. Selecting the real solution, the eigenvalue fixes the primal refinement $\boldsymbol{h}$, up to a normalisation. The eigenvector fixes the dual refinement, again up to a normalisation. The resulting primal scaling and wavelet functions are depicted in Figure 7.6.

The symmetric CDF wavelets with less dissimilar lengths are popular in image processing. The representative with four primal and four dual vanishing

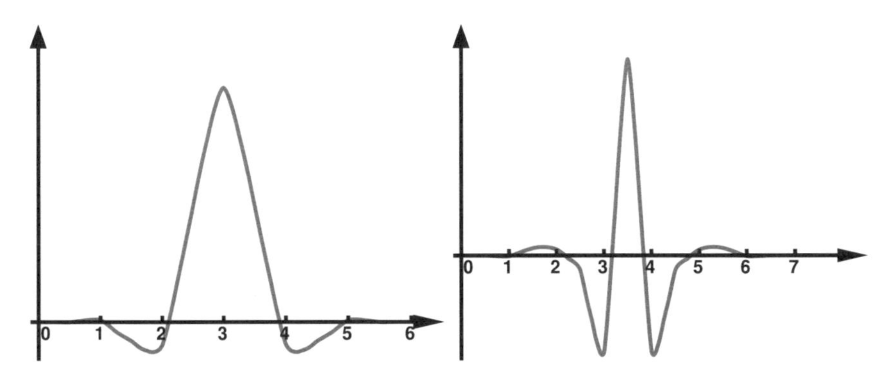

Figure 7.6
Primal father function $\varphi(x)$ (left) and mother function $\psi(x)$ (right) of the Cohen-Daubechies-Feauveau wavelets with four primal and four dual vanishing moments from the family with less dissimilar lengths.

moments, plotted in Figure 7.6, is used for compression in the JPEG-2000 compression standard.

The family of CDF wavelets with less dissimilar lengths are no spline wavelets. This can be verified by applying a CDF spline wavelet transform to a father function with less dissimilar filter lengths (keeping the same number of dual vanishing moments). It turns out that the detail coefficients will not be identically zero, leading to the conclusion that the father function cannot be expressed as a linear combination of B-splines; hence it is not a spline function.

8

Approximation in a wavelet basis

Before proceeding to more advanced multiscale transforms, we need to take a moment to discuss the approximation in a wavelet basis. The approximation of a function in a wavelet basis is an important component of the bias in the estimation of that function from noisy observations.

8.1 Linear approximation in a multiscale basis

As discussed in Chapter 1 and summarised in Section 1.4, a wavelet analysis is of particular interest for a nonlinear approximation of data that come from piecewise smooth functions. In order to understand how a multiresolution analysis of piecewise smooth data works, it is important to explore the approximation of the smooth part first.

Let $f(x)$ be defined on $[0, 1]$, with at least $\widetilde{p}$ continuous and bounded derivatives[1]. On a dyadic and equidistant set of $n = 2^J$ knots $x_{J,k} = k2^{-J}$, the projection $f_J(x) = \Phi_J(x)\boldsymbol{s}_J$ with $\boldsymbol{s}_J$ as in (2.79) can be proven to have a pointwise approximation error $|f(x) - f_J(x)| = \mathcal{O}\left(2^{-J\widetilde{p}}\right)$ (Sweldens and Piessens, 1994, (3.5) and (3.9)). The same order of magnitude holds, also in the equidistant case, for the integrated squared error (Strang and Nguyen, 1996, Theorem 7.5, page 230)

$$\|f - f_J\|_2 = \left[\int_0^1 |f(x) - f_J(x)|^2 dx\right]^{1/2} = \mathcal{O}\left(2^{-J\widetilde{p}}\right). \tag{8.1}$$

Thanks to the smoothness of the function $f(x)$, a resolution of $1/n_J = 2^{-J}$ is enough to approximate the function with an accuracy of $2^{-J\widetilde{p}}$. With $n_J = 2^J$ scaling coefficients, the approximation rate is

$$\|f - f_J\|_2 = \mathcal{O}\left(n_J^{-\widetilde{p}}\right).$$

[1]This can be relaxed to a function which is uniformly Lipschitz-α continuous, with $\alpha \geq \widetilde{p}$, see Definitions 6.2.13 and 6.2.14.

DOI: 10.1201/9781003265375-8

8.2 Nonlinear approximation

In order to illustrate how a wavelet approximation of a piecewise continuous function works, consider first a square integrable, Lipschitz $\widetilde{p}$ continuous function $f(x)$, except at a finite number of unknown but isolated locations. The wavelet basis functions are supposed to have bounded support. In the subsequent discussion the number of singular points is not important, so we further concentrate on a function with a single singularity at $x_0 \in (0,1)$. The function has at least $\widetilde{p}-1$ continuous derivatives elsewhere. We further restrict discussion to the case where all left and right derivatives at x_0 are finite. At least one pair of left and right derivatives is unequal. For $q = 0, 1, \ldots, \widetilde{p}-1$ define the coefficients

$$a_q = \lim_{x \to x_0+} f^{(q)}(x) - \lim_{x \to x_0-} f^{(q)}(x),$$

and the function $f_0(x) = 0$ for $x < x_0$ and

$$f_0(x) = \sum_{q=0}^{\widetilde{p}-1} \frac{a_q}{q!}(x - x_0)^q \text{ for } x \geq x_0.$$

Then $g(x) = f(x) - f_0(x)$ has at least $\widetilde{p}-1$ continuous and bounded derivatives, allowing a linear approximation with integrated squared error $\|g - g_L\|_2 = \mathcal{O}\left(2^{-L\widetilde{p}}\right)$, where $g_L(x)$ is a projection onto a coarse scaling basis, as in (2.79).

The detail coefficients in a wavelet decomposition at scales finer than L are used for the location and approximation of the singularity. In particular, suppose that an oracle observes both $g(x)$ and $f_0(x)$; then it knows which values of

$$d_{0;j,k} = \int_0^1 \widetilde{\psi}_{j,k}(x) f_0(x) dx$$

are nonzero. Those are the coefficients that interfere with the singularity. Indeed, $f_0(x)$ is a piecewise polynomial of degree $\widetilde{p}-1$. As a result, the detail $d_{0;j,k}$ is nonzero only if x_0 is within the support of $\widetilde{\psi}_{j,k}(x)$.

On an equidistant grid, the supports at level j are twice as wide as those at level $j+1$, which is compensated by the fact that there are only half as many basis functions. As a result, thanks to the subsampling, the number of nonzeros is a constant, say n_{11} at each resolution level[2]. For an approximation of $f_0(x)$ at level J it is sufficient to keep the $(J-L)n_{11}$ nonzero detail coefficients, along with the $n_L = 2^L$ coarse scaling coefficients.

[2]In a continuous wavelet transform, where there is no subsampling, the impact of a singularity stretches over a wider range at coarse scales. The widening range at coarse levels is referred to as the cone of influence, as discussed in Chapter 9.

By the definition of a multiresolution analysis in Section 5.5.1, the expansion of the approximation error converges in $L_2([0,1])$, thus leading to the expression

$$f_0(x) - f_{0,J}(x) = \sum_{j=J}^{\infty} \sum_k d_{0;j,k} \psi_{j,k}(x).$$

The norm of the approximation error can be bounded by

$$\|f_0 - f_{0,J}\|_2^2 \leq \frac{1}{\gamma} \sum_{j=J}^{\infty} \sum_k |d_{0;j,k}|^2,$$

where γ is a Riesz constant, as in (5.42), Definition 5.4.5. Assuming $|f_0(x)| \leq A$ for all $x \in [0,1]$, we have, at each level j, n_{11} nonzeros bounded by

$$|d_{0;j,k}| \leq A \int_0^1 |\widetilde{\psi}_{j,k}(x)| dx \leq A 2^{-j} 2^{j/2} \int_0^1 |\widetilde{\psi}(x)| dx = \mathcal{O}\left(2^{-j/2}\right),$$

leading to

$$\|f_0 - f_{0,J}\|_2^2 \leq \frac{A^2}{\gamma} \left(\sum_{j=J}^{\infty} n_{11} 2^{-j/2} \right) = \frac{A^2 n_{11}}{\gamma} 2^{-J}. \tag{8.2}$$

A nonlinear approximation of a piecewise polynomial in a wavelet basis needs just $\mathcal{O}(J)$ terms for a rate of $\mathcal{O}\left(2^{-J/2}\right)$. Taking $J = \lceil 2L\widetilde{p} \rceil$ in the superposition of the piecewise polynomial $f_0(x)$ and the smooth function $g(x)$, the oracle identifies the $(\lceil 2L\widetilde{p} \rceil - L)n_{11}$ detail coefficients that are necessary to approximate the singularity with the same accuracy as the smooth part, where $\|g - g_L\|_2^2 = \mathcal{O}\left(2^{-2L\widetilde{p}}\right)$. We then have

$$\begin{aligned} n_1 &= (\lceil 2L\widetilde{p} \rceil - L)n_{11} + 2^L = n_L + \lceil \log_2(n_L)(2\widetilde{p} - 1)n_{11} \rceil \\ &= \mathcal{O}\left(n_L + \log_2(n_L)\right) \end{aligned} \tag{8.3}$$

coefficients for an approximation error of $\mathcal{O}\left(n_L^{-\widetilde{p}}\right) = \mathcal{O}\left(n_1^{-\widetilde{p}}\right)$.

As a conclusion, the general idea of a nonlinear wavelet approximation is to select first a coarse scale so that the approximation error $\mathcal{O}(2^{-L\widetilde{p}})$ of the smooth part achieves the premised rate, and then to use an oracle that knows where to add detail coefficients at fine scales, in order to capture the singularities. In practice, the oracle is well mimicked by a selection of the largest detail coefficients. The largest details thus provide a well working automatic location of the singularities.

8.3 Function classes for data with singularities

In order to quantify the performance of a nonlinear approximation and estimation in a multiresolution basis, it is interesting to define function classes whose members are typically piecewise smooth as well as classes of sparse coefficient sequences. If membership of the function class can be linked in an unconditional way to the sequence of coefficients belonging to the sparse coefficient class, then performances can be analysed on the coefficient sequences.

By definition, multiresolution bases are Riesz bases, i.e., unconditional bases for square integrable functions. This property guarantees that linear or nonlinear operations on the coefficients have a bounded effect on the reconstruction, at least measured in the Euclidean norm. The unconditional membership of smoothness classes adds further information that can be retrieved from the coefficients in a multiscale decomposition.

8.3.1 Sobolev spaces

The definitions 6.2.13 and 6.2.14 of Lipschitz continuity in Section 6.2.4 concern pointwise and uniform smoothness. Uniform Lipschitz smoothness over an interval I can be expressed in terms of a bounded supremum. More precisely, with

$$\|f\|_{L^\alpha} = \max\left(\max_{0\leq k\leq \lceil\alpha\rceil-1} \sup_{x\in I} |f^{(k)}(x)|, \sup_{x_1,x_2\in I} \frac{|f(x_2)-f(x_1)|}{|x_2-x_1|^{\alpha+1-\lceil\alpha\rceil}} \right),$$

then $f(x)$ is uniformly Lipschitz α continuous if $\|f\|_{L^\alpha} \leq L < \infty$. The expression $\|f\|_{L^\alpha}$ is a seminorm, meaning that it is a norm, except for the fact that nonzero functions may have a zero norm. Uniform Lipschitz smoothness does not allow a single singularity. The decay of the Fourier transform can be used to check it.

For functions that are Lipschitz α continuous except for a few singularities, the supremum and maximum can be replaced by integrals and sums. Taking integer values for the smoothness, i.e., restricting discussion to derivatives, we define the Sobolev norm as

$$\|f\|_{W_p^k} = \left(\sum_{l=0}^{k} \|f^{(l)}\|_p^p \right)^{1/p}, \tag{8.4}$$

where $\|f^{(l)}\|_p$ is the L_p norm of $f^{(l)}(x)$, as defined in (5.48). The Banach space $W_p^k(I)$ of all functions with finite Sobolev norm is termed a Sobolev space. For $p = 2$, the space is often denoted as $H^k(I)$. In that case, the Sobolev norm is induced by an inner product, based on the L_2 inner products of the derivatives,

$$\langle f, g \rangle = \sum_{l=0}^{k} \langle f^{(l)}, g^{(l)} \rangle_2,$$

making the Sobolev space a Hilbert space. For $p = \infty$, the norm becomes

$$\|f\|_{W^k_\infty} = \max_{0 \leq l \leq k} \|f^{(l)}\|_\infty,$$

where $\|g\|_\infty = \operatorname{ess\,sup}_{x \in I} |g(x)|$, the essential supremum of $|g(x)|$ on I. The essential supremum of a function $h(x)$ is the supremum except for a set of points with measure zero, typically a finite or countable set of isolated points in which a larger, outlying value is tolerated. Hence $W^k_\infty(I)$ is the space of all functions that are Lipschitz-k continuous on I, except for a finite or countable set of isolated points in which a discontinuity is accepted.

Membership of the Hilbert space $H^k(I) = W^k_2(I)$ puts quite some restrictions on the smoothness of a squared integrable function $f(x)$. For instance, being a member of $H_1(I)$ means that $f(x)$ and $f'(x)$ are square integrable. Square integrability is a weak condition when it comes to simple jump or cusp discontinuities, but it is far stricter when it applies to the derivative of a sharp transition. This can be illustrated with the family of functions

$$f_a(x) = [1 + \tanh(\operatorname{sign}(x - x_0)|x - x_0|^a / a)] / 2,$$

on $I = [0, 1]$, depicted, along with their derivatives, in Figure 8.1, for $x_0 = 0.345$ and for $a \in \{3/4, 2/3, 1/2, 1/3, 1/4\}$.

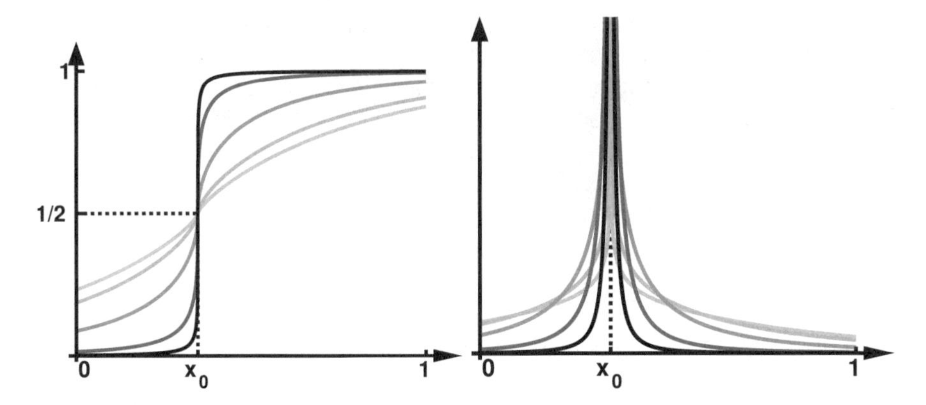

Figure 8.1
Left panel: family of functions with sharp transition in x_0. Right panel: the corresponding derivatives. If the transition is too sharp, the function is not in the Sobolev space $W^1_2(I) = H^1(I)$.

The derivative of $f_a(x)$ is given by

$$f'_a(x) = \left[1 - \tanh^2(\operatorname{sign}(x - x_0)|x - x_0|^a / a)\right] / 2 \cdot |x - x_0|^{a-1}.$$

The first factor in $f_a'(x)$ tends to $1/2$ when $x \to x_0$, but the second factor is unbounded. Moreover,

$$\int_0^1 \left[|x - x_0|^{a-1}\right]^2 dx < \infty \Leftrightarrow 2(a-1) > -1 \Leftrightarrow a > 1/2.$$

This means that only the two smoothest functions plotted in Figure 8.1 are in $H^1(I)$. As discussed in Section 8.3.5, these functions do not have a sparse wavelet decomposition. As a matter of fact, the spaces $H^1(I)$ are well characterised by decay of the coefficients in a Fourier series. Indeed, using version (6.20) of Parseval's theorem, we have on $I = [0, 1]$

$$\|f\|_2^2 + \|f'\|_2^2 = \sum_{k=-\infty}^{\infty} \left(|a_k|^2 + |b_k|^2\right),$$

where the $\boldsymbol{a}$ and $\boldsymbol{b}$ denote the Fourier series coefficients of $f(x)$ and $f'(x)$. Using partial integration on the formula for these coefficients (1.4), we find that $b_k = [f(1) - f(0)] + i(2\pi)ka_k$. Assuming periodic boundary conditions $f(0) = f(1)$, we arrive at

$$\|f\|_2^2 + \|f'\|_2^2 = \sum_{k=-\infty}^{\infty} (1 + (2\pi)^2 k^2)|a_k|^2,$$

which imposes a rate of decay upon $\boldsymbol{a}$. Also note that discontinuities are not allowed, not even in the boundary points upon periodic extension, as otherwise the norm of the derivative function cannot possibly be finite. That explains the periodic boundary conditions.

In general, Sobolev spaces combine smoothness with square integrability. The interesting Sobolev spaces, from a multiscale analysis that is, have p smaller than two. These spaces are larger, as on a bounded I it holds that $L_p(I) \subset L_q(I)$ when $q < p$. In Section 8.3.5, it is discussed that the multiscale analysis of functions in those classes shows sparse coefficient sequences.

8.3.2 The use of Sobolev norms in applications

When two objects are close to each other in a Sobolev norm, they have similar values, but also similar smoothness, derivatives that is[3]. In applications like denoising, Sobolev or other smoothness spaces may be useful to describe the objects to be estimated, but the observed version of that object is far from smooth. The Sobolev norm of the difference between the observed and the estimated objects is large. Therefore, Sobolev norms cannot be used to express closeness of fit w.r.t. the observation. For that part of the estimation or smoothing procedure we need the L_2 distance. Working with L_2

[3]As a Sobolev distance is an integrated expression, it is tolerant to isolated singularities, so two objects close in a Sobolev norm may still have cusps at different locations. Therefore these objects may look different. See also Section 8.3.5.

distances, however, puts additional constraints onto the multiscale basis. For instance, hierarchical bases cannot be used, as explained in Section 3.1.1. In other applications, such as the numerical solution of differential equations, we may look for approximations that are close to each other in value but also in derivatives. In that case, Sobolev norms provide good distance measures.

The discussion is illustrated by the construction of a nontrivial expansion of the zero function in a hierarchical basis, using (3.1) in Section 3.1.1. In Figure 8.2, the construction is applied to a hierarchical basis of cubic B-splines on nested sets of knots. The coarse scale knots are those from Figure 1.8. Unlike Figure 1.8 the corresponding basis does not include the boundary splines, leaving us with $n_L = 7$ basis functions on n_l knots. The initial coefficients are all set to one. Thanks to the partition of unity in (4.43), the resulting function, $f_L(x) = \Phi_L(x)\mathbf{1}_L$ is a constant function, except near the boundaries, where the missing boundary B-splines lead to a decay of $f_L(x)$. This function is depicted in the solid black line in the top panel and, for comparison, in the dotted black line in the next panels of Figure 8.2. The next panels show successive refinements, by inserting a knot between each pair of existing knots. The refinement knots are depicted in grey, along with the corresponding B-splines at that level. The refinement B-splines act as detail functions $\Psi_j(x)$ in the hierarchical basis. Linear combinations of the refinement B-splines define a refinement of the initial function $f_L(x)$,

$$f_J(x) = f_L(x) + \sum_{j=L}^{J-1} \Psi_j(x)\boldsymbol{d}_j.$$

As explained in Section 3.1.1, taking for the coefficients $\boldsymbol{d}_j$ the values found by an orthogonal projection as in (3.1) minimises the L_2-norm of $f_J(x)$. The functions $f_J(x)$ converge to the zero function in the L_2-norm. The sequence does not converge in a Sobolev norm, as it becomes more and more wiggly. This can be understood in an intuitive way because the coefficients $\boldsymbol{d}_j$ correspond to the odd *half* of the basis functions in $\Phi_{j+1}(x)$. As a result, at each scale, an optimisation takes place over half of the basis functions, causing heavily fluctuating behaviour between the two halves of the basis functions. The experiment confirms that hierarchical bases are useful for the study of objects in Sobolev spaces, where distances are measured by Sobolev norms, but not in situations where distances are measured by L_2-norms.

8.3.3 Functions of bounded variation

The Sobolev spaces, discussed in Section 8.3.1, are somehow tolerant to singularities, especially when parameter p is smaller than two, when the singularity is a cusp, or an isolated outlying value where left and right limit are the same and could be used in the definition of a continuous function. For instance, in the Sobolev space $W_1^1(I)$, the function $f(x) = 1 + x$ on $I = [0, 1]$, except for $f(1/2) = 0$, poses no problem. Also the cusp at $x = 1/2$

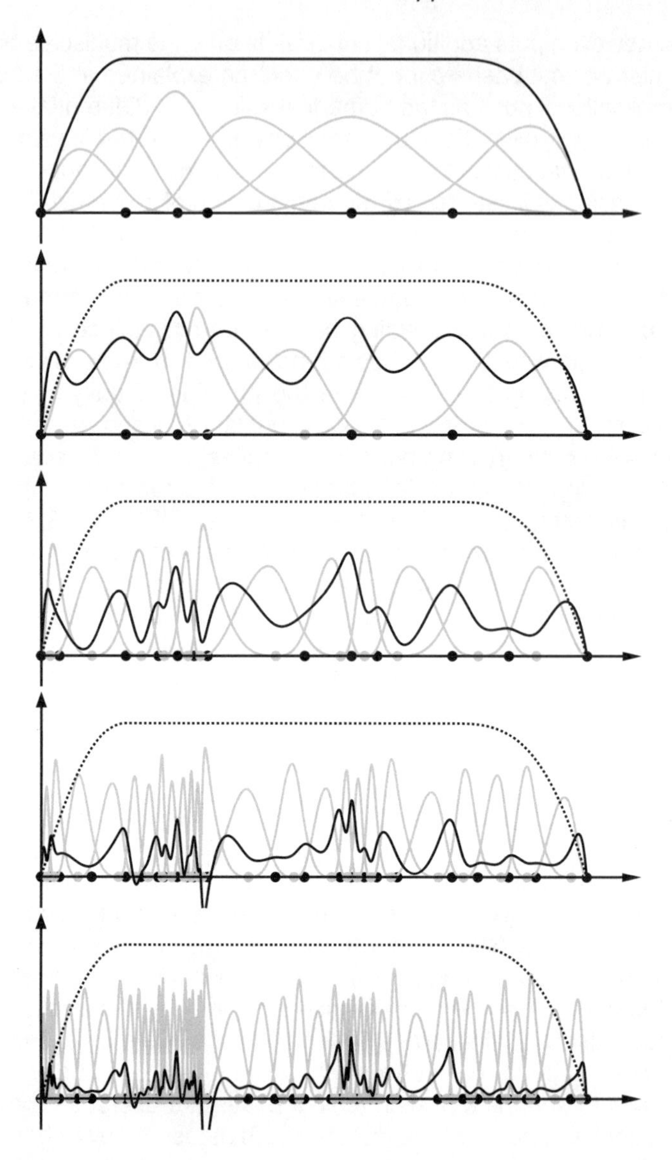

Figure 8.2
Construction of a nontrivial expansion of the zero function in a hierarchical basis of cubic B-splines. The approximation at each level is depicted in the solid black line. The detail basis at each level is plotted in grey. See text for more details.

in $f(x) = |x - 1/2|$ is accepted. The limiting case in Figure 8.1, the step function $f(x) = \chi_{[x_0,1]}(x)$, however, does not belong to $W_1^1(I)$, because we need to have a derivative of $f(x)$. Even in the weak sense, there is no function $Df(x)$ so that for all infinitely differentiable functions $g(x)$,

$$\int_I f(x)g'(x)\,dx = -\int_I g(x)Df(x)\,dx.$$

Hence, except for isolated outliers, all functions in $W_1^1(I)$ are continuous[4]

The Sobolev norm $\|f\|_{W_1^1}$ on I contains a term

$$\mathrm{TV}(f) = \int_I |f'(x)|dx,$$

named the **total variation** of the $W_1^1(I)$ function. This value can be expressed in terms of the local extrema of $f(x)$. More precisely, assuming that the set of points $\{x_i^*\}$ where $f'(x_i^*) = 0$ is countable, we have

$$\mathrm{TV}(f) = \sum_i |f(x_i^*) - f(x_{i-1}^*)|.$$

This expression can be adopted without imposing differentiability. The step function $f(x) = \chi_{[x_0,1]}(x)$ does not have a countable set of local extrema. For these cases, further generalisation comes from replacing the local extrema by a supremum over any countable subset of I. Hence, let $\boldsymbol{S}_n$ denote an ordered vector of $n+1$ points $x_0 < x_2 < \ldots < x_n$ in I; then

$$\mathrm{TV}(f) = \sup_{n\in\mathbb{N}} \sup_{\boldsymbol{S}_n} \sum_{i=1}^n |f(x_i) - f(x_{i-1})|.$$

Applied to the functions in Figure 8.1, we see that they all have the same value of $\mathrm{TV}(f) = 1$. Unlike the Sobolev norm, the total variation of a function does not depend on how steep the transition is. This is an interesting property in image processing, where blurred and sharp edges may have different visual quality, without however disqualifying one or the other as a rightful image. Functions, typically images, with finite total variation are said to have **bounded variation**.

The total variation is tolerant to jumps if they are isolated and not too numerous. Highly oscillating functions such as $\sin(1/x)$ on $I = [0,1]$ do not have bounded variation. Attenuated oscillations, such as $x^r \sin(1/x)$, may add up to a finite total variation, but then r should be strictly larger than one. Noise also causes the total variation to be large. The Gibbs oscillations in a Fourier analysis of a function with discontinuities lead to approximations that do not have bounded variation. Indeed, the example of the sawtooth function in Section 1.3.3 saw an oscillating approximation with local extrema

[4]Functions in $W_1^1(I)$ are almost everywhere absolutely continuous, a notion slightly stronger than just continuous.

$x^*_{n,i} = \frac{i}{2n+1}$ in the error function $R_n(x)$. The total variation on the interval $[0, n/(2n+1)] \subset [0, 1/2]$ is then

$$\mathrm{TV}(R_n) = \sum_{i=1}^{n} \left| \int_{(i-1)/(2n+1)}^{i/(2n+1)} R_n'(u)du \right| = \sum_{i=1}^{n} \left| \int_{i-1}^{i} \frac{\sin(\pi v)}{\sin(\pi v/(2n+1))} \frac{dv}{(2n+1)} \right|.$$

Using the same limit transition as in Section 1.3.3, we find

$$\lim_{n\to\infty} \mathrm{TV}(R_n) = \lim_{n\to\infty} \sum_{i=1}^{n} \left| \int_{i-1}^{i} \frac{\sin(\pi v)}{\pi v} dv \right| = \infty.$$

Fourier analysis is not the appropriate tool for the approximation of bounded variation functions.

Wavelets are a much better option in this function class. The success of wavelets is explained by the combination of two facts. First, the class of bounded variations functions on I, $\mathrm{BV}(I)$, is comprised between two other function classes, in the sense (Choi and Baraniuk, 1999) that

$$B^1_{1,1}(I) \subset \mathrm{BV}(I) \subset B^1_{1,\infty}(I).$$

The $B^1_{1,1}(I)$ and $B^1_{1,\infty}(I)$ are two Besov spaces, introduced in Section 8.3.4. The second reason for the success of wavelets in the class of bounded variation functions is the fact that membership of Besov spaces can be established from the wavelet decomposition of a function. This is developed in Section 8.3.5.

8.3.4 Besov spaces

The smoothness parameter k in the Sobolev space $W^k_p(I)$ admits integer values only. There are several ways of extending the family, including fractional Sobolev spaces (Di Nezza et al., 2012) and Triebel spaces (Peetre, 1975). This section develops Besov norms as an important extension.

The definition of Besov norms requires the introduction of a few additional notions. First, the k**-th difference** of a function $f(x)$ is given by the recursion

$$\begin{aligned} \Delta_h^{(1)} f(x) &= f(x+h) - f(x) \\ \Delta_h^{(k)} f(x) &= \Delta_h^{(k-1)} f(x+h) - \Delta_h^{(k-1)} f(x), \end{aligned}$$

which amounts to

$$\Delta_h^{(k)} f(x) = \sum_{l=0}^{r} (-1)^l \binom{k}{l} f(x+lh). \tag{8.5}$$

Intuitively, the difference could be seen as a coarse scale measure of the smoothness of the function. At scale s, and with $I = [0,1]$, the **k-th modulus of smoothness** is defined by

$$\nu_{k,p}(f;s) = \sup_{h \leq s} \left(\int_0^{1-kh} \left| \Delta_h^{(k)} f(x) \right|^p dx \right)^{1/p}. \tag{8.6}$$

The k-th modulus of smoothness can be interpreted as a multiscale (for all $h \leq s$, that is) analysis of the differences. At finer scales s, the modulus of smoothness should tend to zero sufficiently fast, so that in a global way (except for occasional exceptions) the k-th differences can be divided by s^k. The multiscale sufficiently fast decay is formalised by an integral expression, known as the **Besov semi-norm**,

$$|f|_{B_{p,q}^{\alpha}} = \left[\int_0^1 \left(\frac{\nu_{k,p}(f;s)}{s^{\alpha}} \right)^q \frac{ds}{s} \right]^{1/q}, \tag{8.7}$$

where $k-1 \leq \alpha < k$. Since $\displaystyle\int_0^1 \frac{ds}{s} = \infty$, the Besov semi-norm may only be finite if $\nu_{k,p}(f;s)/s^{\alpha} \to 0$ when $s \to 0$. As the Besov semi-norm may vanish for essentially nonzero functions, the actual Besov norm adds an L_p-norm,

$$\|f\|_{B_{p,q}^{\alpha}} = \|f\|_{L_p(I)} + |f|_{B_{p,q}^{\alpha}}. \tag{8.8}$$

The Besov space $B_{p,q}^{\alpha}(I)$ is then defined by all functions $f \in L_p(I)$ for which $|f|_{B_{p,q}^{\alpha}} < \infty$.

As a follow up to the discussion in Section 8.3.2, it should be noted that in applications such as smoothing or denoising, Besov norms are primarily used for the classification of the smoothness class of the objective functions, rather than for measuring the distances between the input and the output. In those applications, it is common practice to work on Besov *balls*, i.e., sets of functions for which the Besov norms are bounded by a finite value, $\|f\|_{B_{p,q}^{\alpha}} \leq R < \infty$.

8.3.5 Sparse sequences

It can be proven that a multiresolution analysis provides an unconditional basis for Besov spaces. More precisely, if the wavelet analysis has at least $\lceil \alpha \rceil$ dual vanishing moments, then $f(x) \in B_{p,q}^{\alpha}(I)$ if and only if $\boldsymbol{w} \in b_{p,q}^{\alpha}$. Here $\boldsymbol{w}$ is the sequence of wavelet coefficients $\boldsymbol{w} = (\boldsymbol{s}_L, \boldsymbol{d}_L, \boldsymbol{d}_{L+1}, \ldots)$ in the expansion

$$f(x) = \Phi_L(x)\boldsymbol{s}_L + \sum_{j=L}^{\infty} \Psi_j(x)\boldsymbol{d}_j.$$

The sequence space $b^{\alpha}_{p,q}$ is defined by all square summable sequences for which the sequence norm

$$\|\boldsymbol{w}\|_{b^{\alpha}_{p,q}} = \left(\sum_{k=0}^{n_L-1} |s_{L,k}|^p\right)^{1/p} + \left[\sum_{j=L}^{\infty} 2^{j\beta q}\left(\sum_{k=0}^{n'_j-1} |d_{j,k}|^p\right)^{q/p}\right]^{1/q}, \quad (8.9)$$

where $\beta = \alpha + 1/2 - 1/p$. For $q = \infty$, this is:

$$\|\boldsymbol{w}\|_{b^{\alpha}_{p,\infty}} = \left(\sum_{k=0}^{n_L-1} |s_{L,k}|^p\right)^{1/p} + \sup_{j\geq L}\left[2^{j\beta}\left(\sum_{k=0}^{n'_j-1} |d_{j,k}|^p\right)^{1/p}\right]. \quad (8.10)$$

The multiresolution basis is unconditional, meaning that there exist constants c and C, not depending on f, so that

$$c\|f\|_{B^{\alpha}_{p,q}} \leq \|\boldsymbol{w}\|_{b^{\alpha}_{p,q}} \leq C\|f\|_{B^{\alpha}_{p,q}}. \quad (8.11)$$

The Sobolev spaces $H^k(I)$, for which we have seen that they contain smooth functions, are nothing but the Besov spaces $B^k_{2,2}$. The Besov sequence norm (8.9) is given by

$$\|\boldsymbol{w}\|_{b^{k}_{2,2}} = \|\boldsymbol{s}_L\|_2 + \left[\sum_{j=L}^{\infty} 2^{jk2}\|\boldsymbol{d}_j\|_2^2\right]^{1/2}.$$

Functions in $H^k(I)$ have a finite Besov sequence norm. This condition imposes the norms of $\boldsymbol{d}_j$ to show sufficiently fast decay, but no sparsity. Indeed, the sizes of the vectors $\boldsymbol{d}_j$ are measured by their ℓ_2 norms, meaning that all vectors on a Euclidean sphere have the same size. In order to promote sparsity, the size of $\boldsymbol{d}_j$ is measured by its ℓ_p norm with $p < 2$. This is illustrated in Figure 8.3, depicting ℓ_p balls for two different sizes and for different values of p. Vectors that have the same size in the ℓ_2 norm lie on a sphere, which in $\mathbb{R}^2$ is a circle. Taking $p < 2$, these vectors on a circle are different in the ℓ_p norm, the smallest norms occurring on the axes, where one component is zero. Points on the circle with maximum ℓ_p norm for $p < 2$ have components with equal magnitudes; hence they are not sparse in the given basis.

In a random model of sparse vectors, the ℓ_p spheres are contour lines of the joint density of an independent sample of random variables whose absolute values follow a generalised Gamma density[5]. In particular, we have

$$f_X(x) = \frac{1}{K}\exp\left[-(a|x|)^p\right],$$

[5] The generalised Gamma density function has the form $f_X(x) = p\lambda^{\alpha}x^{\alpha-1}\exp\left[-(\lambda x)^p\right]/\Gamma(\alpha/p)$ for $x \geq 0$. Here we have $\lambda = a$ and $\alpha = 1$

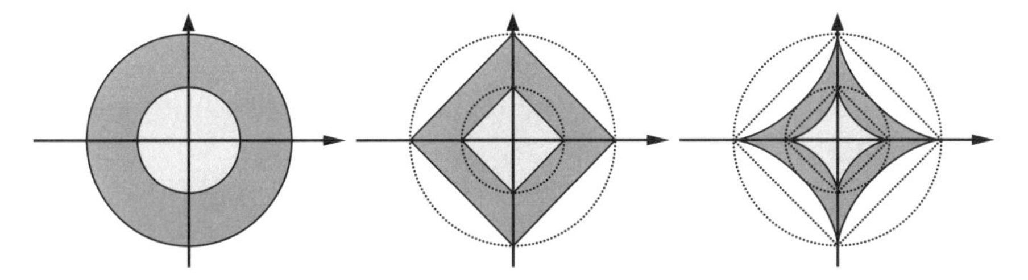

Figure 8.3
ℓ_p-balls in $\mathbb{R}^2$ with radii $r = 1$ and $r = 1/2$, i.e., a set of vectors $\boldsymbol{x}$ for which $\|\boldsymbol{x}\|_p = (|x_1|^p + |x_2|^p)^{1/p} = r$. From left to right: $p = 2, 1, 2/3$. The dotted lines depict all balls with higher p and the same r. Comparison of the balls with the same radius and different p illustrate that any point on an ℓ_p ball lies on an ℓ_q ball with a higher radius if $q < p$, *except* for the points in the corners, i.e., those points where one component is zero. This is why ℓ_p balls with $p < 2$ promote sparsity.

where

$$K = 2\int_0^\infty \exp\left[-(at)^p\right] dt = \frac{2}{pa}\int_0^\infty \exp(-u)u^{1/p-1}du = \frac{2}{pa}\Gamma\left(\frac{1}{p}\right).$$

For $p = 2$, this is the normal density, while for $p = 1$, this is the Laplace (or double exponential) density. The smaller the p, the heavier the tail. In a joint distribution of independent heavy tailed variables, the densities tend to concentrate around the axes, thus promoting sparsity: with $\boldsymbol{X} \in \mathbb{R}^n$ an independent sample, the region $R \subset \mathbb{R}^n$ maximising the probability $P(\boldsymbol{X} \in R)$ for a given area is an ℓ_p ball.

Taking $p < 2$, the Besov sequence norms measure sparsity at each resolution level with a decay across the resolution levels. As a result, they express **multiscale sparsity**. Being norms over a vector space, Besov sequence norms impose no structured sparsity, in the sense that a few large coefficients may occur at arbitrary locations at each scale. The Besov norm is not able to impose groups, clusters, or blocks of large coefficients within or across scales. Modelling that kind of structure is beyond the scope of vector spaces and vector norms. A few cases will be discussed in Chapter 12.

8.4 Preprocessing observed data for use in multiscale decompositions

8.4.1 Situation of the problem

A wavelet transform maps fine scaling coefficients onto a multiscale decomposition, as represented in (1.30). The transform assumes that the fine scaling coefficients are available, but finding or estimating these coefficients is not always a trivial task.

In theory, if the function under consideration $f(x)$ is known by its analytic form on $[0,1]$, an approximation $\Phi_J(x)\boldsymbol{s}_J$ in the scaling basis at level J can be constructed using the orthogonal projection

$$\boldsymbol{s}_J = \left(\int_0^1 \Phi_J^\top(x)\Phi_J(x)dx\right)^{-1}\int_0^1 \Phi_J^\top(x)f(x)dx,$$

where we assume that all functions in $\Phi_J(x)$ are in the Hilbert space of square integrable functions on $[0,1]$. Orthogonal projection is just a special case, taking

$$\widetilde{\Phi}_J(x) = \left(\int_0^1 \Phi_J^\top(x)\Phi_J(x)dx\right)^{-1}\Phi_J^\top(x),$$

of the projection using a dual scaling basis as in (2.79). The dual scaling basis is actually used for the calculation of coarse scaling coefficients from fine scaling coefficients, as in (2.75). In principle, the projection of $f(x)$ onto the fine scaling functions need not use the same dual basis as the further multiscale decomposition.

In most practical situations, the function $f(x)$ is available in a limited number of knots, with possibly noisy observations $Y_i = f(x_i) + \sigma Z_i$. This leaves us with the question of how to find good values of the fine scale coefficients, without introducing too much additional bias. The objective is not to reduce the noise (the variance, that is) at this stage, since the subsequent multiscale decomposition provides a better tool for that task.

A frequently used approach is to simply take the observations as finest scaling coefficients, i.e., $s_{J,k} = Y_k$. This section investigates if and when this simple approach is acceptable from the approximation point of view. At least taking function values as fine scaling coefficients does not reduce any of the noise, so no bias or blur is introduced from that perspective.

8.4.2 Linear approximation with function values at finest scale

Let $\Phi_J(x)$ be an L_2 normalised, dyadic scaling basis with $\widetilde{p}$ dual vanishing moments, defined on an equidistant set of knots, i.e., the basis functions are given by $\varphi_{J,k}(x) = 2^{J/2}\varphi(2^J x - k)$. The objective is to find a linear multiscale approximation for $f(x)$, which has at least $\widetilde{p}$ continuous and bounded derivatives. As discussed in Section 8.1, the approximation using the projection (2.79) admits an error rate of $\mathcal{O}\left(2^{-J\widetilde{p}}\right)$, both for the pointwise and integrated squared errors. When the projection coefficients are replaced by function values in equidistant, dyadic knots $x_{J,k} = 2^{-J}k$,

$$\overline{f}_J(x) = \sum_{k=0}^{2^J-1} f(x_{J,k})2^{-J/2}\varphi_{J,k}(x) = \sum_{k=0}^{2^J-1} f(2^{-J}k)\varphi(2^J x - k), \tag{8.12}$$

then the approximation still converges to $f(x)$, but at a much slower rate, ignoring all benefits from the dual vanishing moments. More precisely, the following theorem (Sweldens and Piessens, 1994, Theorem 2.4) states that the approximation with function values $\overline{f}_J(x)$ lies within a small integrated squared distance from a slowly converging, blurred version of the function $f(x)$.

Theorem 8.4.1 *Let $\varphi(x) \in L_2([0,1])$ be the father function of a multiresolution analysis with $\widetilde{p}$ dual vanishing moments, normalised so that $\int_0^1 \varphi(x)dx = 1$. Let $f(x)$ have at least $\widetilde{p}$ continuous and bounded derivatives. Then for the function $\overline{f}_J(x)$ as defined in (8.12), it holds that $\overline{f}_J(x) = \overline{\overline{f}}_J(x) + \mathcal{O}\left(2^{-J\widetilde{p}}\right)$, where the blurred function $\overline{\overline{f}}_J(x)$ is defined by*

$$\overline{\overline{f}}_J(x) = \sum_{k=1}^{2^J} \varphi(k) f(x - 2^{-J}k). \tag{8.13}$$

Proof. A Taylor expansion of $f(2^{-J}k)$ around x and the equidistance of the knots are used to write,

$$\begin{aligned}
\overline{f}_J(x) &= \sum_{k=0}^{2^J-1}\sum_{q=0}^{\widetilde{p}-1} \frac{f^{(q)}(x)}{q!}(2^{-J}k - x)^q \varphi(2^J x - k) + \mathcal{O}\left((2^{-J}k - x)^{\widetilde{p}}\right) \\
&= \sum_{q=0}^{\widetilde{p}-1} \frac{f^{(q)}(x)}{q!} \sum_{k=0}^{2^J-1} (-1)^q (x - 2^{-J}k)^q \varphi(2^J x - k) + \mathcal{O}\left(2^{-J\widetilde{p}}\right)
\end{aligned}$$

The Poisson summation formula, stated in Theorem 6.2.16, can be used in

$$\sum_{k=0}^{2^J-1} (x - 2^{-J}k)^q \varphi\left(2^J(x - 2^{-J}k)\right) = 2\pi 2^J \sum_{k\in\mathbb{Z}} G_q\left(2\pi 2^J k\right) \exp(i2\pi 2^J kx),$$

with $G_q(\omega)$ the Fourier transform of $g_q(x) = x^q\varphi(2^J x)$. The Fourier transform of $\varphi(2^J x)$ is given by $2^{-J}\Phi(2^{-J}\omega)$, with $\Phi(\omega)$ the Fourier transform of $\varphi(x)$. The Fourier transform of $g_q(x)$ can be verified to be $2^{-2J}\Phi^{(q)}\left(2^{-J}\omega\right)$ We thus arrive at

$$2^J \sum_{k=0}^{2^J-1} (x - 2^{-J}k)^q \varphi(2^J x - k) = \sum_{k\in\mathbb{Z}} \Phi^{(q)}\left(2\pi k\right). \tag{8.14}$$

Because of the vanishing moment conditions (6.15), we have $\Phi^{(q)}(\pi) = 0$. From the Fourier two-scale equation, it then follows that $\Phi^{(q)}(2\pi) = 0$, $\Phi^{(q)}(4\pi) = 0$, and so on. This leaves us with just one nonzero term in the summation over k in (8.14), for $k = 0$, that is. As a result, the left hand side does not depend on x. Fixing $x = 1$ leads to

$$\sum_{k=0}^{2^J-1} (x - 2^{-J}k)^q \varphi(2^J x - k) = \sum_{k=0}^{2^J-1} (1 - 2^{-J}k)^q \varphi(2^J - k) = \sum_{\ell=1}^{2^J} 2^{-Jq}\ell^q \varphi(\ell).$$

This can be substituted into the expression for $\overline{f}_J(x)$,

$$\begin{aligned}
\overline{f}_J(x) &= \sum_{q=0}^{\widetilde{p}-1} \frac{f^{(q)}(x)}{q!}(-1)^q \sum_{\ell=1}^{2^J} (2^{-J}\ell)^q \varphi(\ell) + \mathcal{O}\left(2^{-J\widetilde{p}}\right) \\
&= \sum_{\ell=1}^{2^J} \varphi(\ell) \sum_{q=0}^{\widetilde{p}-1} \frac{f^{(q)}(x)}{q!}(-2^{-J}\ell)^q + \mathcal{O}\left(2^{-J\widetilde{p}}\right) \\
&= \sum_{\ell=1}^{2^J} \varphi(\ell) f(x - 2^{-J}\ell) + \mathcal{O}\left(2^{-J\widetilde{p}}\right).
\end{aligned}$$

The last step is again based on a Taylor expansion, this time of $f(x - 2^{-J}\ell)$ around x. □

8.4.2.1 Discussion on Theorem 8.4.1 – The wavelet crime

When the scale grows finer, the blur converges to zero at a rate which is given by $\|\overline{\overline{f}}_J - f\|_2 = \mathcal{O}(2^{-J})$ in the general case. This rate is slow and does not take any benefit from the vanishing moments in the basis.

The implicit blurring from the use of function values as fine scaling coefficients was termed **the wavelet crime** (Strang and Nguyen, 1996). The wavelet crime is closely related to the bias in local constant kernel fitting, mentioned in Fan and Gijbels (1996, p. 17) as a compelling argument for at least using local linear estimation, thereby quoting Strang (1993). The close connection between the two phenomena is confirmed by the form of the approximation in (8.12), which has the appearance of a kernel smoother, the father function acting as the kernel function.

There are a few exceptions, though. First, when the projection coefficients are close to the function values as in

$$s_{J,k} = \int_0^1 f(x)\widetilde{\varphi}_{J,k}(x)dx = f(x_{J,k}) + \mathcal{O}\left(2^{-J\widetilde{p}}\right),$$

then

$$\begin{aligned}\left|f_J(x) - \overline{f}_J(x)\right| &\leq \sum_{k=0}^{n_J-1} |s_{J,k} - f(x_{J,k})| \cdot |\varphi_{J,k}(x)| \\ &= \mathcal{O}\left(2^{-J\widetilde{p}}\right) \sum_{k=0}^{n_J-1} |\varphi_{J,k}(x)| = \mathcal{O}\left(2^{-J\widetilde{p}}\right).\end{aligned}$$

The projection coefficients are close to the function values if the function is smooth and if the dual scaling basis satisfies

$$\int_0^1 x^q \widetilde{\varphi}_{J,k}(x)dx = 0,$$

for all $q \in \{1, \ldots, \widetilde{p} - 1\}$. Imposing this condition in the framework of an orthogonal basis, so with $\widetilde{\varphi}_{J,k}(x) = \varphi_{J,k}(x)$, leads to the class of **coiflets**, presented in Section 7.1.3.

Another exception to the wavelet crime, already mentioned in Section 3.2.2, holds for **interpolating scaling bases**, such as scaling bases associated to a Deslauriers-Dubuc refinement. Indeed, if $\varphi_{J,0}(x_{J,k}) = \delta_k$, then $\overline{\overline{f}}_J(x) = f(x)$. Note that the first exception is stated in terms of the dual basis, while the second exception is due to properties of the primal scaling basis.

Since the proof of Theorem 8.4.1 hinges on the Poisson summation formula and equidistant refinement, there is no hope that a result similar to that of the theorem, as weak as it is, could hold for multiscale analyses on irregular sets of knots. In other words, there is no guarantee that the use of function values as scaling coefficients in a non-equidistant multiscale analysis would even converge to the original function.

8.4.3 Nonlinear approximation using function values at finest scale

The wavelet crime is about the loss of the linear approximation power provided by the dual vanishing moments. The question is whether the phenomenon has any impact on a nonlinear approximation scheme as described in Section 8.2.

8.4.3.1 The case of equispaced knots

We first consider the case of an equidistant, dyadic multiscale analysis. The oracular nonlinear scheme has a linear component, in the sense that all coefficients at scales L or lower are preserved. The smooth part of the function $f(x)$ is then approximated with an error of $\mathcal{O}(2^{-L\widetilde{p}})$. The singularities are approximated with the same accuracy if all coefficients near the singularities up to a scale $J > L\widetilde{p}$ are preserved, thereby locating the position of the singularities up to a precision of $\mathcal{O}(2^{-J})$.

According to Theorem 8.4.1, the use of function values instead of projection coefficients at the fine scale J introduces an error of $\mathcal{O}(2^{-J})$. This is of the same order as the error in the location of the singularities, and it is smaller than the error from thresholding away all the small coefficients up to scale L, at least if $J > L\widetilde{p}$.

Moreover, on equidistant knots, the error introduced in the smooth part at fine scale J is not inflated through the subsequent thresholding process. This is because, on equidistant knots, the scaling coefficients from a polynomial are recognised as such, thanks to the result in Theorem 5.5.15. As a consequence, the detail coefficients computed by a wavelet transform applied to a vector of function values are zero if and only if the detail coefficients from a vector of projection coefficients are zero. Thresholding the small coefficients therefore has no further impact on the order of approximation.

As a conclusion, a nonlinear selection of wavelet coefficients on an equidistant data set works fine if it is initiated with unpreprocessed function values instead of proper projection. In this specific situation, a common practice in the literature on wavelet based smoothing and noise reduction, the wavelet crime is not an issue.

8.4.3.2 The case of non-equispaced knots

The situation is totally different on a set of non-equispaced knots. As Theorem 5.5.15 holds specifically for equispaced scaling functions, the wavelet crime may induce additional errors after applying a wavelet transform and thresholding the coefficients. Indeed, observations from a polynomial (of degree $\widetilde{p}-1$) are not identified by the wavelet transform as coefficients from a polynomial, thus leading to nonzero detail coefficients. The application of a threshold to these coefficients then introduces additional errors, up to the order of $\mathcal{O}(\Delta_L)$, where Δ_L is the maximal interknot distance $x_{L,k} - x_{L,k-1}$. Again, the wavelet crime neutralises all benefits from the dual vanishing moments. The propagation of fine scale approximation errors through the thresholding scheme also leads to visually unpleasant and unsmooth reconstructions. Indeed, the small coefficients in a non-dyadic wavelet transform of a vector of polynomial observations are a combined effect of the lack of equidistance and the incorrect use of observations as fine scaling coefficients. Taking away these small coefficients makes the irregularity of the knots become visible in the reconstruction, pretty much in a similar way as if the transform itself would ignore the irregularity of the knots by straightforwardly applying the dyadic form for analysis and reconstruction.

8.4.4 Preprocessing the fine scaling coefficients

In many applications the observations are available through a model $\boldsymbol{Y} = \boldsymbol{\mu} + \boldsymbol{\varepsilon} = \boldsymbol{\Phi}_J \boldsymbol{s}_J + \boldsymbol{\varepsilon}$. In this model $\boldsymbol{\Phi}_J$ denotes the matrix with elements $\Phi_{J;k,l} = \varphi_{J,l}(x_{J,k})$. The goal is to find unbiased estimators of the scaling coefficients $\boldsymbol{s}_J$. As for now, the variance of these estimators is not a primary objective, since the search for a balance between a controlled bias and a reduced variance is the topic of the subsequent multiscale analysis. In this sense, the fine scale preprocessing should be seen as complementary to the control of the variance propagation throughout the actual multiscale transform, as discussed in Section 3.1.5. Whereas the variance control plays a crucial role throughout the transform, the bias requires special attention in the preprocessing step.

We need to find $\widehat{\boldsymbol{s}}_J$ so that $E(\widehat{\boldsymbol{s}}_J) \approx \boldsymbol{s}_J$. Moreover, if the noise-free observations $\boldsymbol{\mu}$ come from a polynomial, then $E(\widehat{\boldsymbol{s}}_J)$ should be as close as possible to a form that is recognised by the wavelet transform as polynomial coefficients. For the sake of superposition, the estimator is taken to be linear, $\widehat{\boldsymbol{s}}_J = \mathbf{S}_J \boldsymbol{Y}$. It is tempting to impose that any vector of observations is exactly reconstructed,

$$\boldsymbol{\Phi}_J \mathbf{S}_J \boldsymbol{Y} = \boldsymbol{Y}, \tag{8.15}$$

meaning that $\mathbf{S}_J$ is the inverse or a pseudo-inverse of $\boldsymbol{\Phi}_J$. In most situations, this solution has unpleasant effects. First, as the scaling functions typically have compact support, the matrix $\boldsymbol{\Phi}_J$ has a band structure. Its inverse, however, cannot be expected to have such a structure. The calculation of a single scaling coefficient involves all observations, which is a non-local and computationally expensive operation. Moreover, the matrix $\boldsymbol{\Phi}_J$ cannot be expected to have small singular values, leading to uncontrolled variance propagation. As an alternative, the matrix $\mathbf{S}_J$ can be designed to be as close as possible in Frobenius norm to the identity matrix, under the constraint that (8.15) holds when the observations are exact polynomial values, i.e., $Y_i = x_{J,i}^q$ with $q = 0, 1, \ldots, \widetilde{p} - 1$. The case of spline scaling coefficients is developed in (Jansen, 2016).

9

Overcomplete wavelet transforms

9.1 The nondecimated wavelet transform

The wavelet transforms discussed in Chapter 7 are not **shift or translation invariant**. Indeed, adding one covariate and one response value in front of the existing pairs of covariates and responses changes all coefficients in the multiscale decomposition. In a lifting scheme, this is because in the split phase all evens and odds switch roles. In a classic filterbank implementation on equispaced data, the decimation that comes after the convolution sums in (5.13) and (5.14) keeps the complementary half of the intermediate result.

From the perspective of a statistical analysis, as well as in other applications, a translation invariant decomposition would be appreciated. If apart from possible local boundary effects, shifted inputs yield shifted but otherwise identical outputs, a translation invariant analysis avoids a simple shift to lead to different conclusions. The shift invariance argument is quite similar to the motivation for symmetric filters on equidistant knots, or symmetric constructions on irregular knots, such as B-splines or Deslauriers-Dubuc schemes. These schemes are all reflection or flip invariant. Unfortunately, unlike flip invariance, shift invariance cannot be realised by choosing the right filters.

This motivates an alternative multiresolution analysis, which has emerged quite rapidly and simultaneously in different contributions in the literature. The alternative analysis has been given several names, all reflecting the viewpoints of the contributions. The same multiresolution scheme can thus be referred to as the *translation invariant wavelet transform* (Coifman and Donoho, 1995; Berkner and Wells, 2002), the *redundant wavelet transform*, the *stationary wavelet transform* (Nason and Silverman, 1995), the *maximal overlap* discrete wavelet transform (Percival and Walden, 2008), *cycle spinning*. For reasons to be explained, some papers adopt the term of an *à trous* scheme (Shensa, 1992), although this should not be confused with the à trous scheme of an interpolating wavelet transform, discussed in Section 5.3.1.

The translation invariant scheme is developed throughout the subsequent sections. Next to the obvious translation invariance, it also exhibits superior behaviour when it comes to variance propagation. This is investigated in Section 9.1.7.

DOI: 10.1201/9781003265375-9

9.1.1 Subsampling and translation invariance

The forward fast wavelet transform includes an operation that reduces the size of the vector of scaling coefficients when proceeding from fine to coarse scales. In the lifting implementation, the size reduction is realised by the split into evens and odds. The filterbank version of Section 2.2 operates with rectangular matrices $\widetilde{\mathbf{H}}_j^\top$ and $\widetilde{\mathbf{G}}_j^\top$ mapping a scaling vector of length n_{j+1} onto a scaling vector of length n_j and a detail vector of complementary length $n'_j = n_{j+1} - n_j$. These operations can be formally represented as $\widetilde{\mathbf{H}}_j^\top = \widetilde{\mathbf{J}}_j^\top \widetilde{\boldsymbol{\mathcal{H}}}_j^\top$, and $\widetilde{\mathbf{G}}_j^\top = \widetilde{\mathbf{J}}_j^{o\top} \widetilde{\boldsymbol{\mathcal{G}}}_j^\top$, where $\widetilde{\mathbf{J}}_j^\top$ is the $n_j \times n_{j+1}$ subsampling matrix, introduced in (2.2), $\widetilde{\mathbf{J}}_j^{o\top}$ is its complement, as in Section 2.2.2, while $\widetilde{\boldsymbol{\mathcal{H}}}_j^\top$ and $\widetilde{\boldsymbol{\mathcal{G}}}_j^\top$ are square $n_{j+1} \times n_{j+1}$ operations. These matrices are not unique, because only half of their elements are specified by the subsampled counterparts $\widetilde{\mathbf{H}}_j^\top$ and $\widetilde{\mathbf{G}}_j^\top$, leaving the complementary half of the rows in $\widetilde{\boldsymbol{\mathcal{H}}}_j^\top$ and $\widetilde{\boldsymbol{\mathcal{G}}}_j^\top$ to be filled in.

Formally, and focussing on $\widetilde{\boldsymbol{\mathcal{H}}}_j^\top$ throughout the subsequent discussion, we have the freedom to define the complementary lines $\widetilde{\mathbf{H}}_j^{[1]\top}$ in $\widetilde{\boldsymbol{\mathcal{H}}}_j^\top$

$$\widetilde{\boldsymbol{\mathcal{H}}}_j^\top = \widetilde{\mathbf{J}}_j \widetilde{\mathbf{H}}_j^\top + \widetilde{\mathbf{J}}_j^o \widetilde{\mathbf{H}}_j^{[1]\top}. \tag{9.1}$$

The matrices $\widetilde{\mathbf{J}}_j$ and $\widetilde{\mathbf{J}}_j^o$ are upsampling operations, placing the outcome of the the matrix operations with $\widetilde{\mathbf{H}}_j^\top$ and $\widetilde{\mathbf{H}}_j^{[1]\top}$ into the even and odd rows respectively.

The complementary refinement matrix $\widetilde{\mathbf{H}}_j^{[1]\top}$ can be filled in by a procedure referred to as **cycle spinning** in some literature. Cycle spinning switches the roles of evens and odds. It is implemented by defining $\mathbf{S}_j$, a specific version of the **shift or translation matrix** encountered earlier, in (5.7). More precisely, the $(n_j + 1) \times n_j$ matrix $\mathbf{S}_j$ has zeros everywhere, except for the very first element, $S_{j;0,0}$, and the elements on the first upper diagonal, $S_{j;k,k+1}$, which are all equal to one. It then holds that $(\mathbf{S}_j \boldsymbol{s}_j)_0 = s_{j,0}$ and $(\mathbf{S}_j \boldsymbol{s}_j)_k = s_{j,k-1}$. The complementary refinement is then constructed by a lifting scheme on the shifted grid of knots $\mathbf{S}_j \boldsymbol{x}_j$, followed by omission of the lines and columns in the lifting matrices corresponding to the newly inserted first element $(\mathbf{S}_j \boldsymbol{x}_j)_1$.

On equidistant knots, and with the appropriate boundary handling, the matrices $\widetilde{\boldsymbol{\mathcal{H}}}_j^\top$ and $\widetilde{\boldsymbol{\mathcal{G}}}_j^\top$ can be completed to have a Toeplitz[1] structure. The Toeplitz matrices commute with a simple translation,

$$\widetilde{\boldsymbol{\mathcal{H}}}_j^\top \mathbf{S}_{j+1} = \mathbf{S}_{j+1} \widetilde{\boldsymbol{\mathcal{H}}}_j^\top \tag{9.2}$$

[1] A **Toeplitz matrix** $\mathbf{T}$ is a matrix with constant main- and off-diagonals, i.e., $A_{i+k,j+k} = A_{i,j}$ for any combination of i, j, k fitting into the not necessarily square dimensions of the matrix.

On infinite or periodically extended equidistant grids, the Toeplitz matrices turn into convolution sums, as in (5.13) and (5.14) without decimation steps.

On nonequidistant knots, the translation invariance of the matrices $\widetilde{\mathcal{H}}_j^\top$ and $\widetilde{\mathcal{G}}_j^\top$ includes their construction. The matrices themselves do not have a Toeplitz structure, but the translation of the grid of covariates induces a shift in the matrices, leading to mere translation of the output coefficients.

9.1.2 Translation invariant decompositions

The use of the nondecimated operations $\widetilde{\mathcal{H}}_j^\top$ and $\widetilde{\mathcal{G}}_j^\top$ in a translation invariant multiresolution scheme requires further development.

The first step of the forward nondecimated wavelet transform is straightforward: the nondecimated matrices have nondecimated scaling and wavelet coefficients as outputs. On an infinite equidistant grid, the expressions can be derived from the decimated versions in (5.13) and (5.14) by taking out the dyadic subsampling operation,

$$\boldsymbol{s}_{J-1} = \overline{\boldsymbol{h}} * \boldsymbol{s}_J = \widetilde{\mathcal{H}}_{J-1}^\top \boldsymbol{s}_J \text{ and } \boldsymbol{d}_{J-1} = \overline{\boldsymbol{g}} * \boldsymbol{s}_J = \widetilde{\mathcal{G}}_{J-1}^\top \boldsymbol{s}_J.$$

From now on, and for the rest of the section on nondecimated transforms, the vectors $\boldsymbol{s}_j$ and $\boldsymbol{d}_j$ stand for vectors of nondecimated coefficients. The decimated coefficients are retrieved from subsampling $\boldsymbol{s}_{J-1,e} = \widetilde{\mathbf{J}}_{J-1}^\top \boldsymbol{s}_{J-1}$, and $\boldsymbol{d}_{J-1,e} = \widetilde{\mathbf{J}}_{J-1}^{o\top} \boldsymbol{d}_{J-1}$. The complementary half of the nondecimated scaling coefficients at level $J-1$ is given by the vector $\boldsymbol{s}_{J-1,o} = \widetilde{\mathbf{J}}_{J-1}^{o\top} \boldsymbol{s}_{J-1}$.

At the next level, the $n_{J-1} \times n_{J-1}$ matrix $\widetilde{\mathcal{H}}_{J-2}^\top$ has been designed to transform a vector of length n_{J-1}. In particular, it operates on the decimated scaling vector $\boldsymbol{s}_{J-1,e} = \widetilde{\mathbf{J}}_{J-1}^\top \boldsymbol{s}_{J-1}$, producing a vector of length n_{J-1} at level $J-2$. Within this vector of length n_{J-1}, we have n_{J-2} coefficients of the decimated transform at level $J-2$, along with a complement corresponding to a translation of the vector $\boldsymbol{s}_{J-1}$. For the coefficients in $\boldsymbol{s}_{J-1,o} = \widetilde{\mathbf{J}}_{J-1}^{o\top} \boldsymbol{s}_{J-1}$, we need to construct a separate $n'_J \times n'_J$ nondecimated, forward transform matrix $\widetilde{\mathcal{H}}_{J-2}^{[1]\top}$, using the locations in $\boldsymbol{x}_{J,o} = \widetilde{\mathbf{J}}_{J-1}^{o\top} \boldsymbol{x}_J$. We thus arrive at a nondecimated scaling vector $\boldsymbol{s}_{J-2}$ of length n_J,

$$\begin{aligned} \boldsymbol{s}_{J-2} &= \widetilde{\mathbf{J}}_{J-1}(\widetilde{\mathbf{J}}_{J-1}^\top \boldsymbol{s}_{J-2}) + \widetilde{\mathbf{J}}_{J-1}^{o}(\widetilde{\mathbf{J}}_{J-1}^{o\top} \boldsymbol{s}_{J-2}) \\ &= (\widetilde{\mathbf{J}}_{J-1} \widetilde{\mathcal{H}}_{J-2}^\top \widetilde{\mathbf{J}}_{J-1}^\top + \widetilde{\mathbf{J}}_{J-1}^{o} \widetilde{\mathcal{H}}_{J-2}^{[1]\top} \widetilde{\mathbf{J}}_{J-1}^{o\top}) \boldsymbol{s}_{J-1}. \end{aligned}$$

The expression is understood as follows. The large $n_J \times n_J$ matrix that maps $\boldsymbol{s}_{J-1}$ onto $\boldsymbol{s}_{J-2}$ contains two smaller squared matrices, $\widetilde{\mathcal{H}}_{J-2}^\top$ and $\widetilde{\mathcal{H}}_{J-2}^{[1]\top}$. The matrices $\widetilde{\mathbf{J}}_{J-1}^\top$ and $\widetilde{\mathbf{J}}_{J-1}^{o\top}$ fill up the large matrix with intermediate zeros. On equidistant knots, the matrices $\widetilde{\mathcal{H}}_{J-2}^\top$ and $\widetilde{\mathcal{H}}_{J-2}^{[1]\top}$ have the same elements.

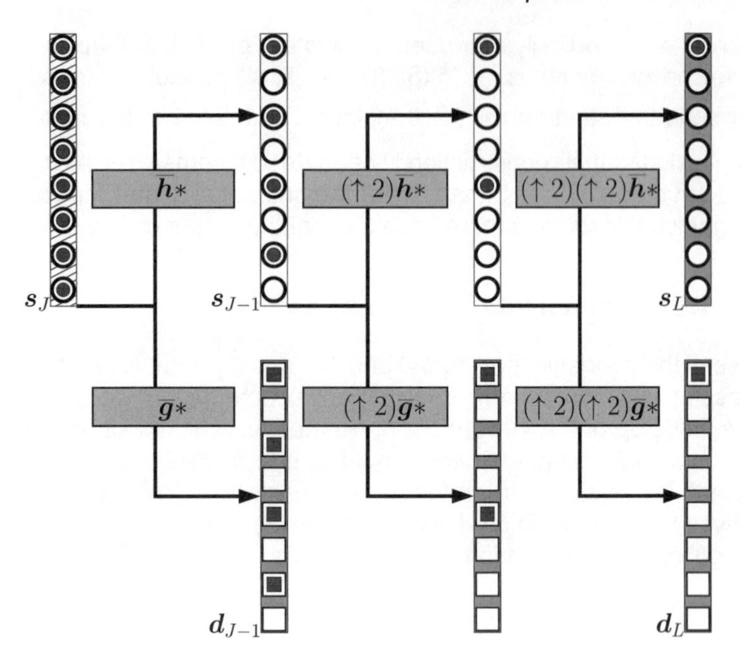

Figure 9.1
Diagram of the equispaced nondecimated wavelet transform. The fine scale input of length $n = 8$ is represented by the eight circles with black centres in the vector s_J. At the second scale, we have two vectors of length n, with scaling and detail coefficients. Half of the coefficients, marked with a black centre, come from the fast, decimated transform. The other half comes from a fast transform on a shifted input. In order to keep the decimated values among the coefficients at the next level, we need to make sure that the black centred values are not mixed with values from the white centred half. Therefore the filters $\overline{h}$ and $\overline{g}$ are upsampled, i.e., interlaced with zeros, leading to an *à trous* filtering. The upsampling is repeated in every subsequent level.

The large matrix then has a Toeplitz structured with the elements from the decimated filter $\overline{h}$, interlaced with zeros. This is the filter $(\uparrow 2)\overline{h}$ in Figure 9.1. The resulting transform is an *à trous* algorithm. Both $\widetilde{\mathcal{H}}_{J-2}$ and $\widetilde{\mathcal{H}}^{[1]}_{J-2}$ consist of two refinement matrices. For $\widetilde{\mathcal{H}}_{J-2}$, these are $\widetilde{\mathbf{H}}^{[1]}_{J-2}$ and the original $\widetilde{\mathbf{H}}_{J-2}$, which is also denoted as $\widetilde{\mathbf{H}}^{[0]}_{J-2}$. The two refinement matrices in $\widetilde{\mathcal{H}}^{[1]}_{J-2}$ are then $\widetilde{\mathbf{H}}^{[2]}_{J-2}$ and $\widetilde{\mathbf{H}}^{[3]}_{J-2}$.

At level j, the square, nondecimated matrix $\widetilde{\mathcal{H}}_j^{\top}$ has size $n_{j+1} \times n_{j+1}$. The matrix is constructed on $\boldsymbol{x}_{j+1}$, which is given by $\boldsymbol{x}_{j+1} = \widetilde{\mathbf{J}}^{[0]^{\top}}_{J-1,o(j+1)}\boldsymbol{x}_J$. This expression adopts the notation of the subsampling matrix $\widetilde{\mathbf{J}}^{[q]^{\top}}_{J-1,o}$ which selects all components $q + ko$ for fixed q and o and $k = 0, 1, \ldots, \lfloor (n_J - q)/o \rfloor$. In our case, the rate o is given by $o(j+1) = 2^{J-1-j}$. The matrix $\widetilde{\mathbf{J}}^{[0]^\top}_{J-1,o(j+1)}$ follows from repeated dyadic subsampling

$$\widetilde{\mathbf{J}}^{[0]^\top}_{J-1,o(j+1)} = \widetilde{\mathbf{J}}^\top_{j+1} \widetilde{\mathbf{J}}^\top_{j+2} \ldots \widetilde{\mathbf{J}}^\top_{J-1}.$$

The matrix $\widetilde{\boldsymbol{\mathcal{H}}}_j^\top$ is accompanied by $2^{J-1-j} - 1$ shifted versions $\widetilde{\boldsymbol{\mathcal{H}}}_j^{[q]^\top}$ constructed on adjacent subsamples $\widetilde{\mathbf{J}}^{[q]^\top}_{J-1,o(j+1)} \boldsymbol{x}_J$. The $n_J \times n_J$ matrices mapping $\boldsymbol{s}_{j+1}$ onto $\boldsymbol{s}_j$ and $\boldsymbol{d}_j$ are then given by

$$\boldsymbol{s}_j = \sum_{q=0}^{o(j+1)-1} \left(\widetilde{\mathbf{J}}^{[q]}_{J-1,o(j+1)} \widetilde{\boldsymbol{\mathcal{H}}}_j^{[q]^\top} \widetilde{\mathbf{J}}^{[q]^\top}_{J-1,o(j+1)} \right) \boldsymbol{s}_{j+1} \tag{9.3}$$

$$\boldsymbol{d}_j = \sum_{q=0}^{o(j+1)-1} \left(\widetilde{\mathbf{J}}^{[q]}_{J-1,o(j+1)} \widetilde{\boldsymbol{\mathcal{G}}}_j^{[q]^\top} \widetilde{\mathbf{J}}^{[q]^\top}_{J-1,o(j+1)} \right) \boldsymbol{s}_{j+1} \tag{9.4}$$

These expressions read as follows: for the calculation of $\boldsymbol{s}_j$, construct for all shifts $q = 0, 1, \ldots, o(j+1) - 1$ the forward transform matrices $\widetilde{\boldsymbol{\mathcal{H}}}_j^{[q]^\top}$, apply them to the corresponding subsampled shifted data $\widetilde{\mathbf{J}}^{[q]^\top}_{J-1,o(j+1)} \boldsymbol{s}_{j+1}$, and store the result into the corresponding subsamples of the output $\widetilde{\mathbf{J}}^{[q]^\top}_{J-1,o(j+1)} \boldsymbol{s}_j$.

On an equidistant grid, the overall transform matrix takes again a Toeplitz structure, with on each row the elements of $\overline{\boldsymbol{h}}$, separated by $o(j+1) - 1$ zeros.

On all grids, the nondecimated matrices $\widetilde{\boldsymbol{\mathcal{H}}}_j^{[q]}$ consist of two refinement matrices

$$\widetilde{\mathbf{H}}_j^{[2q]} = \widetilde{\boldsymbol{\mathcal{H}}}_j^{[q]} \widetilde{\mathbf{J}}_j \text{ and } \widetilde{\mathbf{H}}_j^{[2q+1]} = \widetilde{\boldsymbol{\mathcal{H}}}_j^{[q]} \widetilde{\mathbf{J}}_j^{[1]} \tag{9.5}$$

so that (9.1) is generalised to

$$\widetilde{\boldsymbol{\mathcal{H}}}_j^{[q]} = \widetilde{\mathbf{H}}_j^{[2q]} \widetilde{\mathbf{J}}_j^\top + \widetilde{\mathbf{H}}_j^{[2q+1]^\top} \widetilde{\mathbf{J}}_j^{[1]^\top} \tag{9.6}$$

for $j = L, L+1, \ldots, J-1$ and for $q = 0, 1, \ldots, 2^{J-j-1} - 1$. The same expressions hold, mutatis mutandis, for the the detail matrices $\widetilde{\boldsymbol{\mathcal{H}}}_j^{[q]}$ and for the reconstruction matrices $\boldsymbol{\mathcal{H}}_j^{[q]}$ and $\boldsymbol{\mathcal{G}}_j^{[q]}$.

9.1.3 Reconstruction from a redundant wavelet transform

The perfect reconstruction condition in (2.59) can be reformulated in terms of the nondecimated matrices,

$$\boldsymbol{\mathcal{H}}_j \widetilde{\mathbf{J}}_j \widetilde{\mathbf{J}}_j^\top \widetilde{\boldsymbol{\mathcal{H}}}_j^\top + \boldsymbol{\mathcal{G}}_j \widetilde{\mathbf{J}}_j^o \widetilde{\mathbf{J}}_j^{o^\top} \widetilde{\boldsymbol{\mathcal{G}}}_j^\top = \mathbf{I}_{j+1}.$$

This version of the perfect reconstruction condition includes explicit decimation operations. Hence, if $\boldsymbol{s}_j^{[0,1]} = \widetilde{\boldsymbol{\mathcal{H}}}_j^\top \boldsymbol{s}_{j+1}^{[0]}$ and $\boldsymbol{d}_j^{[0,1]} = \widetilde{\boldsymbol{\mathcal{G}}}_j^\top \boldsymbol{s}_{j+1}^{[0]}$ are the non-decimated coarse scaling and detail coefficients, then the fine scale $\boldsymbol{s}_{j+1}^{[0]}$ can be reconstructed from the decimated coarse coefficients $\widetilde{\mathbf{J}}_j^\top \boldsymbol{s}_j^{[0,1]}$ and $\widetilde{\mathbf{J}}_j^{o\top} \boldsymbol{d}_j^{[0,1]}$,

$$\boldsymbol{s}_{j+1}^{[0]} = \boldsymbol{\mathcal{H}}_j \widetilde{\mathbf{J}}_j \widetilde{\mathbf{J}}_j^\top \boldsymbol{s}_j^{[0,1]} + \boldsymbol{\mathcal{G}}_j \widetilde{\mathbf{J}}_j^o \widetilde{\mathbf{J}}_j^{o\top} \boldsymbol{d}_j^{[0,1]} = \mathbf{H}_j^\top \boldsymbol{s}_{j,e}^{[0,1]} + \mathbf{G}_j^\top \boldsymbol{d}_{j,o}^{[0,1]}.$$

The perfect reconstruction holds for the complementary decimation too,

$$\boldsymbol{\mathcal{H}}_j \widetilde{\mathbf{J}}_j^o \widetilde{\mathbf{J}}_j^{o\top} \widetilde{\boldsymbol{\mathcal{H}}}_j^\top + \boldsymbol{\mathcal{G}}_j \widetilde{\mathbf{J}}_j \widetilde{\mathbf{J}}_j^\top \widetilde{\boldsymbol{\mathcal{G}}}_j^\top = \mathbf{I}_{j+1},$$

so the same vector $\boldsymbol{s}_{j+1}^{[0]}$ can be reconstructed from the complementary coarse scale values,

$$\boldsymbol{s}_{j+1}^{[0]} = \boldsymbol{\mathcal{H}}_j \widetilde{\mathbf{J}}_j^o \widetilde{\mathbf{J}}_j^{o\top} \boldsymbol{s}_j^{[0,1]} + \boldsymbol{\mathcal{G}}_j \widetilde{\mathbf{J}}_j \widetilde{\mathbf{J}}_j^\top \boldsymbol{d}_j^{[0,1]} = \mathbf{H}_j^{[1]^\top} \boldsymbol{s}_{j,o}^{[0,1]} + \mathbf{G}_j^{[1]^\top} \boldsymbol{d}_{j,e}^{[0,1]}.$$

Identifying the fine scale vector $\boldsymbol{s}_{j+1}^{[0]}$ as the subsampled vector $\widetilde{\mathbf{J}}_{J-1,o(j+1)}^{[0]^\top} \boldsymbol{s}_{j+1}$, similar reconstructions can be set up for adjacent subsamples $\boldsymbol{s}_{j+1}^{[q]} = \widetilde{\mathbf{J}}_{J-1,o(j+1)}^{[q]^\top} \boldsymbol{s}_{j+1}$ and the transformed vectors $\boldsymbol{s}_j^{[2q,2q+1]} = \widetilde{\boldsymbol{\mathcal{H}}}_j^{[q]^\top} \boldsymbol{s}_{j+1}^{[q]}$ and $\boldsymbol{d}_j^{[2q,2q+1]} = \widetilde{\boldsymbol{\mathcal{G}}}_j^{[q]^\top} \boldsymbol{s}_{j+1}^{[q]}$. The transformed vectors are split for further processing, i.e.,

$$\boldsymbol{s}_j^{[2q]} = \boldsymbol{s}_{j,e}^{[2q,2q+1]} \text{ and } \boldsymbol{s}_j^{[2q+1]} = \boldsymbol{s}_{j,o}^{[2q,2q+1]},$$

but also for the construction of the two independent inverse transforms. The existence of two independent reconstructions for each of these subsamples is of course due to the redundancy in the nondecimated analysis[2]. When the coefficients $\boldsymbol{d}_j^{[2q,2q+1]}$ are processed for denoising, the two reconstructions most probably yield a different output. An averaged reconstruction

$$\begin{aligned}
\boldsymbol{s}_{j+1}^{[q]} &= \alpha_{j,q} \left(\mathbf{H}_j^{[2q]} \boldsymbol{s}_{j,e}^{[2q,2q+1]} + \mathbf{G}_j^{[2q]} \boldsymbol{d}_{j,o}^{[2q,2q+1]} \right) \\
&\quad + (1 - \alpha_{j,q}) \left(\mathbf{H}_j^{[2q+1]} \boldsymbol{s}_{j,o}^{[2q,2q+1]} + \mathbf{G}_j^{[2q+1]} \boldsymbol{d}_{j,e}^{[2q,2q+1]} \right) \\
&= \alpha_{j,q} \left(\mathbf{H}_j^{[2q]} \boldsymbol{s}_j^{[2q]} + \mathbf{G}_j^{[2q]} \boldsymbol{d}_j^{[2q+1]} \right) \\
&\quad + (1 - \alpha_{j,q}) \left(\mathbf{H}_j^{[2q+1]} \boldsymbol{s}_j^{[2q+1]} + \mathbf{G}_j^{[2q+1]} \boldsymbol{d}_j^{[2q]} \right),
\end{aligned} \tag{9.7}$$

then induces additional smoothing. In most routines, the weights are taken to be constants, $\alpha_{j,q} = 1/2$, thereby simplifying the reconstruction to

$$\boldsymbol{s}_{j+1}^{[q]} = \frac{1}{2} \left(\boldsymbol{\mathcal{H}}_j^{[q]} \boldsymbol{s}_j^{[2q,2q+1]} + \boldsymbol{\mathcal{G}}_j^{[q]} \boldsymbol{d}_j^{[2q,2q+1]} \right). \tag{9.8}$$

[2]Although $\widetilde{\boldsymbol{\mathcal{H}}}_j$ and $\widetilde{\boldsymbol{\mathcal{G}}}_j$ are square matrices, they are (close to) singular. Therefore, using their inverses for reconstruction of the fine scaling coefficients is not really an option.

With all subsamples at once, this becomes

$$\begin{aligned} \boldsymbol{s}_{j+1} &= \frac{1}{2}\left[\sum_{q=0}^{o(j+1)-1}\left(\widetilde{\mathbf{J}}^{[q]}_{J-1,o(j+1)}\boldsymbol{\mathcal{H}}^{[q]}_{j}\widetilde{\mathbf{J}}^{[q]\top}_{J-1,o(j+1)}\right)\boldsymbol{s}_j\right. \\ &\left.+\sum_{q=0}^{o(j+1)-1}\left(\widetilde{\mathbf{J}}^{[q]}_{J-1,o(j+1)}\boldsymbol{\mathcal{G}}^{[q]}_{j}\widetilde{\mathbf{J}}^{[q]\top}_{J-1,o(j+1)}\right)\boldsymbol{d}_j\right] \end{aligned} \tag{9.9}$$

The variance reducing effect of the averaged reconstruction is further explored in Section 9.1.7.

Exercise 9.1.1 *Prove (9.8) from (9.7), using the identity (2.54).*

Exercise 9.1.2 *Let* $[\mathbf{H}_j^{[2q]}\ \mathbf{G}_j^{[2q]}]$ *and* $[\mathbf{H}_j^{[2q+1]}\ \mathbf{G}_j^{[2q+1]}]$ *be two orthogonal transforms; then show that (9.8) is the least squares solution to the problem*

$$\left\{\begin{array}{rcl} \widetilde{\mathbf{H}}_j^{[2q]\top}\boldsymbol{s}_{j+1} &=& \boldsymbol{s}_j^{[2q]} \\ \widetilde{\mathbf{G}}_j^{[2q]\top}\boldsymbol{s}_{j+1} &=& \boldsymbol{d}_j^{[2q]} \\ \widetilde{\mathbf{H}}_j^{[2q+1]\top}\boldsymbol{s}_{j+1} &=& \boldsymbol{s}_j^{[2q+1]} \\ \widetilde{\mathbf{G}}_j^{[2q+1]\top}\boldsymbol{s}_{j+1} &=& \boldsymbol{d}_j^{[2q+1]} \end{array}\right.$$

9.1.4 Redundant representation

The reconstruction in (9.9) writes $\boldsymbol{s}_{j+1}$ as a linear transform of $\boldsymbol{s}_j$ and $\boldsymbol{d}_j$, involving two $n_J \times n_J$ *à trous* matrices. Refinement from $\boldsymbol{s}_L$ up to $\boldsymbol{s}_J$ is an inverse transform where all detail vectors are zero. On equidistant knots, taking $\alpha_{j,q} = 1/2$, all refinement matrices have an $n_J \times n_J$ Toeplitz structure. As a result, the refinement from the canonical vector $\boldsymbol{e}_{J,k}$ is nothing but a shift of the refinement from $\boldsymbol{e}_{J,0}$. Subdivision, i.e., infinite refinement, then leads to scaling functions $\varphi_{j,k}(x)$ that are given by

$$\varphi_{j,k}(x) = \varphi_{j,0}(x - 2^{-J}k) = 2^{j/2}\varphi(2^j x - 2^{j-J}k).$$

The scaling functions are more numerous than in the critically downsampled case, where $\varphi_{j,k}(x) = \varphi(2^j x - k)$. In the nondecimated case, each level has n_J scaling functions and n_J wavelet functions

$$\psi_{j,k}(x) = 2^{j/2}\psi(2^j x - 2^{j-J}k).$$

Similar definitions hold for dual scaling and wavelet functions.

As can be seen in Figure 9.1, in a nondecimated wavelet transform, each level has n_J detail functions $\psi_{j,k}(x)$, along with n_J coefficients. In a decimated version, that would be $n_j = \lceil n_J/2^{J-j}\rceil$, while the number of levels is bounded by $\lceil\log_2(n_J)\rceil$. Keeping the same number of levels in a nondecimated wavelet transform, we see that the number of coefficients is of the order $\mathcal{O}\left(n\log(n)\right)$. This is also the computational complexity of a nondecimated wavelet transform. In most applications, the logarithmic redundancy factor in computations and coefficient vector size poses no problem at all.

9.1.5 Overcomplete families of dilations and translations

Just as in the case of the decimated wavelet transform, the nondecimated version maps the coefficient vector $\boldsymbol{s}_J$ in the decomposition

$$f_J(x) = \Phi_J(x)\boldsymbol{s}_J$$

onto the decomposition

$$f_J(x) = \Phi_L(x)\boldsymbol{s}_L + \sum_{j=L}^{J-1} \Psi_j(x)\boldsymbol{d}_j.$$

In contrast to the fine scale decomposition and unlike the decimated decomposition, the union of low resolution scaling functions $\Phi_L(x)$ and detail functions $\Psi_j(x)$ is overcomplete: these functions do not constitute a basis[3]. Nevertheless, there are still dual functions so that

$$\boldsymbol{s}_L = \int_{-\infty}^{\infty} f_J(x)\widetilde{\Phi}_L^\top(x)dx \text{ and } \boldsymbol{d}_j = \int_{-\infty}^{\infty} f_J(x)\widetilde{\Psi}_j^\top(x)dx.$$

As $\{\Phi_L, \Psi_L, \Psi_{L+1}, \ldots, \Psi_{J-1}\}$ is overcomplete, this is not the only possible analysis[4]. For the mathematical description of the analysis and synthesis in overcomplete sets of functions, the notion of frames is introduced in the following section.

9.1.6 Frames

Although the analysis in an overcomplete family of functions is not unique, the unconditional characterisation of a vector in a Banach space by its expansion can be extended. In a Hilbert space $\mathcal{H}$ the notion of unconditional bases was further refined in the Definition 5.4.5 of Riesz bases, using the ℓ_2 sequence norm of the coefficients. According to Theorem 5.4.7, the characterisation by the coefficients can be replaced by a characterisation by inner products with the basis vectors. This idea is now extended introducing the notion of a frame.

Definition 9.1.3 *(frame) A sequence*[5] $\boldsymbol{\Phi} = (\varphi_0, \varphi_1, \ldots)$ *of vectors* $\varphi_k \in \mathcal{H}$

[3]Even more, the scaling functions at any coarse scale $j < L$ alone do not constitute a basis. More precisely, the nondecimated $\Phi_j(x)$ is the union of 2^{J-j} shifts of the scaling basis in the corresponding decimated transform. As all shifts reproduce the constant function, it is straightforward to find a nontrivial representation of the zero function in the union of the bases.

[4]In Section 9.1.3, the overcompleteness was characterised by multiple reconstructions from a given decomposition. Here, we consider multiple decompositions in a given set of elementary functions.

[5]A frame is a *sequence* of vectors, whereas a basis is a *set* (Heil, 2011). This is because a frame may contain duplicates of the same vector. Although, strictly speaking, it is incorrect to claim that a Riesz basis is a frame, this sort of statement should be understood by thinking of a sequence associated to a set.

in a separable real Hilbert space is a frame if there exist strictly positive and finite constants such that for all $\boldsymbol{f} \in \mathcal{H}$,

$$\gamma \sum_{k=1}^{\infty} \langle \boldsymbol{f}, \varphi_k \rangle^2 \leq \|\boldsymbol{f}\|_{\mathcal{H}}^2 \leq \Gamma \sum_{k=1}^{\infty} \langle \boldsymbol{f}, \varphi_k \rangle^2. \tag{9.10}$$

The tightest values of γ and Γ for which (9.10) holds are termed the ***frame constants****.*

Expression (9.10) is exactly the same as the one in Theorem 5.4.7 for Riesz bases. As a consequence, the elements of a Riesz basis also constitute a frame, but the converse is not necessarily true, since Definition 9.1.3 does not impose the uniqueness of an expansion in a basis. For instance, a frame may contain the zero vector. The presence of this vector has no effect on the frame constants. Another example of a frame is a sequence $\boldsymbol{\Phi}$ constructed with all the elements from $\boldsymbol{\Phi}_1 \cup \boldsymbol{\Phi}_2$, where $\boldsymbol{\Phi}_1$ and $\boldsymbol{\Phi}_2$ are two Riesz bases with Riesz constants γ_1, Γ_1 and γ_2, Γ_2. The frame constants of $\boldsymbol{\Phi}$ are then bounded by $\gamma_1 + \gamma_2 \leq \gamma \leq \Gamma \leq \Gamma_1 + \Gamma_2$. A basis whose vectors are not quasi-normalised as in (5.43) is not a frame, even if the basis is orthogonal. For instance, the basis e_n/n with e_n the canonical basis of $\ell_2(\mathbb{R})$ is not a frame. In a finite dimensional vector space, each spanning set of vectors is a frame.

Definition 9.1.4 *(tight frame) A tight frame is a frame with equal frame constants, i.e., $\gamma = \Gamma$ in (9.10), so, in a real Hilbert space $\mathcal{H}$,*

$$c\|\boldsymbol{f}\|_{\mathcal{H}}^2 = \sum_{k=1}^{\infty} \langle \boldsymbol{f}, \varphi_k \rangle^2. \tag{9.11}$$

Example 9.1.5 *In $\mathbb{R}^2$, a set of $n > 2$ equiangular normalised vectors φ_i, with $i = 0, 1, \ldots, n-1$, is illustrated in Figure 9.2. It can be shown that for any $\boldsymbol{f} \in \mathbb{R}^2$,*

$$\sum_{k=0}^{n-1} \langle \boldsymbol{f}, \varphi_k \rangle^2 = \frac{n}{2} \|\boldsymbol{f}\|_{\mathcal{H}}^2.$$

The notion of tight frames generalises orthogonal bases, for which $\gamma = 1$. Not all tight frames with frame constants 1 are orthogonal bases (Heil, 2011, Example 8.1, Remark 8.5). For instance, let $\varphi_{1,i} \in \boldsymbol{\Phi}_1$ and $\varphi_{2,i} \in \boldsymbol{\Phi}_2$ be vectors from orthogonal bases. Then the elements $\varphi_{1,i}/\sqrt{2}$ and $\varphi_{2,i}/\sqrt{2}$ constitute a tight frame, but obviously not an orthogonal basis. We do have the following result though,

Theorem 9.1.6 *Let $\boldsymbol{\Phi}$ be a tight frame with frame constants equal to one, consisting of normalised vectors. Then these vectors constitute an orthonormal basis.*

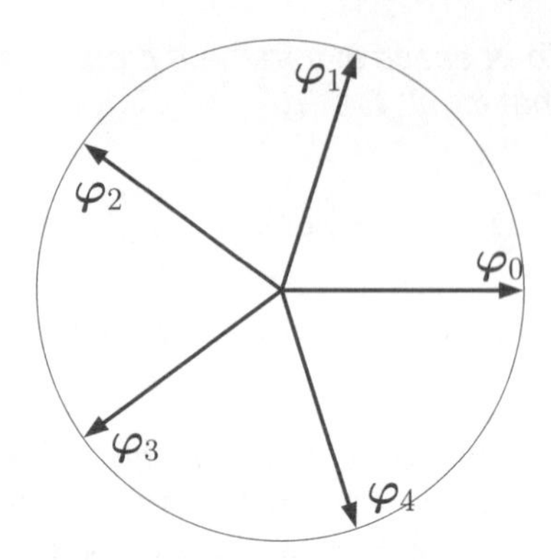

Figure 9.2
An example of a tight frame in $\mathbb{R}^2$. The frame constants are $5/2$.

Proof. Being a frame, the vectors span the Hilbert space. Linear independence follows from orthogonality. The vectors are orthogonal to each other because (9.11) with $c = 1$ and $\boldsymbol{f} = \varphi_i$ reads

$$\sum_{k=1}^{\infty} \langle \varphi_i, \varphi_k \rangle^2 = \|\varphi_i\|_{\mathcal{H}}^2 = 1 \Rightarrow \sum_{k=1, k\neq i}^{\infty} \langle \varphi_i, \varphi_k \rangle^2 = 0 \Rightarrow \langle \varphi_i, \varphi_k \rangle = 0 \text{ for } k \neq i.$$

□

9.1.7 Variance propagation in a nondecimated analysis

As in Section 3.1.5, the control of variance inflation throughout a nondecimated multiscale analysis and reconstruction is assessed by considering the linear operation that reconstructs $\widehat{\boldsymbol{s}}_J$ from the coarse scaling coefficients $\boldsymbol{s}_L$, replacing all intermediate details by zero.

Lemma 9.1.7 *In a nondecimated wavelet transform, the scaling coefficients at low level* L *are obtained from the scaling coefficients at level* J *by* $\boldsymbol{s}_L = \widetilde{\boldsymbol{\mathcal{V}}}_L \boldsymbol{s}_J$, *where* $\widetilde{\boldsymbol{\mathcal{V}}}_L$ *is an aggregate of translated forward transforms,*

$$\widetilde{\boldsymbol{\mathcal{V}}}_L = \sum_{q=0}^{o(L)-1} \widetilde{\mathbf{J}}_{J-1,o(L)}^{[q]} \widetilde{\mathbf{V}}_L^{[q]}, \tag{9.12}$$

with $o(L) = 2^{J-L}$ *as before. The shifted forward projections are defined recursively* $\widetilde{\mathbf{V}}_L^{[q]} = \widetilde{\mathbf{H}}_L^{[q]^\top} \widetilde{\mathbf{V}}_{L+1}^{[q_1]}$, *with* $q_1 = \mod(q, 2^{J-L-1})$, *using* $\mod(a,b) = a - b\lfloor a/b \rfloor$ *for the remainder after Euclidean division.*

Proof. Expression (9.12) is proven by recursion for $j = J-1, J-1, \ldots$, starting off from

$$\widetilde{\boldsymbol{\mathcal{V}}}_{J-1} \boldsymbol{s}_J = \left(\widetilde{\mathbf{J}}_{J-1,2}^{[0]} \widetilde{\mathbf{H}}_{J-1}^{[0]^\top} + \widetilde{\mathbf{J}}_{J-1,2}^{[1]} \widetilde{\mathbf{H}}_{J-1}^{[1]^\top} \right) \boldsymbol{s}_J = \widetilde{\boldsymbol{\mathcal{H}}}_{J-1}^\top \boldsymbol{s}_J = \boldsymbol{s}_{J-1},$$

where we plugged in (9.1) with $j = J-1$, $\widetilde{\mathbf{H}}_{J-1}^{[0]^\top} = \widetilde{\mathbf{H}}_{J-1}$, $\widetilde{\mathbf{J}}_{J-1,2}^{[1]} = \widetilde{\mathbf{J}}_{J-1}^{o}$, and $\widetilde{\mathbf{J}}_{J-1,2}^{[0]} = \widetilde{\mathbf{J}}_{J-1}$.

For the proof of the induction step, we first insert the definition (9.6) $\widetilde{\mathcal{H}}_j^{[q]^\top} = \widetilde{\mathbf{J}}_j \widetilde{\mathbf{H}}_j^{[2q]^\top} + \widetilde{\mathbf{J}}_j^{[1]} \widetilde{\mathbf{H}}_j^{[2q+1]^\top}$ into (9.3), leading to

$$\boldsymbol{s}_j = \sum_{q=0}^{o(j)-1} \left(\widetilde{\mathbf{J}}_{J-1,o(j)}^{[q]} \widetilde{\mathbf{H}}_j^{[q]^\top} \widetilde{\mathbf{J}}_{J-1,o(j+1)}^{[q_1]^\top} \right) \boldsymbol{s}_{j+1}$$

with $q_1 = \mod(q, o(j+1))$. Then, by the induction hypothesis, we have $\boldsymbol{s}_{j+1} = \widetilde{\mathcal{V}}_{j+1} \boldsymbol{s}_J$, with $\widetilde{\mathcal{V}}_{j+1}$ given by (9.12) taking $L = j+1$. Substitution of (9.12) into the expression above and using the fact that the shifted subsamplings satisfy

$$\widetilde{\mathbf{J}}_{J-1,o(j)}^{[q]^\top} \widetilde{\mathbf{J}}_{J-1,o(j)}^{[q']} = \begin{cases} \mathbf{I}_{o(j)} & \text{if } q = q', \\ \mathbf{0} & \text{otherwise,} \end{cases} \tag{9.13}$$

completes the induction step, i.e., it leads to (9.12) for $L = j$. □

In a similar way, the reconstruction $\widehat{\boldsymbol{s}}_J$ at fine scale from the coarse scale approximation $\boldsymbol{s}_L$ with all details set to zero is represented by the matrix $\mathcal{V}_L$, which is an average of $o(L) = 2^{J-L}$ linearly independent reconstructions,

$$\mathcal{V}_L = \frac{1}{o(L)} \sum_{q=0}^{o(L)-1} \mathbf{V}_L^{[q]} \widetilde{\mathbf{J}}_{J-1,o(L)}^{[q]^\top},$$

where each shifted reconstruction is defined by the multiresolution recursion $\mathbf{V}_L^{[q]} = \mathbf{V}_{L+1}^{[q_1]} \mathbf{H}_L^{[q]}$, again with $q_1 = \mod(q, 2^{J-L-1})$.

Again using (9.13), we then find the matrix $\mathcal{P}_L$ that maps a fine scaling coefficient $\boldsymbol{s}_J$ onto a reconstruction $\widehat{\boldsymbol{s}}_J$ from a wavelet analysis after thresholding all the details,

$$\widetilde{\mathcal{P}}_L = \frac{1}{o(L)} \sum_{q=0}^{o(L)-1} \widetilde{\mathbf{P}}_L^{[q]}, \tag{9.14}$$

with $\widetilde{\mathbf{P}}_L^{[q]} = \mathbf{V}_L^{[q]} \widetilde{\mathbf{V}}_L^{[q]}$. The matrix $\widetilde{\mathcal{P}}_L$, being an average of projection matrices, is not a projection matrix itself. It is not idempotent. For instance, taking $L = J-1$, we have

$$\begin{aligned} \widetilde{\mathcal{P}}_{J-1} \widetilde{\mathcal{P}}_{J-1} &= \frac{1}{2} \left(\widetilde{\mathbf{P}}_{J-1}^{[0]} + \widetilde{\mathbf{P}}_{J-1}^{[1]} \right) \frac{1}{2} \left(\widetilde{\mathbf{P}}_{J-1}^{[0]} + \widetilde{\mathbf{P}}_{J-1}^{[1]} \right) \\ &= \frac{1}{4} \left(\widetilde{\mathbf{P}}_{J-1}^{[0]} + \widetilde{\mathbf{P}}_{J-1}^{[1]} + \widetilde{\mathbf{P}}_{J-1}^{[0]} \widetilde{\mathbf{P}}_{J-1}^{[1]} + \widetilde{\mathbf{P}}_{J-1}^{[1]} \widetilde{\mathbf{P}}_{J-1}^{[0]} \right). \end{aligned}$$

The singular values of $\widetilde{\mathcal{P}}_L$ are not all zero or larger than one. This is illustrated in Figure 9.3, which depicts the singular values of $\widetilde{\mathbf{P}}_{J-1}$ in a cubic spline wavelet transform with two dual vanishing moments on a grid of $n_J = n = 64$

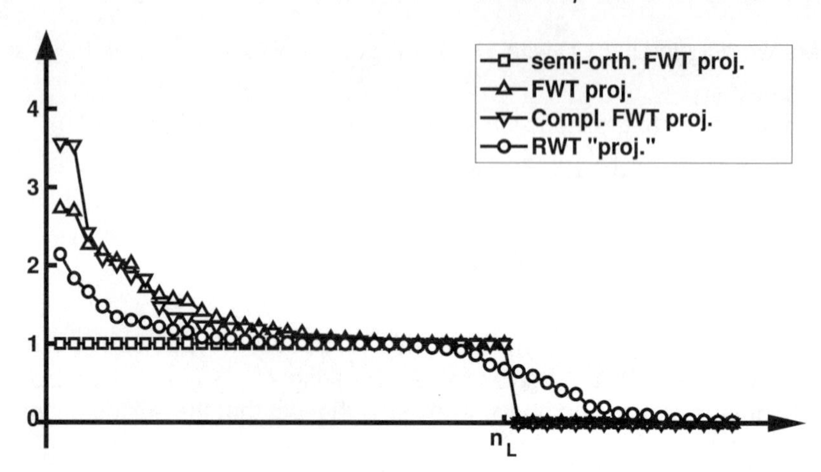

Figure 9.3
Singular values in descending order for projections $\widetilde{\mathbf{P}}_{J-1} = \mathbf{V}_{J-1}\widetilde{\mathbf{V}}_{J-1}$ in decimated spline wavelet transforms, along with the singular values of the nondecimated version $\widetilde{\boldsymbol{\mathcal{P}}}_{J-1} = \boldsymbol{\mathcal{V}}_{J-1}\widetilde{\boldsymbol{\mathcal{V}}}_{J-1}$. The decimated wavelet transforms include the cubic spline transform with two vanishing moments and its complementary shift in the nondecimated transform. The decimated transforms also include the semi-orthogonal version of the cubic spline wavelet transform, depicted here as a benchmark. It can be seen that the nondecimated transform, which is merely an interlaced combination of two decimated transforms, performs better than the individual decimated transforms, with lower singular values and hence less variance inflation.

uniformly distributed knots. The plot confirms the intuition that the averaging reconstruction 9.9 has an important effect on variance control. Indeed, the nondecimated transform has lower singular values than the best of the two shifted decimated transforms of which it combines the coefficients.

9.2 The continuous wavelet transform

On an equidistant grid, the expression for a nondecimated wavelet coefficient using the dual basis functions reads

$$d_{j,k} = \int_{-\infty}^{\infty} f(x) 2^{j/2} \widetilde{\psi}\left(2^j x - 2^{j-J} k\right) dx.$$

With $a = 2^{-j}$ and $b = k2^{-J}$, this can be seen as a discretisation of

$$d(a,b) = \frac{1}{\sqrt{a}} \int_{-\infty}^{\infty} f(x) \widetilde{\psi}\left(\frac{x-b}{a}\right) dx, \tag{9.15}$$

which is the **continuous wavelet transform** (CWT). Historically, the continuous wavelet transform came first, as a way to analyse the time-frequency contents of geophysical signals (Goupillaud et al., 1984; Grossmann and Morlet, 1984).

9.2.1 The continuous wavelet transform and its inverse

Whereas the discussion above concluded that the fast and overcomplete discrete wavelet transforms can be considered as discretisations of a continuous wavelet transform, the continuous version extends far beyond the framework of multiresolution analysis, as defined in Section 5.5.1 or the two-scale equations defined by the lifting scheme.

As a matter of fact, the continuous wavelet transform does not rely on refinability. There are no scaling functions and no dual wavelet functions involved. As the wavelet functions $\psi(x)$ can be complex, the classical definition of a continuous wavelet transform $\mathcal{W}\{f\} = d_f(a,b)$ reads

$$d_f(a,b) = \langle f, \psi_{a,b} \rangle = \int_{-\infty}^{\infty} f(x) \overline{\psi_{a,b}(x)} dx = \frac{1}{\sqrt{a}} \int_{-\infty}^{\infty} f(x) \overline{\psi\left(\frac{x-b}{a}\right)} dx, \tag{9.16}$$

where

$$\psi_{a,b}(x) = \frac{1}{\sqrt{a}} \psi\left(\frac{x-b}{a}\right),$$

with scale $a \in \mathbb{R}^+$, and translation $b \in \mathbb{R}$. The function $f(x)$ is supposed to be squared integrable, i.e., $f(x) \in L_2(\mathbb{R})$. As the continuous wavelet transform operates on functions of continuous variables, its properties can be easily analysed in the Fourier domain. Let

$$\Psi(\omega) = \mathcal{F}\{\psi(x)\} = \frac{1}{2\pi} \int_{-\infty}^{\infty} \psi(x) \exp(-i\omega x) dx,$$

be the Fourier transform of $\psi(x)$; then it is straightforward to check that

$$\mathcal{F}\left\{\psi\left(\frac{x-b}{a}\right)\right\} = a \exp(-i\omega b)\, \Psi(a\omega).$$

Application of Parseval's Theorem 6.2.1 then leads to the calculation of the CWT from the Fourier transformed functions,

$$d_f(a,b) = 2\pi\sqrt{a} \int_{-\infty}^{\infty} F(\omega) \exp(i\omega b) \overline{\Psi(a\omega)} d\omega. \tag{9.17}$$

From the Fourier analysis, we can prove the following reconstruction formula:

Theorem 9.2.1 *(inverse CWT) Let $d_f(a,b)$, with $a \in \mathbb{R}^+$ and $b \in \mathbb{R}$, be the CWT of an $L_2(\mathbb{R})$ function $f(x)$. Then the function can be recovered from*

$$f(x) = \frac{1}{C_\psi} \int_0^\infty \int_{-\infty}^\infty d_f(a,b) \frac{1}{\sqrt{a}} \psi\left(\frac{x-b}{a}\right) db \frac{da}{a^2}, \tag{9.18}$$

*provided that $\psi(x)$ satisfies the **admissibility condition**, meaning that*

$$C_\psi = (2\pi)^2 \int_0^\infty \frac{|\Psi(\omega)|^2}{\omega} d\omega \tag{9.19}$$

is finite.

Proof. The following lines provide a proof in a weak L_2-sense. Pointwise proofs can be found, for instance in Koornwinder (1993).

Consider two functions $f(x), g(x) \in L_2(\mathbb{R})$. By introducing the function

$$\widetilde{f}_a(b) = \int_{-\infty}^\infty F(\omega) \exp(i\omega b) \overline{\Psi(a\omega)} d\omega, \tag{9.20}$$

Expression (9.17) takes the form $d_f(a,b) = 2\pi\sqrt{a}\widetilde{f}_a(b)$, while (9.20) itself takes the form of an inverse Fourier transform (1.11), meaning that

$$\mathcal{F}(\widetilde{f}_a) = F(\omega) \overline{\Psi(a\omega)}.$$

With a similar definition for $\widetilde{g}_a(b)$, and by application of Parseval's Theorem 6.2.1, we find

$$\begin{aligned}
&\int_0^\infty \int_{-\infty}^\infty d_f(a,b) \overline{d_g(a,b)} db \frac{da}{a^2} \\
&= (2\pi)^2 \int_0^\infty \int_{-\infty}^\infty \widetilde{f}_a(b) \overline{\widetilde{g}_a(b)} db \frac{da}{a} \\
&= (2\pi)^3 \int_0^\infty \int_{-\infty}^\infty F(\omega) \overline{\Psi(a\omega)} \overline{G(\omega)} \Psi(a\omega) d\omega \frac{da}{a} \\
&= (2\pi)^3 \int_{-\infty}^\infty F(\omega) \overline{G(\omega)} \int_0^\infty |\Psi(a\omega)|^2 \frac{da}{a} d\omega.
\end{aligned}$$

Substitution $a\omega = \xi$ yields $\int_0^\infty |\Psi(a\omega)|^2 \frac{da}{a} = \int_0^\infty |\Psi(\xi)|^2 \frac{d\xi}{\xi} = \frac{C_\psi}{(2\pi)^2}$.
One more application of Parseval's Theorem then leads to

$$\begin{aligned}
C_\psi \int_{-\infty}^\infty f(x)\overline{g(x)} dx &= \int_0^\infty \int_{-\infty}^\infty d_f(a,b) \overline{d_g(a,b)} db \frac{da}{a^2} \\
&= \int_0^\infty \int_{-\infty}^\infty d_f(a,b) \int_{-\infty}^\infty \overline{g(x)} \psi\left(\frac{x-b}{a}\right) dx db \frac{da}{a^2}.
\end{aligned}$$

As this holds for any function $g(x) \in L_2(\mathbb{R})$, this leads, in a weak sense, to (9.18). □

9.2.2 Admissible wavelets

The reconstruction from a continuous wavelet transform in Theorem 9.2.1 does not involve any scaling functions, dual basis, or multiresolution. Any function $\psi(x)$ for which the admissibility condition holds can be used as a wavelet. Admissible wavelets have $C_\psi < \infty$ with C_ψ as in (9.19). For a continuous $\Psi(\omega)$ the integral in (9.19) can only be finite if

$$\Psi(0) = \int_{-\infty}^{\infty} \psi(x)dx = 0. \tag{9.21}$$

As all absolute integrable functions have uniformly continuous Fourier transforms (see Theorem 6.2.12), this means that an absolute integrable function cannot possibly be an admissible wavelet unless its integral vanishes. Although the zero integral in (9.21) is a necessary condition for admissibility, it is not so far from being sufficient (Daubechies, 1992; Mallat, 2001).

As a conclusion, more or less all absolute integrable functions with a zero integral may be used as a wavelet. This corresponds precisely to the notion of a wavelet as a function that fluctuates above and below zero in a damped way or on a bounded interval. The set of admissible wavelets includes all wavelets encountered in the discrete wavelet transform, but also others that do not fit into the multiresolution framework. An important class of wavelets is derived from the Gaussian bell curve (the normal density function, that is). The popularity of the Gaussian bell curve in location-frequency representations is due to the fact that it reaches the minimum set by Heisenberg's uncertainty principle.

The earliest example is the **Morlet wavelet** $\psi(x) = e^{-x^2/2}(e^{i\omega_0 x} - c)$, with c chosen so that the function has zero integral. This function is an enveloped complex oscillation, leading to a transform which is closely connected to windowed Fourier analysis.

Examples of real wavelets derived from the Gaussian bell curve are the first derivative $\psi(x) = xe^{-x^2/2}$ and the second derivative $\psi(x) = (1 - x^2)e^{-x^2/2}$. The latter is well known as the **Mexican hat wavelet**.

9.2.3 Examples of continuous wavelet analyses

The continuous wavelet transform is highly redundant, as it maps a one dimensional function onto a two dimensional function. The values of $d_f(a, b)$ can be represented as a grey level image, as in Figures 9.4, 9.5, and 9.6, where b is on the horizontal axis, while the vertical axis has $\log_2(a)$. The grey values $g(a, b)$ range from 0 (black) to 255 (white). The value in the middle (127) represents a zero, i.e., $g(a, b) = 127.5 \Leftrightarrow d_f(a, b) = 0$, while the other values are computed as

$$g(d) = 127.5 \left[1 - \mathrm{sign}(d) \left(\frac{|d|}{|d|_{\max}}\right)^{\kappa}\right],$$

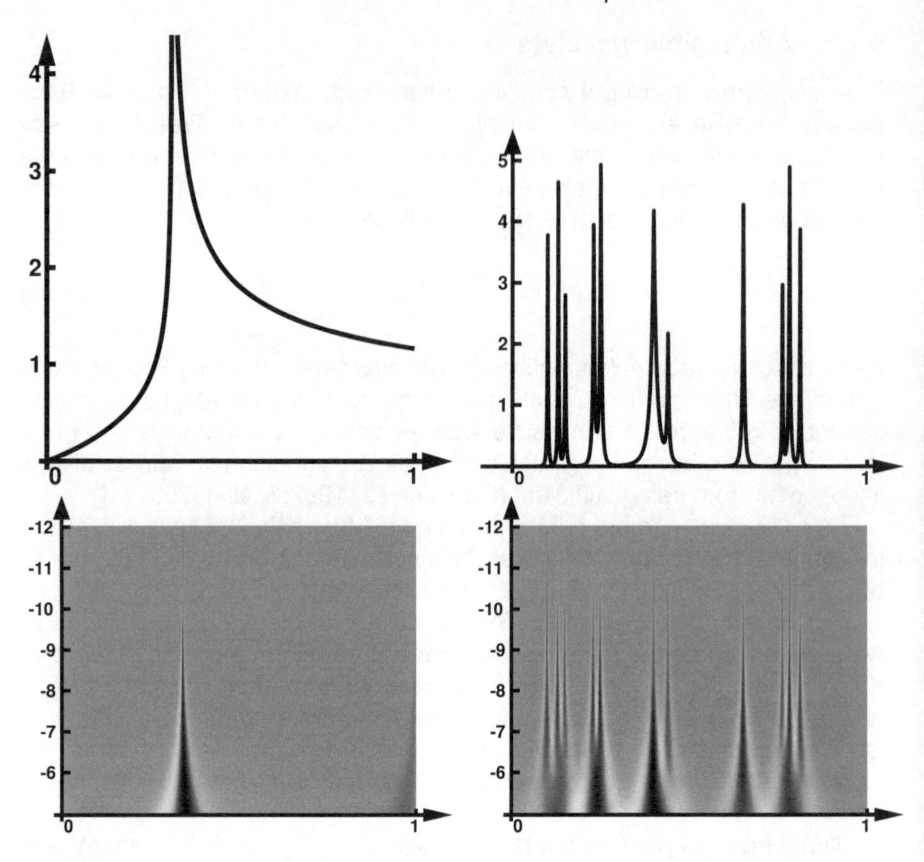

Figure 9.4
Top figures: test signals with bumps. Bottom figures: corresponding continuous wavelet transforms, using the Mexican hat wavelet. On the horizontal axes, the location parameter b. On the vertical axes, the logarithmic scale parameter $\log_2(a)$. See text for more details.

where κ is a contrast parameter, for which the figures adopt the value of 0.7.

The test signal in the left panel of Figure 9.4 is defined by $f(x) = x(x_0 - x)^\alpha$ for $0 < x < x_0$ and $f(x) = (x - x_0)^\beta$ for $x_0 < x < 1$, where we set $x_0 = 0.345$, $\alpha = -0.4$ and $\beta = -0.35$. The test signal in the right panel of that figure is the well known "bumps", defined by

$$f(x) = \sum_{\ell=1}^{B} f_\ell/(1 + |x - b_\ell|/a_\ell)^4,$$

with $\boldsymbol{b} = \begin{bmatrix} 10 & 13 & 15 & 23 & 25 & 40 & 44 & 65 & 76 & 78 & 81 \end{bmatrix}/100,$

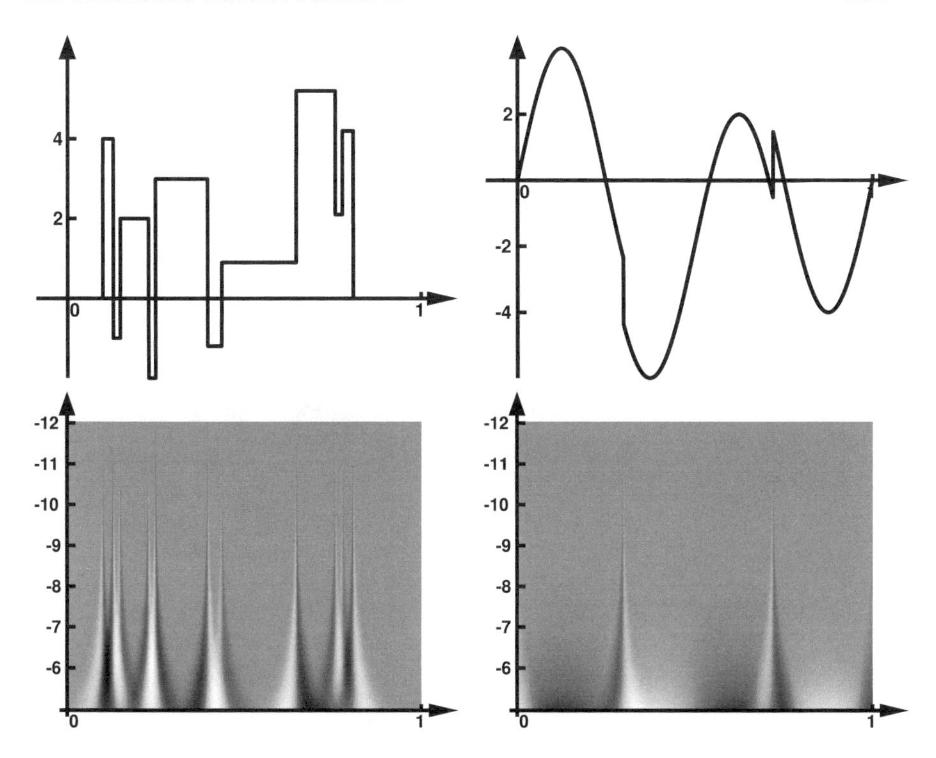

Figure 9.5
Top figures: test signals with jumps. Bottom figures: corresponding continuous wavelet transforms, using the Mexican hat wavelet. On the horizontal axes, the location parameter b. On the vertical axes, the logarithmic scale parameter $\log_2(a)$. See text for more details.

$\boldsymbol{f} = \begin{bmatrix} 4 & 5 & 3 & 4 & 5 & 4.2 & 2.1 & 4.3 & 3.1 & 5.1 & 4.2 \end{bmatrix}$,
$\boldsymbol{a} = \begin{bmatrix} 5 & 5 & 6 & 10 & 10 & 30 & 10 & 10 & 5 & 8 & 5 \end{bmatrix} /1000.$

Two other well known test signals appear in Figure 9.5. These two functions have jump singularities. The "blocks" signal on the left is defined by

$$f(x) = \sum_{\ell=0}^{B} f_\ell \varphi_\ell(x),$$

where $\varphi_\ell(x)$ are Haar scaling functions on the intervals $[b_\ell, b_{\ell+1}]$ with b_ℓ as in the bumps example (and $b_0 = 0$) and
$\boldsymbol{f} = \begin{bmatrix} 4 & -1 & 2 & -2 & 3 & -1.2 & 0.9 & 5.2 & 2.1 & 4.2 \end{bmatrix}$.
The "heavisine" on the right is given by $f(x) = 4\sin(4\pi x) - 2I(x > 0.3) + 2I(x > 0.72)$.

The oscillating function in Figure 9.6 is given by $f(x) = \cos(1/(x - x_0))$,

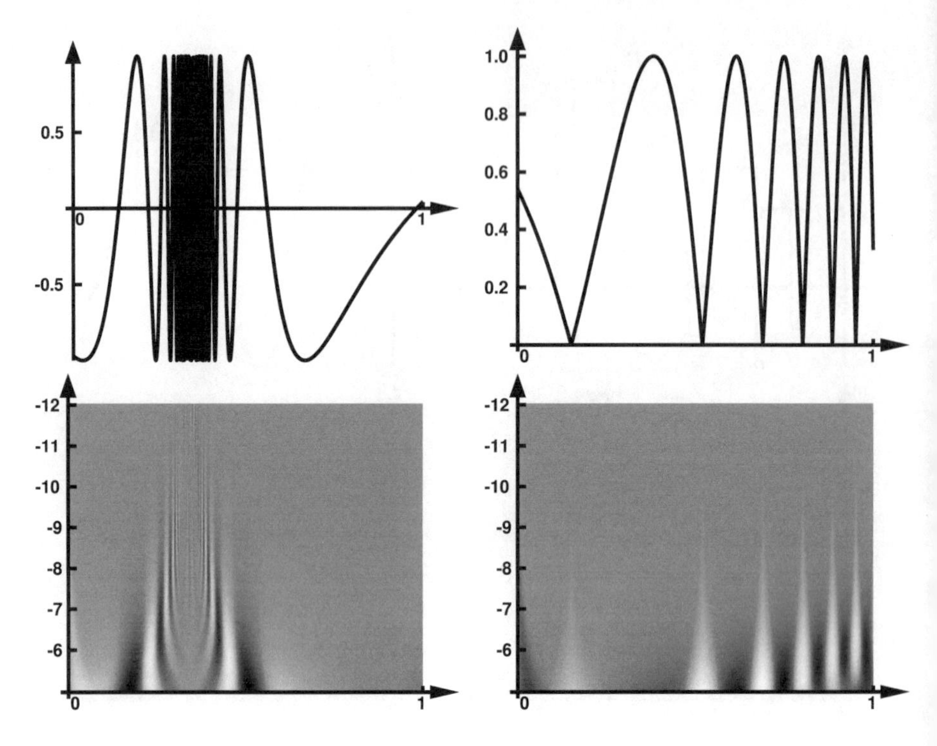

Figure 9.6
Top figures, left panel: test signal with oscillating singularity. Right panel: test signal with cusps. Bottom figures: corresponding continuous wavelet transforms, using the Mexican hat wavelet. On the horizontal axes, the location parameter b. On the vertical axes, the logarithmic scale parameter $\log_2(a)$. See text for more details.

where $x_0 = 0.345$. The function with the cusp singularities in the right panel of the same figure is given by $f(x) = |\cos(\exp(3x))|$.

The wavelet function used in all analyses is the Mexican hat wavelet.

9.2.4 Local regularity analysis using wavelets

In Section 6.2.4, the decay of the Fourier transform was linked to the global smoothness of a function. Given that wavelets have a compact support or at least are clearly located, it is no surprise that a wavelet analysis carries more local and more balanced information on the smoothness of a function.

The information provided by the magnitude of a wavelet coefficient depends on two main features of the wavelet function. The first feature is the width of the support, or, in the case where that support is unbounded, the asymptotic rate of the decay of the function. This feature has influence on whether or how much a wavelet coefficient $d_f(a,b)$ is affected by the presence of a singularity. Since the continuous wavelet transform has no subsampling, the range of coefficients affected by a point singularity widens at coarse scales. This is clearly visible in Figures 9.4, 9.5, and 9.6, where each point singularity has its own **cone of influence** (Jaffard, 1991) in which the wavelet coefficients $d_f(a,b)$ at given scale a reach a local maximum as a function of b. In the case of an **oscillating singularity**, the coefficients have two local maxima at each scale left and right of the location of the singularity (Arneodo et al., 1998; Mallat, 2001). An intuitive explanation for this observation starts from the Riemann-Lebesgue Lemma, which states that the Fourier transform of a square integrable function has always asymptotic decay towards zero. In other terms,

$$\lim_{|\omega|\to\infty} \int_{-\infty}^{\infty} \exp(-i\omega x) f(x) dx = 0.$$

In this expression, the oscillation $\exp(-i\omega x)$ with $|\omega| \to \infty$ can be used to approximate the oscillating singularity. Taking for $f(x)$ in this expression the wavelet function $\psi_{a,b}(x)$ with b at the location of the singularity, the integral is determined by the immediate neighbourhood of b, as $\psi_{a,b}(x)$ has compact support or fast decay. In that neighbourhood the fast oscillations lead to small integrals. The local maximum is produced by an oscillation a bit away from the singularity, so outside the cone of influence of that singularity. The oscillation producing the maximum is the one that has the highest inner product or overlap with the wavelet function at that scale. The effect can be seen as a sort of resonance.

The information provided by a continuous wavelet coefficient also depends on a second feature, namely the **vanishing moments** of the wavelet function. A wavelet function has p vanishing moments if

$$\int_{-\infty}^{\infty} \psi(x) x^q dx = 0,$$

for $q = 0, 1, \ldots, p-1$. Unlike in the discrete wavelet transform, vanishing moments do not come in a dual and primal pair, nor do they show up as an approximation order of a projection onto a scaling basis. The following theorem (Jaffard, 1991) states that the **local** smoothness of a function can be measured from the decay of its continuous wavelet coefficients across scales, provided that the number of vanishing moments is higher than the degree of smoothness to be measured[6].

Theorem 9.2.2 *If $\psi(x)$ is a wavelet function with p vanishing moments and*

[6]Note that the results in this chapter require a number of vanishing moments higher than the smoothness, so that the smoothness can be measured from the evolution of wavelet coefficients across scales. This is in contrast to Chapter 8 where we impose that the Lipschitz smoothness is higher than the number of vanishing moments, so that the vanishing moments can be fully exploited.

$f(x)$ is Lipschitz α at x_0 with $\alpha \leq p$, then

$$|d_f(a,b)| = \mathcal{O}\left(a^{\alpha+1/2}\left(1+\left|\frac{b-x_0}{a}\right|^{\alpha}\right)\right).$$

In particular, for b within the ***cone of influence*** *around x_0 defined by the inequality $|b-x_0| \leq Ca$, this becomes*

$$|d_f(a,b)| = \mathcal{O}\left(a^{\alpha+1/2}\right).$$

Nearly converse versions of this theorem, with some subtle modifications in the precise formulation, exist as well (Jaffard, 1991; Mallat, 2001). The smoother the function at a point x_0, the faster the decay of the wavelet coefficients due to that smoothness. In the analysis of a function from its wavelet decay, the points of singularity play a central role. At each scale a, the modulus of the wavelet transform $|d_f(a,b)|$ is expected to reach a local maximum within the cone of influence around each singularity. The local maxima of a wavelet transform carry the essential information, allowing even the reconstruction of the original function from those maxima (Mallat and Zhong, 1992; Carmona, 1995).

If the function $f(x)$ is available in a finite number of locations only, and even if these observations are noisy, a continuous wavelet analysis allows us, through Theorem 9.2.2, to measure the regularity of that function *up to the observational resolution*, or, in the presence of noise, almost up to that resolution, *as far as we can see*, that is.

Whereas the **discrete** decimated wavelet transform is a popular tool in (typically nonlinear) data **processing** (non-parametric smoothing, enhancement and so on) and in data **compression**, the **continuous wavelet transform** is mostly used in local data analysis. In a statistical context, this may amount to a **parametric inference** problem, such as the estimation or testing of the parameters of a **time series model**. The nondecimated wavelet transform (often called maximal overlap transform in publications on time series) takes an intermediate position. In the framework of a continuous wavelet transform, vanishing moments and the cone of influence near singularities play a central role in the analysis of the decay of local maxima of the wavelet transform modulus. The analysis may proceed up to a finite scale if the data are available at a finite number of observations. The analysis then reveals the degree of smoothness, not at the theoretically required infinitely fine scale, but **up to the resolution of the observations**.

9.3 Wavelet packets

9.3.1 Multiscale analysis of multiscale coefficients

In some applications, such as sound data, the detail coefficients of a wavelet decomposition still show structure for which a further multiscale analysis seems to be appropriate (Coifman et al., 1992). Whereas the scaling coefficients $\boldsymbol{s}_j$ are associated with the "even" subset of the fine scale knots, i.e., $\boldsymbol{x}_j = \boldsymbol{x}_{j+1,e}$, the details $\boldsymbol{d}_j$ can be associated with the odd subset $\boldsymbol{x}_{j+1,o}$. The next step of the transform of $\boldsymbol{d}_j$ is constructed on the knots in $\boldsymbol{x}_{j+1,o}$.

The dual refinement and detail matrices applied to the detail coefficients $\boldsymbol{d}_{j+1}$ at scale $j+1$, with $j = J-2, J-3, \ldots, L$ are given by $\widetilde{\mathbf{H}}_j^{[2q]}$ and $\widetilde{\mathbf{G}}_j^{[2q]}$, defined in (9.6), where $q = 2^{J-j-2}$. This is because the knots in $\boldsymbol{x}_{j+1,o}$ are given by the shifted subsample $\widetilde{\mathbf{J}}_{J-1,o(j+1)}^{[q]^\top} \boldsymbol{x}_J$. Defining $\boldsymbol{d}_j^{[0]} = \boldsymbol{s}_j$ and $\boldsymbol{d}_j^{[1]} = \boldsymbol{d}_j$, we can find all further decompositions of the detail coefficients

$$\boldsymbol{d}_j^{[2t]} = \widetilde{\mathbf{H}}_j^{[2q(j,t)]^\top} \boldsymbol{d}_{j+1}^{[t]} \quad (9.22)$$

$$\boldsymbol{d}_j^{[2t+1]} = \widetilde{\mathbf{G}}_j^{[2q(j,t)]^\top} \boldsymbol{d}_{j+1}^{[t]}, \quad (9.23)$$

taking $q(j,t) = 2q(j+1,t)$ and $q(j, t+2^{J-j-2}) = q(j,t)+1$ for $t \in \{0, 1, \ldots, 2^{J-j-2}\}$. This is illustrated for $j = J-1, J-2, J-3$ in Figure 9.7.

9.3.2 Libraries or dictionaries of basis functions

The diagram of a wavelet packet transform takes the form of a binary tree. This tree can be pruned at any branch. For instance, if a multiscale analysis is considered to be unnecessary, the first branch can be pruned, leaving a uniscale representation $\Phi_J(x)\boldsymbol{s}_J$. Secondly, all branches leading away from a fast wavelet transform can be cut off. When a branch outside the fast wavelet transform is kept, the details at that scale, $\Psi_j(x)\boldsymbol{d}_j$, are further decomposed into a coarser scale, using basis functions that follow from

$$\Psi_j^{[2t]}(x) = \Psi_{j+1}^{[t]}(x)\mathbf{H}_j^{[2q(j,t)]} \quad (9.24)$$

$$\Psi_j^{[2t+1]}(x) = \Psi_{j+1}^{[t]}(x)\mathbf{G}_j^{[2q(j,t)]}, \quad (9.25)$$

where $\Psi_j^{[0]}(x) = \Phi_j(x)$ are the only members of the family that have a nonzero integral. Figure 9.8 depicts a few members of the cubic B-spline wavelet packet family $\Psi_L^{[t]}(x)$, at a fixed, coarse level L. The functions have two primal vanishing moments, while the knots come from a uniform random distribution. All basis functions at this single scale L have roughly the

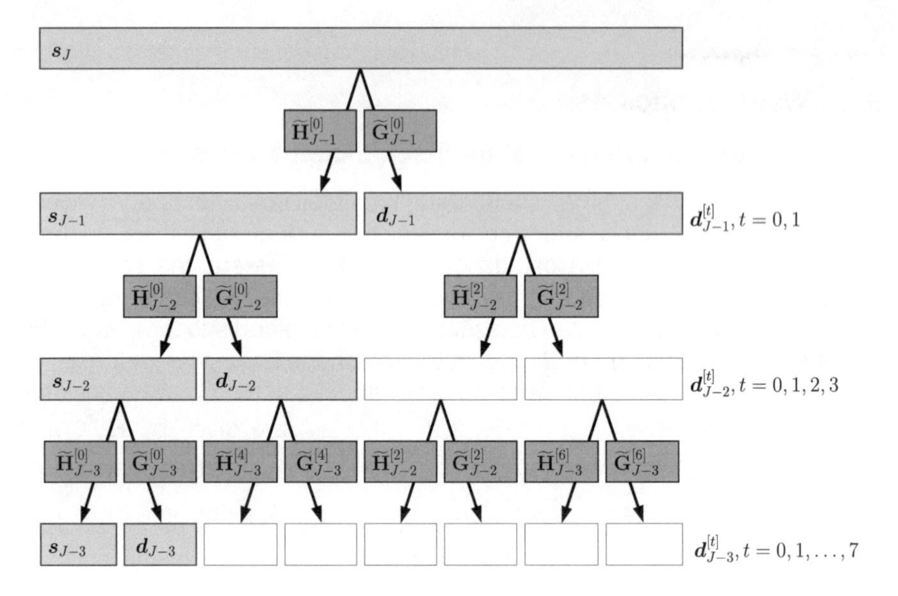

Figure 9.7
Diagram of a wavelet packet transform. The light-shaded coefficients are those from the fast wavelet transform. The set of dual refinement and details matrices is the non-cycle spun half of the set used in the nondecimated wavelet transform in Figure 9.1.

same size of support, where differences in support size are due to boundary effects and to the non-equispaced knot locations. The oscillations within the support reflect the scale of the detail function from which the function in question descends by further multiscale decomposition. The computational complexity of the full transform up to scale L is $\mathcal{O}\left((J-L)n\right)$, where $J-L$ is typically $\mathcal{O}(\log(n))$. When the cubic B-spline basis is replaced by the set of characteristic functions on each interval between the knots and all refinements proceed as in the Haar transform, then the full transform up to a coarse level is referred to as the **Hadamard, Rademacher, or Walsh** transform. It can be understood as a blocky version of a sort of discrete Fourier transform.

Whereas the collection of functions $\Psi^{[t]}_j(x)$ for $j = J, J-1, \ldots$ is overcomplete, the pruned wavelet packet decomposition leads to a selected basis representation in the leaves of the tree. The overcomplete set of functions $\Psi^{[t]}_j(x)$ is referred to as a **dictionary** or **library** of functions from which words, i.e., the basis functions, can be borrowed to construct sentences, i.e., valid basis decompositions.

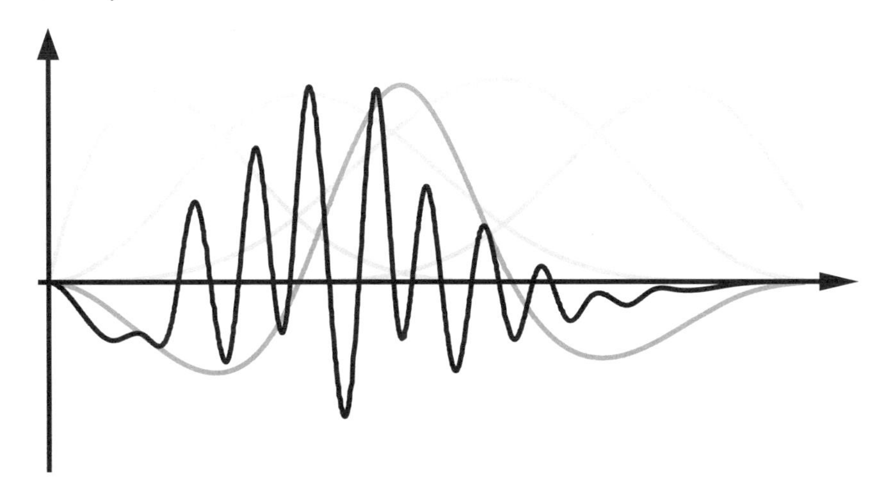

Figure 9.8
A few members of the cubic B-spline wavelet packet family with two primal vanishing moments at a given coarse level L. The functions in light grey are the members of $\Psi_L^{[0]}(x) = \Phi_L(x)$, i.e., the scaling functions. The function in dark grey is a member of $\Psi_L^{[0]}(x) = \Psi_L(x)$; i.e., it is an original detail function from the fast wavelet transform. The function in black comes from the multiscale decomposition of a detail function at finer scale. It combines the coarse scale support with the finer scale oscillations.

9.3.3 Dictionaries of nondecimated wavelet functions

A dictionary of basis functions can also be constructed within the framework of the nondecimated wavelet transform. Indeed, consider in (9.7) binary values of $\alpha_{j,q} \in \{0,1\}$. Then, at all levels j and for all shifts q, we have either

$$\boldsymbol{s}_{j+1}^{[q]} = \mathbf{H}_j^{[2q]} \boldsymbol{s}_{j,e}^{[2q,2q+1]} + \mathbf{G}_j^{[2q]} \boldsymbol{d}_{j,o}^{[2q,2q+1]}$$

or

$$\boldsymbol{s}_{j+1}^{[q]} = \mathbf{H}_j^{[2q+1]} \boldsymbol{s}_{j,o}^{[2q,2q+1]} + \mathbf{G}_j^{[2q+1]} \boldsymbol{d}_{j,e}^{[2q,2q+1]}.$$

Each step of the reconstruction uses only half of the available overcomplete vectors of coefficients. This implies that the other half need not be computed in the forward transform. As a result, the forward transform and reconstruction operate in a selected subset, defined by the binary values of $\alpha_{j,q}$. The selected subset of coefficients, having the same size as the outcome of a fast wavelet transform, is a basis representation. The averaged reconstruction in a frame is replaced by a selected reconstruction in a dictionary of basis functions. In statistical terms, this is model selection instead of model averaging.

9.4 Best basis selection

The selection of a basis from an overcomplete set of functions can be made data-adaptive, leading to a model selection procedure. This procedure may also include a selection of significant coefficients within the selected basis. The search can be steered by the tree structure of the dictionary. Among the early procedures are *Best Orthogonal Basis* (Coifman and Wickerhauser, 1992) and extensions (Cohen et al., 1997a,b). *Matching Pursuit* (Mallat and Zhang, 1993) is a greedy approach. In each iteration step, it selects the function from the dictionary that has the largest inner product with the current residual, i.e., the part of the data not yet represented by the current selection. *Basis pursuit* (Chen et al., 1998) formulates the selection as a convex optimisation problem minimising the sum of squared residuals regularised by the ℓ_1 norm of the selected coefficients, where $\|\boldsymbol{d}\|_{\ell_1} = \sum_i |d_i|$. The ℓ_1 norm can be considered as a measure of entropy, low entropy meaning a well structured and sparse representation. Nowadays, basis pursuit is better known as the lasso (Tibshirani, 1996), for which a group version (Yuan and Lin, 2007; Simon et al., 2013) performs structured selection. In the analysis of multivariate functions, basis selection procedures are even more prominent, for reasons explained in Section 10.4.

10

Two-dimensional wavelet transforms

10.1 There is more in 2D than in twice 1D: $2\text{D} > (1\text{D})^2$

When the observations come from a two-dimensional function, i.e., $Y_i = f(x_1, x_2) + \sigma Z_i$, with covariates x_1 and x_2, the analysis or estimation of $f(x_1, x_2)$ requires more specialised approaches. On the one hand, the two-dimensional background often adds helpful information. In particular, one observation is now surrounded not by just one left and one right neighbour. Instead many adjacent observations in horizontal, vertical, and diagonal directions may now contribute to the estimation in a given knot. On the other hand, as observations have more directions in which to diverge from each other as compared to the 1D case, distances between observations become larger. As a result, to maintain the same degree of accuracy, the number of observations has to increase, roughly from n to n^2. This phenomenon, known as the **curse of dimensionality**, is even more outspoken in higher dimensions, where we need n^d observations in d dimensions to achieve the same order of accuracy as n observations in one dimension.

A second challenge in two dimensions is the **notion of adjacency**. In order to construct lifting schemes of two-dimensional data, we need a routine for splitting the data into evens and odds. When the data are scattered over a two-dimensional domain, there is no straightforward natural ordering of the data on which the binary partitioning can be based. Moreover, the actual lifting steps require information on which knots can be considered as immediate neighbours of a given knot. This information is used in defining predictions and updates.

A third problem lies in the dimension of the singularities. In one dimension, isolated point singularities have no dimension. In two dimensions, singularities may take the form of a line or a curve. This has important consequences, further discussed in Section 10.4.

DOI: 10.1201/9781003265375-10

10.2 Rectangular and square 2D wavelet transforms

When observations of $f(x_1, x_2)$ are available on a Cartesian product of two sets of knots, $\mathcal{X} = \mathcal{X}_1 \times \mathcal{X}_2$, then the observations can be structured as a data matrix $\boldsymbol{Y}$ with elements

$$Y_{k_1,k_2} = f(x_{k_1,1}, x_{k_2,2}) + \varepsilon_{k_1,k_2}.$$

With n_1 and n_2 the sizes of the two sets of knots, $\mathcal{X}_1$ and $\mathcal{X}_2$, a general wavelet transform of the data would require the application of a big $n_1 n_2 \times n_1 n_2$ transform matrix, producing $n_1 n_2$ multiresolution coefficients. A fast wavelet transform would be able to keep the computational complexity at the optimal order of $\mathcal{O}(n_1 n_2)$. Nevertheless, by exploiting the 2D structure, it is possible to work with much smaller matrices, $\widetilde{\mathbf{W}}_1$ and $\widetilde{\mathbf{W}}_2$, defined on the sets $\mathcal{X}_1$ and $\mathcal{X}_2$, respectively. A two dimensional wavelet transform, applied to a matrix of finest scale coefficients s, is then given by

$$\boldsymbol{W} = \widetilde{\mathbf{W}}_1 \boldsymbol{Y} \widetilde{\mathbf{W}}_2^\top. \tag{10.1}$$

This version of a 2D wavelet transform is illustrated in Figure 10.1, where it applied to the image in Figure 1.1. The black bars in Figure 10.1 mark the boundaries of the successive scales on both components x_1 and x_2. The pixels represent contrast enhanced absolute values of the wavelet coefficients. White pixels correspond to zero detail coefficients. This version of the 2D wavelet transform is termed the **rectangular wavelet transform**, because the boundaries of the scales in the matrix of coefficients define mostly rectangular regions, as can be seen in Figure 10.1. In terms of basis functions, the rectangular 2D transform decomposes $f(x_1, x_2)$ into a tensor product basis

$$\begin{aligned}
f(x_1, x_2) &= \sum_{k_1=0}^{n_{1;L_1}-1} \sum_{k_2=0}^{n_{2;L_2}-1} s_{L_1,L_2,0;k_1,k_2} \varphi_{1;L_1,k_1}(x_1) \varphi_{2;L_2,k_2}(x_2) \\
&+ \sum_{j_1=L_1}^{J_1-1} \sum_{k_1=0}^{n'_{1;j_1}-1} \sum_{k_2=0}^{n_{2;L_2}-1} d_{j_1,L_2,1;k_1,k_2} \psi_{1;j_1,k_1}(x_1) \varphi_{2;L_2,k_2}(x_2) \\
&+ \sum_{k_1=0}^{n_{1;L_1}-1} \sum_{j_2=L_2}^{J_2-1} \sum_{k_2=0}^{n'_{2;j_2}-1} d_{L_1,j_2,2;k_1,k_2} \varphi_{1;L_1,k_1}(x_1) \psi_{2;j_2,k_2}(x_2) \\
&+ \sum_{j_1=L_1}^{J_1-1} \sum_{k_1=0}^{n'_{1;j_1}-1} \sum_{j_2=L_2}^{J_2-1} \sum_{k_2=0}^{n'_{2;j_2}-1} d_{j_1,j_2,0;k_1,k_2} \psi_{1;j_1,k_1}(x_1) \psi_{2;j_2,k_2}(x_2).
\end{aligned}$$

In this expression, $n_{1;L_1}$ and $n_{2;L_2}$ are the number of course scale knots in both dimensions, while $n'_{1,j}$ and $n'_{2,j}$ are the number of refinement points in

Figure 10.1
A rectangular wavelet transform (CDF2,2) of the image in Figure 1.1. White pixels correspond to zero coefficients. The rectangular wavelet transform is a variant of the squared wavelet transform in Figure 1.2.

both dimensions, at resolution level j. Using row vectors of basis functions, as in (1.30), the matrix $\boldsymbol{s}_{J_1,J_2}$ of finest scale coefficients is associated to a function in the tensor product space,

$$f(x_1, x_2) = \Phi_{1;J_1}(x_1)\boldsymbol{s}_{J_1,J_2}\Phi_{2;J_2}^\top(x_2), \tag{10.2}$$

which is then decomposed into

$$\begin{aligned} f(x_1, x_2) &= \Phi_{1;L_1}(x_1)\boldsymbol{s}_{L_1,L_2}\Phi_{2;L_2}^\top(x_2) \\ &+ \sum_{j_1=L_1}^{J_1-1} \Psi_{1;j_1}(x_1)\boldsymbol{d}_{j_1,L_2,1}\Phi_{2;L_2}^\top(x_2) \\ &+ \sum_{j_2=L_2}^{J_2-1} \Phi_{1;L_1}(x_1)\boldsymbol{d}_{L_1,j_2,2}\Psi_{2;j_2}^\top(x_2) \\ &+ \sum_{j_1=L_1}^{J_1-1}\sum_{j_2=L_2}^{J_2-1} \Psi_{1;j_1}(x_1)\boldsymbol{d}_{j_1,j_2,0}\Psi_{2;j_2}^\top(x_2). \end{aligned} \tag{10.3}$$

The rectangular 2D wavelet transform arises in a natural way in the computation of the covariance of wavelet coefficients in a 1D transform. Indeed, if $\boldsymbol{W} = \widetilde{\mathbf{W}}\boldsymbol{Y}$, then the covariance matrix of $\boldsymbol{W}$, $\boldsymbol{\Sigma}_{\boldsymbol{W}}$, is given by

$$\boldsymbol{\Sigma}_{\boldsymbol{W}} = \widetilde{\mathbf{W}}\boldsymbol{\Sigma}_{\boldsymbol{Y}}\widetilde{\mathbf{W}}^\top. \tag{10.4}$$

For images, the rectangular wavelet transform (10.1) may not be the best 2D decomposition. Indeed, the wavelet transform depicted in Figure 1.2 adopts an alternative version of a 2D transform, known as the **squared wavelet transform**. In a square[1] wavelet transform, the number of detail resolution levels is the same in the two covariates, x_1 and x_2. For the sake of notation, the resolution levels are labelled so that $J_1 = J_2 = J$, even if[2] $n_1 \neq n_2$. Unlike the rectangular version, the detail coefficients in a squared wavelet transform are situated at a single resolution level. Starting from the finest scaling coefficients, now denoted $\boldsymbol{s}_J$ with a single scale index J instead of $\boldsymbol{s}_{J_1,J_2}$, the transform computes, for $j = J-1, J-2, \ldots, L$,

$$\boldsymbol{s}_j = \widetilde{\mathbf{H}}_{j,1}^\top \boldsymbol{s}_{j+1} \widetilde{\mathbf{H}}_{j,2} \tag{10.5}$$

$$\boldsymbol{d}_{j,1} = \widetilde{\mathbf{G}}_{j,1}^\top \boldsymbol{s}_{j+1} \widetilde{\mathbf{H}}_{j,2} \tag{10.6}$$

$$\boldsymbol{d}_{j,2} = \widetilde{\mathbf{H}}_{j,1}^\top \boldsymbol{s}_{j+1} \widetilde{\mathbf{G}}_{j,2} \tag{10.7}$$

$$\boldsymbol{d}_{j,0} = \widetilde{\mathbf{G}}_{j,1}^\top \boldsymbol{s}_{j+1} \widetilde{\mathbf{G}}_{j,2}, \tag{10.8}$$

leading to the decomposition

$$\begin{aligned} f(x_1, x_2) &= \Phi_{1;L}(x_1) \boldsymbol{s}_L \Phi_{2;L}^\top(x_2) \\ &+ \sum_{j=L}^{J-1} \left(\Psi_{1;j}(x_1) \boldsymbol{d}_{j,1} \Phi_{2;j}^\top(x_2) + \Phi_{1,j}(x_1) \boldsymbol{d}_{j,2} \Psi_{2;j}^\top(x_2) \right. \\ &\left. + \Psi_{1;j}(x_1) \boldsymbol{d}_{j,0} \Psi_{2;j}^\top(x_2) \right). \end{aligned} \tag{10.9}$$

The diagonal blocks in Figure 1.2 contain the coarse scaling coefficients $\boldsymbol{s}_L$ and the detail coefficients $\boldsymbol{d}_{j,0}$. These coefficients are exactly the same as their rectangular counterparts. The other square transform coefficients, $\boldsymbol{d}_{j,1}$ and $\boldsymbol{d}_{j,2}$, could be further transformed into rectangular coefficients, as indeed, the squared decomposition is an incomplete version of the rectangular alternative. It follows that the squared transform is computationally faster than the rectangular one, and yet, its coefficients have a clear directional interpretation. In particular, the coefficients $\boldsymbol{d}_{j,1}$ represent features that are smooth in the direction of x_2 and oscillating in the direction of x_1. In Figure 1.2 these coefficients are found in the upper right corner of the image. These coefficients represent the vertical components at a given scale. Likewise, the coefficients $\boldsymbol{d}_{j,2}$ carry information about horizontal structures, while the coefficients $\boldsymbol{d}_{j,0}$, which are typically much smaller than the other two, correspond to diagonal features.

[1]The regions bounded by the scales in a squared wavelet transform are not really squares, unless the input data $\boldsymbol{Y}$ is a square matrix.

[2]The number of a resolution level is indeed a mere convention. The same convention could be imposed for the rectangular transform, especially if the number of detail levels in both covariates is taken to be the same, i.e., if $J_1 - L_1 = J_2 - L_2$. As the rectangular transform combines details at different resolutions j_1 and j_2, the convention that $J_1 = J_2 = J$ does not lead to simplified notations as in the squared transform.

Obviously, the decomposition into diagonal, horizontal, and vertical components is of little meaning when the data at hand are *anisotropic*, for instance when the two dimensions x_1 and x_2 represent time and distance respectively, or even when x_1 and x_2 are measured at different scales. In that case, the rectangular transform is the natural choice. On the other hand, for images and data observed at squared shaped grids, the square transform is probably the better alternative.

For both the rectangular and the square wavelet transform, a nondecimated version can be implemented. The nondecimated square wavelet transform delivers at every resolution level three matrices of the same size as that of the input, representing the diagonal, horizontal, and vertical components.

10.3 Nonseparable wavelets on scattered knots

10.3.1 Neighbourhoods in two dimensions

One of the challenges faced by a two-dimensional refinement scheme is how to split the available knots into even and odd subsets. When a wavelet coefficient is defined as the offset between an odd indexed fine scaling coefficient and a prediction from its even neighbours, it is crucial to know which coefficients are labelled even and which of those are neighbours to the odd coefficient at hand. In one dimension or on a Cartesian product of two one dimensional sets of knots, there is a natural ordering of knots, leading to a unique definition of the left and right neighbour in one dimension and also the upper and lower neighbour in two dimensions.

Neighbourhoods in arbitrary dimensions can be represented by **graphs**.

Definition 10.3.1 *(Graphs) Let $\mathcal{X}$ be a set of n knots $\boldsymbol{x}_i$, $i = 1, 2, \ldots, n$, termed* ***vertices*** *(singular: vertex) in this context. Then a graph $G(\mathcal{X})$ is a couple of sets $(\mathcal{X}, \mathcal{E}(\mathcal{X})$, where the elements of $\mathcal{E}(\mathcal{X})$ are unordered pairs of vertices, called* ***edges****. If $(\boldsymbol{x}_i, \boldsymbol{x}_j) \in \mathcal{E}(\mathcal{X})$, then $\boldsymbol{x}_i$ and $\boldsymbol{x}_j$ are termed* ***neighbours****.*

One way to construct a graph on scattered data is by triangulation. We first introduce a **boundary** $\partial\Omega(\mathcal{X})$ as a closed polyline without self-intersections circumscribing all points in $\mathcal{X}$. In particular, let $\boldsymbol{b}$ be a circular m-tuple with elements from $\{1, 2, \ldots, n\}$, i.e., $b_1 = b_m$. Then

$$\partial\Omega(\mathcal{X}) = \bigcup_{i=1}^{m-1} S(\boldsymbol{x}; b_i, b_{i+1}),$$

where the **polyline segments** $S(\boldsymbol{x}; b_i, b_{i+1})$ are defined as

$$S(\boldsymbol{x}; r, s) = [\boldsymbol{x}_r \boldsymbol{x}_s] = \{\alpha \boldsymbol{x}_r + (1 - \alpha)\boldsymbol{x}_s; 0 \leq \alpha \leq 1\}.$$

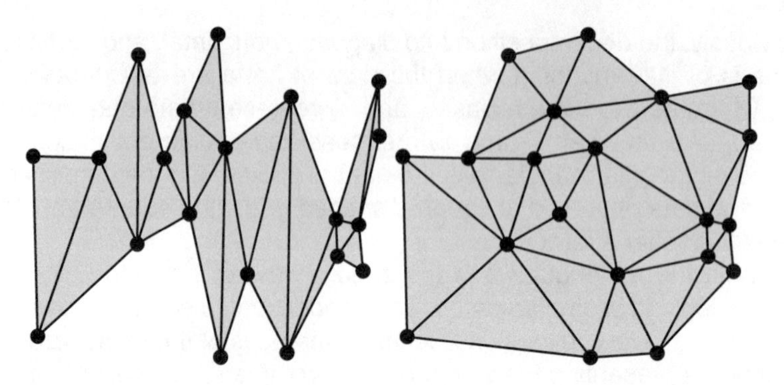

Figure 10.2
Two triangulations of two different polygons defined on the same set of vertices. The polygons are shaded. The figure on the right is a Delaunay triangulation.

For a closed polyline to be a boundary, we impose two conditions. First, it has no self-intersections,

$$S(\boldsymbol{x}; b_i, b_{i+1}) \cap S(\boldsymbol{x}; b_j, b_{j+1}) = \emptyset \text{ for any } i \neq j, 1 \leq i, j \leq m-1.$$

Second, the set of knots is a subset of the closed polygon circumscribed by $\partial\Omega(\mathcal{X})$, i.e., $\mathcal{X} \subset \Omega(\mathcal{X})$.

Definition 10.3.2 *(Triangulation) Let $\Omega \subset \mathbb{R}^2$ be a closed planar polygon with boundary $\partial\Omega$, and let $\mathcal{X} \subset \Omega$ be a set of vertices so that $\partial\Omega$ consists of segments with end points in $\mathcal{X}$. Then a set $\mathcal{T}(\Omega)$ of triangles $T_i \subset \Omega$ is a triangulation or a triangular mesh of Ω if the following conditions are met.*

1. *The triangles T_i can be defined as $T_i = \{\boldsymbol{x} \in \Omega; \boldsymbol{x} = \alpha_1\boldsymbol{x}_1 + \alpha_2\boldsymbol{x}_2 + \alpha_3\boldsymbol{x}_3; 0 \leq \alpha_1, \alpha_2, \alpha_3; \alpha_1 + \alpha_2 + \alpha_3 = 1; \boldsymbol{x}_1, \boldsymbol{x}_2, \boldsymbol{x}_3 \in \mathcal{X}\}$.*
2. $\bigcup_{i=1}^{\#\mathcal{T}(\Omega)} T_i = \Omega$.
3. *For all $T_1, T_2 \in \mathcal{T}(\Omega)$, if $\boldsymbol{x} \in T_1 \cap T_2$, then this point is on a common edge, i.e., there exist vertices $\boldsymbol{x}_1, \boldsymbol{x}_2 \in \mathcal{X}$, and $0 \leq \alpha \leq 1$ so that $\boldsymbol{x} = \alpha\boldsymbol{x}_1 + (1-\alpha)\boldsymbol{x}_2$.*

Figure 10.2 shows two triangulations of two different polygons defined on the same set of vertices. The figure makes clear that the triangles may be long and skinny, and so their edges may not reflect true adjacency. An appropriate triangulation should be based on Euclidian distances. This is realised by the Delaunay triangulation and the underlying Voronoi tesselation.

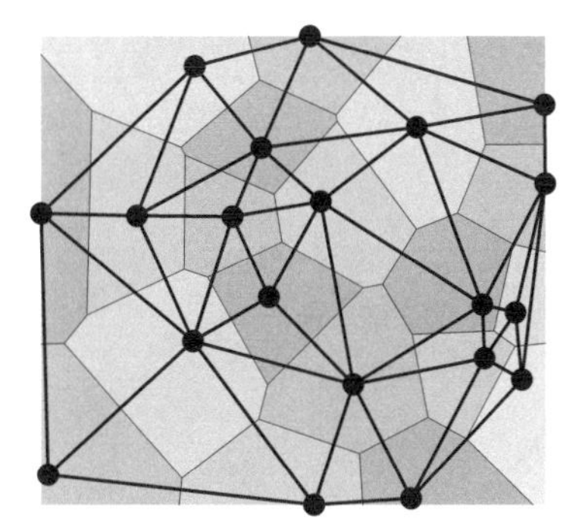

Figure 10.3
Voronoi cells around vertices. Vertices whose Voronoi cells have a common edge are linked by an edge in the corresponding Delaunay triangulation.

Definition 10.3.3 *(Voronoi tesselation) Given a planar region Ω and a set of vertices $\mathcal{X} \subset \Omega$, and a distance function d, then the Voronoi tesselation is a set of subsets $V_i \subset \Omega$, termed Voronoi cells, associated to the vertices $\boldsymbol{x}_i \in \mathcal{X}$ and defined by*

$$V_i = \{\boldsymbol{x} \in \Omega | \forall \boldsymbol{x}_j \in \mathcal{X}, j \neq i : d(\boldsymbol{x}, \boldsymbol{x}_i) \leq d(\boldsymbol{x}, \boldsymbol{x}_j)\}, \tag{10.10}$$

where $d(\cdot, \cdot)$ is a distance function. In other words, the Voronoi cell associated to vertex $\boldsymbol{x}_i$ is the set of points in Ω that are closer to $\boldsymbol{x}_i$ than to any other vertex in $\mathcal{X}$.

Figure 10.3 has an example of a Voronoi tesselation, defined on the entire planar space, $\Omega = \mathbb{R}^2$, based on the Euclidian distance. Based on the Voronoi tesselation, two vertices can be called neighbours if their Voronoi cells share a common border. Where three borders meet, we have a point at equal distance from three vertices. This point is the centre of the circle circumscribing the triangle formed by the three vertices. By construction of the Voronoi cells, that circle cannot possibly contain another vertex. The collection of all triangles circumscribed by these circles constitutes the **Delaunay triangulation**.

Definition 10.3.4 *(Delaunay triangulation) A triangulation $\mathcal{T}(\Omega(\mathcal{X}))$ of a set of vertices $\mathcal{X}$ in a polygon $\Omega(\mathcal{X})$ is the Delaunay triangulation if each triangle $T_i \in \mathcal{T}(\Omega(\mathcal{X}))$ is circumscribed by a circle centred in the intersection point of the three Voronoi cells associated to the three vertices defining the triangle.*

With $\Omega = \mathbb{R}^2$, the edges of the Delaunay triangulation include the convex hull of $\mathcal{X}$. Some of the edges in the convex hull are much longer than interior edges. The Delaunay triangulation in the right panel of Figure 10.2 is obtained by repeatedly taking away the triangles adjacent to a boundary with an obtuse interior angle.

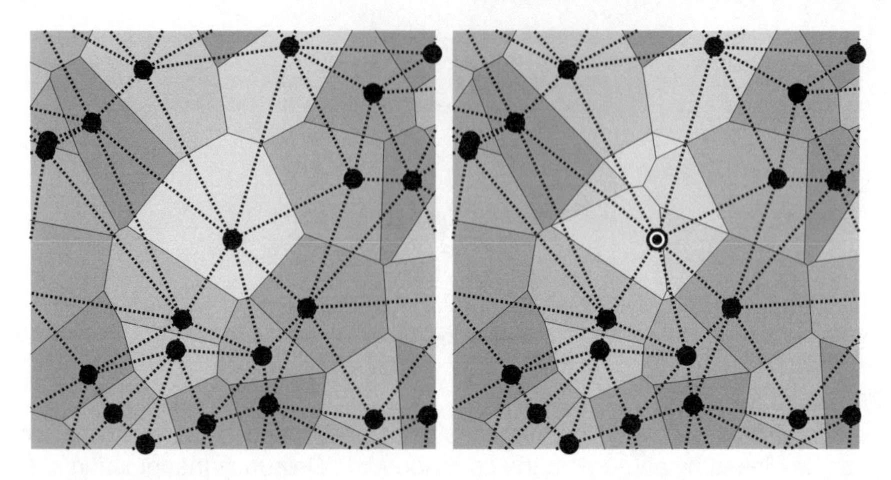

Figure 10.4
Natural interpolation using areas of Voronoi cells in the prediction of an interpolating lifting scheme for scattered data.

10.3.2 Using the Voronoi cells in prediction

The Voronoi cells can be used in the construction of a prediction step in a refinement scheme. Let $\mathcal{X}_{j+1}$ be a set of vertices, assumed to constitute level $j+1$ in a multilevel set $\{\mathcal{X}_L, \mathcal{X}_{L+1}, \ldots \mathcal{X}_J\}$. Consider the Voronoi tesselation and the corresponding Delaunay triangulation of this set, as depicted in the left panel of Figure 10.4. The right panel of the figure illustrates the prediction of a vertex that is taken out when proceeding from resolution level $j+1$ to level j. Let $\boldsymbol{x}_{j+1,o}$ be one of the knots taken out, and let $N_{j+1,o}$ be the set of its neighbours, which are all assumed to remain at level j. Then all points in the Voronoi cell $V_{j+1,o}$ of $\boldsymbol{x}_{j+1,o}$ are reassigned to one of the neighbours, as can be seen in the right panel of Figure 10.4. The overlap between the fine scale Voronoi cell $V_{j+1,o}$ and the coarse scale cells $V_{j,k}$ with $k \in N_{j+1,o}$ can be used in a **natural neighbour interpolation** scheme. In particular, let $\sigma_2(V)$ denote the Lebesgue measure, the area, that is, of $V \subset \mathbb{R}^2$; then a prediction operation can be defined by

$$d_{j,o} = s_{j+1,o} - \sum_{k \in N_{j+1,o}} P_{j;o,k} s_{j+1,k},$$

with

$$P_{j;o,k} = \frac{\sigma_2\left(V_{j,k} \cap V_{j+1,o}\right)}{\sigma_2\left(V_{j+1,o}\right)}. \qquad (10.11)$$

This is known as **Sibson interpolation**. The interpolation has the important property that it reproduces constants and linear functions,

$$\sum_{k \in N_{j+1,o}} P_{j;o,k} = 1 \tag{10.12}$$

$$\sum_{k \in N_{j+1,o}} P_{j;o,k} \boldsymbol{x}_{j+1,k} = \boldsymbol{x}_{j+1,o}. \tag{10.13}$$

In other words, Sibson interpolation has two dual vanishing moments. The proof of the linear reproduction (Sibson, 1980) is based on an integration of the intersections of the Voronoi cells with lines parallel to one the axes. A variant of Sibson interpolation, known as **Laplace interpolation** and also satisfying the constant and linear reproduction properties, measures the length of the boundary between adjacent Voronoi cells and the distance between the corresponding vertices. More precisely, let $\sigma_1 (V_{j+1,o} \cap V_{j+1,k})$ be the length of the edge shared by two adjacent Voronoi cells $V_{j+1,o}$ and $V_{j+1,k}$; then

$$P_{j;o,k} = \frac{1}{C_{j;o}} \frac{\sigma_1 (V_{j+1,o} \cap V_{j+1,k})}{d(\boldsymbol{x}_{j+1,o}, \boldsymbol{x}_{j+1,k})}, \tag{10.14}$$

where $C_{j;o}$ is a normalising constant,

$$C_{j;o} = \sum_{k \in N_{j+1,o}} \frac{\sigma_1 (V_{j+1,o} \cap V_{j+1,k})}{d(\boldsymbol{x}_{j+1,o}, \boldsymbol{x}_{j+1,k})}.$$

10.3.3 Subsampling scattered data

Next to the definition of a natural neighbourhood interpolating prediction, the adjacency information provided by a triangulation can also be used in the partitioning of vertices into "evens" and "odds". For proper implementation of the prediction, two odds should not be neighbours of each other. As illustrated in Figure 10.4, one odd has several even neighbours. From this observation it follows that in the subsampling of triangulated scattered data, the maximal number of odds is typically much smaller than half of the number of vertices at the current resolution level, leading to a slow coarsening process. Moreover, the Voronoi cells of the distinct odds united in one resolution level may have quite different areas, meaning that a single resolution level represents contributions at several scales. All of these arguments motivate us to compute **one coefficient at-a-time**, i.e., to pick at each resolution level just one "odd".

In applications where the location of the grid points is better controlled, refinement can proceed at much higher rates. This holds for images, of course, where the rate is three detail coefficients (vertical, horizontal, diagonal) for a single coarse scaling coefficient, but also for **semiregular grids** as in Figure 10.5, where a coarse scale scattered grid is refined in a regular way.

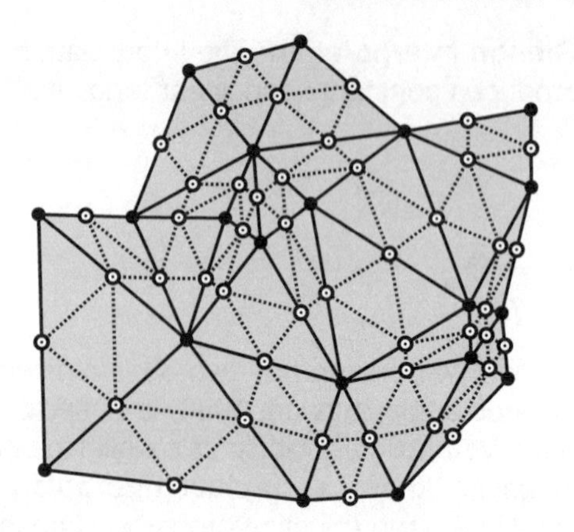

Figure 10.5
Semiregular refinement. The fine scale vertices are found by regular refinement of all edges in the coarse scale triangulation. Prediction by natural neighbourhood interpolation is not possible on this refinement scheme, because odd vertices (depicted by dotted circles) have odd neighbours.

For scattered data at fine scales, faster refinement is possible within the framework of **multiscale local polynomial transforms**, developed in Chapter 11. The multiscale local polynomial transform does not rely on a multilevel triangulation, making it much more flexible to deal with irregularly spaced data.

10.4 Multiscale analyses of data with line singularities

As illustrated in Figure 10.6, bivariate, piecewise smooth functions may feature sets of points of discontinuities taking the form of a long curve. The image in the left panel of the figure represents a simple bivariate piecewise constant function $f(x_1, x_2)$, where a low intensity area is separated from a high intensity area by a singularity curve. Assuming that the domain of the function is $[0,1] \times [0,1]$, the two-dimensional tensor product Fourier series would develop the function as

$$f(x_1, x_2) = \sum_{k=-\infty}^{\infty} \sum_{l=-\infty}^{\infty} a_{k,l} \exp\left[i(2\pi)kx_1\right] \exp\left[i(2\pi)lx_2\right], \tag{10.15}$$

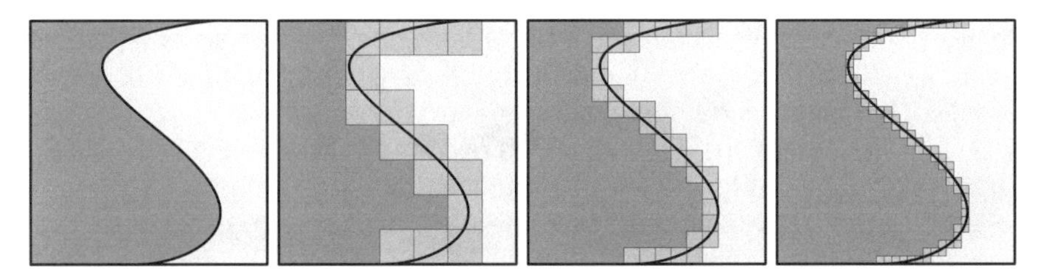

Figure 10.6
Line singularity in an image (or a 2D function). Using tensor product basis functions, $\varphi_{J;k,l}(x_1, x_2) = \varphi_{J,k}(x_1)\varphi_{J,k}(x_2)$, leads to an increasing number of supports with intersections with the singularity at finer scales. This leads to an increasing number of significant coefficients in a multiscale decomposition.

thereby extending (1.2) into two dimensions. The coefficients can be found in a tensor product version of (1.4), which reads

$$a_{k,l} = \int_0^1 \int_0^1 f(x_1, x_2) \exp\left[-i(2\pi)lx_1\right] \exp\left[-i(2\pi)lx_2\right] dx_1 dx_2. \qquad (10.16)$$

In a way similar to the discussion in Section 1.3.3, the coefficients are of the order $a_{k,l} = \mathcal{O}\left(1/|kl|\right)$. For a linear approximation

$$f_{m^2}(x_1, x_2) = \sum_{k=-m}^{m} \sum_{l=-m}^{m} a_{k,l} \exp\left[i(2\pi)kx_1\right] \exp\left[i(2\pi)lx_2\right],$$

the order of approximation, expressed in L_2-error, becomes

$$\|f - f_{m^2}\|_2^2 = \sum_{k=-\infty}^{\infty} \sum_{l=-\infty}^{\infty} |a_{k,l}|^2 [1 - I(|k| \leq m) I(|l| \leq m)] = \mathcal{O}\left(m^{-1}\right),$$

which is $\|f - f_n\|_2 = \mathcal{O}\left(n^{-1/4}\right)$. Comparison with (1.18) reveals that for the same approximation order, the number of coefficients needed in a two-dimensional Fourier transform is the square of the number in a one-dimensional Fourier transform.

In the case of basis functions with compact support, the situation is far more dramatic. The approximation of a piecewise constant function with line singularities is sketched in panels 2, 3, and 4 of Figure 10.6. For the approximation of the singularity in a tensor product basis $\varphi_{J;k,l}(x_1, x_2) = \varphi_{J,k}(x_1)\varphi_{J,l}(x_2)$, we need $\mathcal{O}\left(2^J\right)$ square shaped supports. Within each of these supports, the pointwise approximation error is $\mathcal{O}(1)$, leading to an integrated squared approximation error of

$$\|f - f_J\|_2^2 = \mathcal{O}\left(2^J \cdot 2^{-J} \times 2^{-J}\right) = \mathcal{O}\left(2^{-J}\right).$$

So, with n nonzero coefficients in a nonlinear approximation scheme, we arrive at $\|f - f_n\|_2 = \mathcal{O}\left(n^{-1/2}\right)$, which is definitely better than the Fourier alternative, but far from the one-dimensional exponential result of $\|f - f_J\|_2 = \mathcal{O}\left(2^{-J/2}\right)$ in (8.2). In one dimension, the exponentially decaying error is negligible in comparison to the linear approximation error of the smooth, non-polynomial part. In two dimensions, there are 2^{2L} scaling coefficients at level L for an L_2-approximation error of order $\|f - f_{2^{2L}}\|_2 = \mathcal{O}\left(2^{-L\widetilde{p}}\right)$, provided that there exists a single Lipschitz constant so that the Lipschitz exponent of $f(x_1, x_2)$ as a function of x_1 with fixed x_2 and vice versa is uniformly larger than the number of dual vanishing moments $\widetilde{p}$ of $\varphi_{J,k}(x)$. This means that with n_L coarse scaling coefficients, the L_2 order of approximation is $\|f - f_{n_L}\|_2 = \mathcal{O}\left(n_L^{-\widetilde{p}/2}\right)$. For an equally fine approximation of the singularities, we need the finest scale to be $J = \lceil 2L\widetilde{p} \rceil$, just as in Section 8.2. Due to the line singularities, however, this represents $\mathcal{O}\left(2^{2L\widetilde{p}}\right)$ detail coefficients. A nonlinear approximation of a piecewise smooth bivariate function thus requires a total of $n = 2^{2L\widetilde{p}} + 2^{2L} = \mathcal{O}\left(n_L^{\widetilde{p}}\right)$ significant coefficients for an approximation order of

$$\|f - f_n\|_2 = \mathcal{O}\left(n_L^{-\widetilde{p}/2}\right) = \mathcal{O}\left(n^{-1/2}\right).$$

This implies that, from an asymptotic approximation point of view, all benefits from vanishing moments are lost.

The problem of tensor-product bases in the representation of piecewise smooth data is that the line singularities are approximated in a blocky way by the square basis supports. Better approximations can be constructed when the basis includes functions with anisotropic, long, thin supports, that could be used along the edges. Numerous methods have been proposed, such as curvelets (Candès and Donoho, 2000), contourlets (Do and Vetterli, 2003), wedgelets (Donoho, 1999; Romberg et al., 2003), platelets (Willett and Nowak, 2003), bandelets (Le Pennec and Mallat, 2005), beamlets (Donoho and Huo, 2001), and many others (Dekel and Leviatan, 2005; Shukla et al., 2005; Figueras i Ventura et al., 2006; Jansen et al., 2005). Since there is an infinite number of orientations, lengths, and widths of the supports, these methods typically use a nonlinear and data adaptive decomposition in an overcomplete set of functions, i.e., a frame.

11

The multiscale local polynomial transform

11.1 Combining the benefits of spline and interpolating refinements

The wavelet transforms presented in Chapters 3 and 4 are complementary, both in the way they are constructed and in their properties. For the construction of the B-spline wavelets in Chapter 4, first came the basis functions, the B-splines, which are known to be refinable. From there, and using the factoring into lifting schemes, the actual refinement scheme was established, and on top of that came the full wavelet transform. Because of the order of things, there exist closed expressions for the scaling and wavelet basis functions. With basis functions that are piecewise polynomials, the closed expressions may be a bit heavy to actually write them down; at least we have a full understanding and control of the smoothness of the basis, and so the smoothness of any reconstruction in that basis. This is how the B-spline wavelets extend the scope of spline methods to multiscale problems.

Unfortunately, B-splines are not interpolating. As explained in Section 8.4, a B-spline basis on irregular knots does not recognise power function evaluations as coming from a power function. Evaluations or observations cannot be used as scaling coefficients. For use in a B-spline basis, function values need to be pre-processed, leading to additional computation time, changed statistics, and a possible loss of accuracy.

The Deslauriers-Dubuc scheme of Chapter 3 and other interpolating refinements can take function values as valid input. On the downside, unlike B-splines as in Figure 4.1, the Deslauriers-Dubuc basis functions as in Figure 3.7 are not bounded between zero and one. The B-spline refinement writes a scaling function as a convex combination of scaling functions at finer scale. This convex refinement ensures optimal numerical behaviour. No such guarantee exists for interpolating scaling functions, where irregularly spaced knots may result in highly fluctuating scaling functions. Also, the smoothness of the scaling functions in an interpolating scheme on an irregular grid is often hard to analyse.

In order to reduce the highly fluctuating lobes in the prediction, we now investigate using smoothing instead of interpolation as a prediction operator.

DOI: 10.1201/9781003265375-11

Whereas the definition below adopts local polynomial smoothing, most of the subsequent ideas and problems apply to other smoothing schemes as well.

Definition 11.1.1 *(Local polynomial prediction) Let $\boldsymbol{x}_{j+1}$ be an n_{j+1}-vector of ordered knots at resolution level $j+1$, with associated scaling coefficients $\boldsymbol{s}_{j+1}$, and let e and o be two subsets of the index set $\{0, 1, 2, \ldots, n_j - 1\}$ at level $j+1$, so that the union of both subsets equals the whole index set. The subsets, referred to as evens and odds, do not necessarily correspond to the classical number theoretic notions of even and odd. We leave the option open that evens and odds are not disjoint, meaning that some knots act as both even and odd.*

Then the local polynomial prediction matrix for use in a dual lifting step $\boldsymbol{d}_j = \boldsymbol{s}_{j+1,o} - \mathbf{P}_j \boldsymbol{s}_{j+1,e}$, i.e., (2.23) at level j, is constructed with the following ingredients.

The vector of knots $\boldsymbol{x}_j$ at level j is given by the even subset, $\boldsymbol{x}_j = \boldsymbol{x}_{j+1,e}$.

Each knot $x_{j,l}$ acts as a centre in the weight function $\mathrm{W}_{j;l,l}(x) = K\left(\frac{x - x_{j,l}}{h_j}\right)$, in which $K(u)$ is a positive function, integrating to one, termed the kernel function. In most cases, $K(u)$ is taken to be symmetric, i.e., $K(-u) = K(u)$. The parameter h_j, known as the bandwidth, defines the scale associated to the resolution level j. As a result, scale is user controlled, and does not have to be dyadic.

The weight functions $\mathrm{W}_{j;l,l}(x)$ appear as diagonal elements in the $n_j \times n_j$ diagonal weight matrix $\mathbf{W}_j(x)$.

The construction of the prediction also involves the row vector of functions $\mathrm{X}^{(\widetilde{p})}(x) = [1\, x \ldots x^{\widetilde{p}-1}]$ and the $n_j \times \widetilde{p}$ matrix $\mathbf{X}_j^{(\widetilde{p})}$ with elements $\mathbf{X}^{(\widetilde{p})}_{j;k,r} = x_{j,k}^{r-1}$, where, as before, the parameter $\widetilde{p}$ stands for the number of dual vanishing moments, meaning that for any $\widetilde{p}$-vector $\boldsymbol{a}$, the polynomial $\mathrm{X}^{(\widetilde{p})}(x)\boldsymbol{a}$ has degree $\widetilde{p} - 1$.

Then, writing by $\boldsymbol{e}_l$ is the lth canonical vector, the elements of the local polynomial prediction matrix $\mathbf{P}_j$ are given by the values of a prediction function, i.e., $P_{j;k,l} = \mathrm{P}_{j,l}(x_{j+1,k}; \boldsymbol{x}_j)$ where

$$\mathrm{P}_{j,l}(x; \boldsymbol{x}_j) = \mathrm{X}^{(\widetilde{p})}(x) \left(\mathbf{X}_j^{(\widetilde{p})^\top} \mathbf{W}_j(x) \mathbf{X}_j^{(\widetilde{p})} \right)^{-1} \left(\mathbf{X}_j^{(\widetilde{p})^\top} \mathrm{W}_j(x) \boldsymbol{e}_l \right). \quad (11.1)$$

Unlike in polynomial prediction, the function $\mathrm{P}_{j,l}(x; \boldsymbol{x}_j)$ is not a polynomial. In each point x its value is taken from the evaluation of a polynomial, as can be seen in (11.1), but the weights in that expression depend on x, and so do the coefficients of the polynomial.

With $\widetilde{p} = 1$, the local polynomial smoothing reduces to a locally weighted average, i.e., the Nadaraya-Watson kernel regression in (1.15),

$$\mathrm{P}_{j,l}(x; \boldsymbol{x}_j) = \frac{K\left(\frac{x - x_{j,l}}{h_j}\right)}{\sum_{\ell=0}^{n_j - 1} K\left(\frac{x - x_{j,\ell}}{h_j}\right)}. \quad (11.2)$$

11.2 The Laplacian pyramid

Figure 11.1 investigates what happens when the smoothing prediction of Definition 11.1.1 is inserted into a subdivision scheme (2.69). Subdivision repeatedly refines the scaling coefficients $\boldsymbol{s}_L$ in the expansion $\Phi_L(x)\boldsymbol{s}_L = \Phi_J(x)\mathbf{H}_J\mathbf{H}_{J-1}\ldots\mathbf{H}_L\boldsymbol{s}_L = \Phi_J(x)\boldsymbol{s}_J$. The fine scaling coefficients are then $\boldsymbol{s}_J = \mathbf{H}_J\mathbf{H}_{J-1}\ldots\mathbf{H}_L\boldsymbol{s}_L$. Starting off from a canonical vector at coarse scale, $\boldsymbol{s}_L = \boldsymbol{e}_k$, we find the fine scale coefficients $\boldsymbol{s}_J$ in the approximation of the scaling function $\varphi_{K,k}(x)$. For schemes with just one prediction, we have $\mathbf{H}_j = \widetilde{\mathbf{J}}_j + \widetilde{\mathbf{J}}_j^o\mathbf{P}_j$ according to (2.57). The term $\widetilde{\mathbf{J}}_j$ reads as copying the coarse scaling coefficients into the even fine scale locations, while the other term, $\widetilde{\mathbf{J}}_j^o\mathbf{P}_j$, means that the odd fine scale locations are given a smoothed value.

The refinement scheme as such is still interpolating, from which it follows that the scaling function in Figure 11.1 satisfies $\varphi_{j,l}(x_{j,k}) = \delta_{lk}$, with δ_{lk} the Kronecker delta, as in (3.23). Moreover, the scaling function is nicely bounded between 0 and 1. Unfortunately, it has a highly non-smooth, fractal shape. This is due to plugging in a non-interpolating smoothing prediction into an interpolating scheme. In particular, it holds that

$$\lim_{x\to x_{j,k}} \mathrm{P}_{j,l}(x;\boldsymbol{x}_j) \neq \delta_{j,k},$$

from which it follows that the prediction functions, depicted in grey dashed lines in the first three plots of Figure 11.1, are not continuous at the evens, i.e., the white squares in the figure. Formally this reads

$$\lim_{x\to x_{j,k}} \sum_{l=0}^{n_j-1} \mathrm{P}_{j,l}(x;\boldsymbol{x}_j)s_{j,l} \neq s_{j,k}.$$

The discontinuity in the refinement function persists if, at increasingly finer scales, the further refinement (odd) points come arbitrarily close to the even points, leading to the fractal oscillations in the limiting function in the bottom plot of Figure 11.1. Moreover, the outcome of the refinement process is heavily dependent on the order in which refinement points are added. If a point close to a knot is inserted in the first refinement round, then this results in a sharp slope persisting over the further refinement steps.

In order to deal with the fractal subdivision, the values of the even fine scaling coefficients should take input from the smoothing operation as well, at least when the even coefficient is close to one of its odd neighbours. Therefore, the smoothing prediction operation should be evaluated at the evens as well. This happens automatically when the evens are also kept as odds. The option of overlapping evens and odds is available in Definition 11.1.1. We thus define one step in a simple forward **multiscale local polynomial trans-**

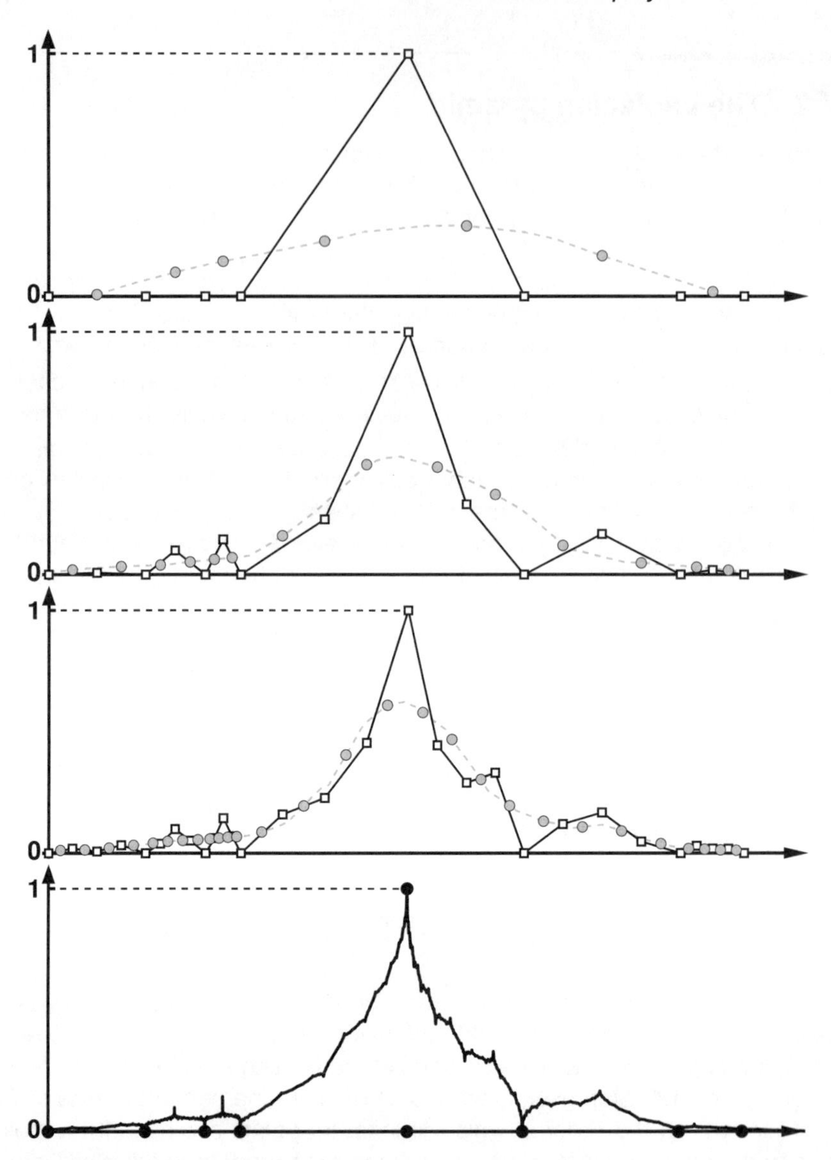

Figure 11.1
Naive kernel smoothing refinement. The top three plots depict the first three steps of the refinement, with the evens in white squares and the odds or refinement points in grey bullets. The refinement points are found as evaluations of a prediction function, plotted in the grey dashed line. The prediction function is found from a smoothing operation in the evens. In this case, the smoothing operation is a Nadaraya-Watson kernel smoothing. The resulting scaling function in the bottom plot, although interpolating and bounded between 0 and 1, has a fractal appearance.

form (MLPT) as

$$
\begin{aligned}
\boldsymbol{s}_j &= \widetilde{\mathbf{J}}_j^\top \boldsymbol{s}_{j+1} = \boldsymbol{s}_{j+1,e} && (11.3)\\
\boldsymbol{d}_j &= \boldsymbol{s}_{j+1} - \mathbf{P}_j\widetilde{\mathbf{J}}_j^\top \boldsymbol{s}_{j+1} = \boldsymbol{s}_{j+1} - \mathbf{P}_j\boldsymbol{s}_{j+1,e} = \boldsymbol{s}_{j+1} - \mathbf{P}_j\boldsymbol{s}_j. && (11.4)
\end{aligned}
$$

Going from coarse to fine scales in a reconstruction, we can write $\boldsymbol{s}_{j+1,e} = \boldsymbol{s}_j$, but also $\boldsymbol{s}_{j+1,e} = \widetilde{\mathbf{J}}_j^\top(\boldsymbol{d}_j + \mathbf{P}_j\boldsymbol{s}_j)$. In general, a reconstruction for the combined even and odd scaling coefficient vector at level $j+1$ can be written as

$$
\boldsymbol{s}_{j+1} = (\mathbf{I}_{j+1} - \mathbf{Q}_{j+1})\widetilde{\mathbf{J}}_j\boldsymbol{s}_j + \mathbf{Q}_{j+1}(\boldsymbol{d}_j + \mathbf{P}_j\boldsymbol{s}_j). \quad (11.5)
$$

In this expression, $\mathbf{Q}_{j+1}$ is an $n_{j+1} \times n_{j+1}$ diagonal matrix with entries that control the balance between a reconstruction by smoothing subdivision and an interpolation of the coarser scale coefficients. Obviously the odd coefficients $\boldsymbol{s}_{j+1,o}$ must be computed by subdivision, so when $i \in o(j+1)$, then $Q_{j+1;ii} = 1$. For the evens, we impose that $Q_{j+1;ii}$ tends to one whenever $x_{j+1,i}$ tends to point in the set of odds $o(j+1)$. The diagonal elements of $\mathbf{Q}_{j+1}$ can be defined through a continuous function $\mathrm{Q}_{j+1}(x; \boldsymbol{x}_{j+1,o})$ interpolating the constant function $y(x) = 1$ at all the odds $\boldsymbol{x}_{j+1,o}$. We then set $Q_{j+1;kk} = \mathrm{Q}_{j+1}(x_{j,k})$. With $\mathrm{Q}_{j+1}(x) = 1$, the reconstruction is simply

$$
\boldsymbol{s}_{j+1} = \boldsymbol{d}_j + \mathbf{P}_j\boldsymbol{s}_j. \quad (11.6)
$$

The reconstruction in (11.6) remains valid when the subsampling $\widetilde{\mathbf{J}}_j^\top$ in the forward transform is replaced by a more sophisticated prefiltering $\widetilde{\mathbf{F}}_j^\top$ to obtain

$$
\begin{aligned}
\boldsymbol{s}_j &= \widetilde{\mathbf{F}}_j^\top \boldsymbol{s}_{j+1} && (11.7)\\
\boldsymbol{d}_j &= \boldsymbol{s}_{j+1} - \mathbf{P}_j\widetilde{\mathbf{F}}_j^\top \boldsymbol{s}_{j+1} = \boldsymbol{s}_{j+1} - \mathbf{P}_j\boldsymbol{s}_j. && (11.8)
\end{aligned}
$$

The **prefilter** matrix has size $n_j \times n_{j+1}$, just like the subsampling matrix. It is termed a **reduce** operation in the signal and image processing literature. The **prediction** matrix $\mathbf{P}_j$ now has size $n_{j+1} \times n_j$. It is referred to as an **expand** operation. In applications with equisampled digital signals and images, the reduce and expand operations are time-invariant filters. The multiresolution scheme of (11.7), (11.8), and its inverse in (11.6) are then known as the **Laplacian pyramid** (Burt and Adelson, 1983). A flowchart of this scheme is depicted in Figure 11.2.

11.3 Local polynomial subdivision in a Laplacian pyramid

Just like a wavelet transform, the multiscale local polynomial transform can be described in terms of scaling and detail basis functions,

$$
\Phi_{j+1}(x)\boldsymbol{s}_{j+1} = \Phi_j(x)\boldsymbol{s}_j + \Psi_j(x)\boldsymbol{d}_j. \quad (11.9)
$$

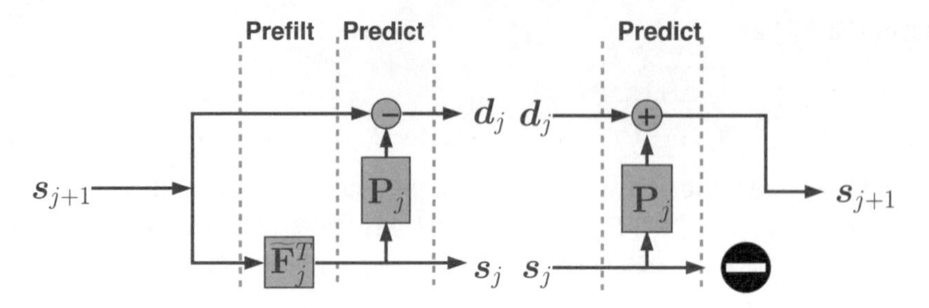

Figure 11.2
Flowchart of the forward and inverse Laplacian pyramid in (11.7), (11.8), and (11.6). Unlike the lifting scheme for wavelet transforms, this scheme has no simple split operation, thus duplicating the fine scaling coefficients in the two branches. Instead, the prefilter $\widetilde{\mathbf{F}}_j^\top$ reduces the size of the scaling vector. The scheme of the Laplacian pyramid, applied to non-equispaced data, can be used for a simple multiscale local polynomial transform.

The scaling functions are found through subdivision or iterated refinement. More precisely, in order to obtain $\phi_{L,k}(x)$, the inverse transform in (11.6) is launched at level L with $\boldsymbol{s}_L = \boldsymbol{e}_k$, the kth canonical vector of length n_L. At all subsequent, finer levels the details are set to zero, $\boldsymbol{d}_j = \mathbf{0}$ for $j = L, L+1, \ldots$, leading to the iterated refinement $\boldsymbol{s}_{j+1} = \mathbf{P}_j\boldsymbol{s}_j$. At some arbitrarily fine level J^*, the scaling coefficients can be interpolated or smoothed to define the basis function. A simple interpolation is to use characteristic functions, $\chi_{J^*,l}(x)$, defined on the interval $[x_{J^*,l}, x_{J^*,l+1})$,

$$\phi_{L,k}(x) = \sum_{l=0}^{n_{J^*}-1} s_{J^*,l}\chi_{J^*,l}(x).$$

As an alternative, the prediction functions $\mathrm{P}_{j^*,l}(x; \boldsymbol{x}_J^*)$ can take over the roles of the characteristic functions, in a way similar to the procedure in Section 3.2.3.

In contrast to the refinement in Figure 11.1, the Laplacian pyramid refinement applies the same prediction to all knots at the fine resolution level. The outcome can be expected to be smooth. The precise form of the basis functions depends on the bandwidths used at each scale. When the refinement inserts new knots in a more or less equidistant way between existing knots, it is reasonable to adopt at each resolution level a bandwidth that is inversely proportional to the number of knots at that level,

$$h_j = h_0/n_j.$$

This way, the number of evens within each interval $[x_{j+1,k} - h_j, x_{j+1,k} + h_j]$ is roughly the same at each resolution level. When the refinement occurs on a given, finite set of non-equispaced, yet uniformly distributed knots, experiments suggest that at fine scales the bandwidth should be a bit larger, as otherwise too many fine scale intervals may have less than a critical minimum of evens for the construction of the prediction (Jansen and Amghar, 2017). A heuristic choice of the bandwidth would then be

$$h_j = h_0 \log(n_j)/n_j.$$

The user is free, however, to apply other bandwidths as well.

The **bandwidth** in a multiscale local polynomial transform plays an essential and double role.

First, it is nothing but a **user controlled notion of scale**. This scale need not be dyadic; it can be considered as a continuous notion, to be discretised in a way that fits best with the vector of knots at hand.

Second, the bandwidth defines **adjacency**, fixing for every refinement point $x_{j+1,k}$ which even knots in $\boldsymbol{x}_j = \boldsymbol{x}_{j+1,e}$ will be used for the prediction in $\boldsymbol{x}_{j+1,k}$. As such, the prediction adapts automatically to local concentrations of knots, keeping the same range or scale for the prediction at all refinement points at a given resolution level.

Figure 11.3 illustrates the importance of the choice of the bandwidths. It displays the scaling bases of two multiscale local polynomial refinements. Both refinements start from the same coarse set of knots as in Figures 2.12, 3.7, and 4.1, representing interpolating and B-spline scaling functions. Both refinements insert new knots in the middle between two existing knots, and adopt an accordingly dyadic choice, i.e., $h_j = 2h_{j+1}$. The only difference between the two refinements is the choice of the coarse scale bandwidth h_0. In the second refinement, the coarse scale bandwidth, from which the bandwidths at all resolution levels follow, is too small, leading to predictions that are often no more than linear interpolation, due to a lack of evens within distance h_j from the refinement points. Coarse scale bandwidths that are too small thus lead to suboptimal smoothness properties and an inferior approximation power, which is a source of bias. On the other hand, the bandwidths at fine scales should not be too large either, as this may induce oversmoothing, another form of bias. At the same time, the bandwidths have an impact on the variance propagation throughout the forward transform, coefficient processing, and reconstruction. The variance propagation is further discussed in Section 11.6.

The user controlled notion of scale marks a fundamental difference between wavelet and multiscale local polynomial transforms. In wavelet transforms, the number of adjacent evens used in a prediction does not depend on the local configuration of the knots. As a result, the scales of the refinements within a single resolution level depend on the location of the fixed number of adjacent evens, leading to different scales within a single resolution level.

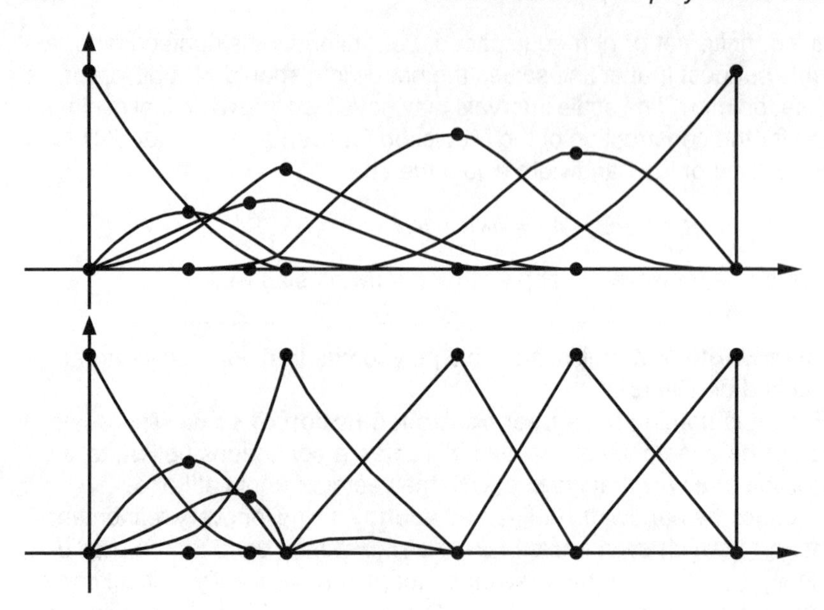

Figure 11.3
Two scaling bases corresponding to multiscale local polynomial transforms with dyadic refinements and dyadic bandwidths, i.e., dyadic scales.

Just like B-splines, the scaling functions associated with an MLPT are all positive. The scaling functions are not interpolating, and yet the function values of a polynomial with degree $q < \widetilde{p}$ appear in the reconstruction of that polynomial, provided that $\Phi_j(x)$ has $\widetilde{p}$ dual vanishing moments. Indeed, subdivision from the values $x^q_{j,k}$ leads to the values $x^q_{J,k}$ at level J, with $J \to \infty$, from which it follows that

$$x^q = \sum_{k=0}^{n_j-1} x^q_{j,k} \varphi_{j,k}(x),$$

just like in Theorem 3.2.1 for interpolating scaling bases.

11.4 Detail basis functions with zero integrals

It is remarkable that the prefilter operation $\widetilde{\mathbf{F}}_j^\top$ in the Laplacian pyramid (11.7) and (11.8) has no impact on the reconstruction in (11.6). The design of the prefilter is completely free, making it an ideal tool for nonlinear and data-adaptive processing. On the other hand, the prefilter cannot be combined with the weighted reconstruction in (11.5), because that reconstruction relies on the availability of the unfiltered even values $s_{j+1,e}$. Also, as the prefilter does not appear in the inverse transform, it has no impact on the basis functions, as indeed, the basis functions follow from the reconstruction. The prefilter cannot be used to impose, for instance, that

$$\int_{-\infty}^{\infty} \Psi_j(x)dx = \mathbf{0}.$$

It is possible, though, to design the prefilter for a sort of *global* vanishing moments, by making the detail coefficients $\boldsymbol{d}_j$ satisfy the condition

$$\int_{-\infty}^{\infty} \Psi_j(x)\boldsymbol{d}_j dx = 0,$$

for any input $\boldsymbol{s}_{j+1}$. For this condition to hold, it suffices to impose that

$$\int_{-\infty}^{\infty} \Phi_{j+1}(x)dx \cdot (\mathbf{I}_{j+1} - \mathbf{P}_j\widetilde{\mathbf{F}}_j^\top) = \mathbf{0}. \quad (11.10)$$

Indeed, by subdivision it holds that $\Psi_j(x) = \Phi_{j+1}(x)$, while on the other hand the forward transform reads as $\boldsymbol{d}_j = (\mathbf{I}_{j+1} - \mathbf{P}_j\widetilde{\mathbf{F}}_j^\top)\boldsymbol{s}_{j+1}$, leading to $\Psi_j(x)\boldsymbol{d}_j = \Phi_{j+1}(x)(\mathbf{I}_{j+1} - \mathbf{P}_j\widetilde{\mathbf{F}}_j^\top)\boldsymbol{s}_{j+1}$. Imposing the global zero integral for any $\boldsymbol{s}_{j+1}$ then yields the condition in (11.10). The prefilter may also be used to control the variance propagation throughout the analysis, and to reduce aliasing effects.

As an alternative to the Laplacian pyramid scheme, the multiscale local polynomial may also adopt slightly different forms. The forward and inverse schemes in Figures 11.4 and 11.5 include an **update** step, taking over most of the functionalities of the prefilter. It also allows for detail functions $\Psi_j(x)$ with zero integrals, which comes in quite handy in applications such as density estimation. The integral of a density estimator, decomposed as in (11.9), remains unaffected by any manipulation of the detail coefficients. The forward multiscale local polynomial transform corresponding to the scheme in Figure 11.4 reads

$$\boldsymbol{d}_j = \boldsymbol{s}_{j+1} - \mathbf{P}_j\widetilde{\mathbf{J}}_j^\top \boldsymbol{s}_{j+1} \quad (11.11)$$

$$\boldsymbol{s}_j = \widetilde{\mathbf{J}}_j^\top \boldsymbol{s}_{j+1} + \mathbf{U}_j\boldsymbol{d}_j. \quad (11.12)$$

The inverse transform represented in the flowchart of Figure 11.5 is given by

$$\boldsymbol{s}_{j+1} = (\mathbf{I}_{j+1} - \mathbf{Q}_{j+1})\widetilde{\mathbf{J}}_j(\boldsymbol{s}_j - \mathbf{U}_j\boldsymbol{d}_j) + \mathbf{Q}_{j+1}(\boldsymbol{d}_j + \mathbf{P}_j\boldsymbol{s}_j - \mathbf{P}_j\mathbf{U}_j\boldsymbol{d}_j). \quad (11.13)$$

11.5 Redundancy

As a forward multiscale local polynomial transform decimates the scaling coefficients but not the detail coefficients, each step produces longer output

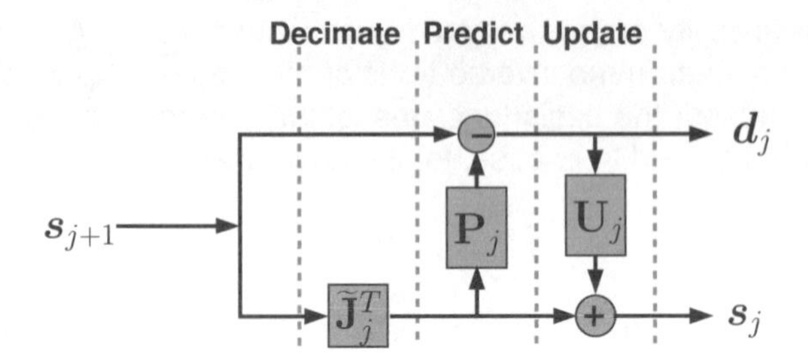

Figure 11.4
Flowchart of the forward multiscale local polynomial transform in (11.11) and (11.12).

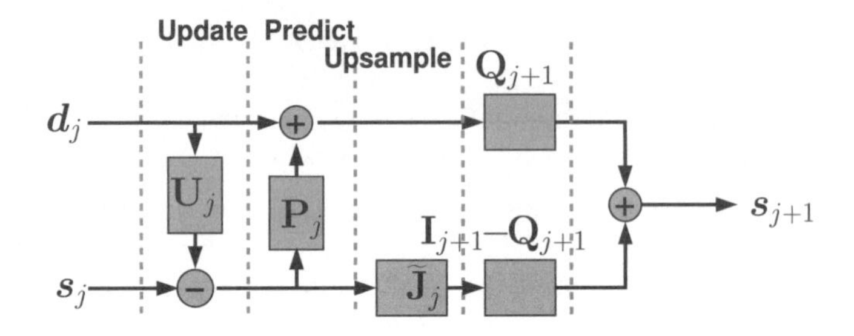

Figure 11.5
Flowchart of the inverse multiscale local polynomial transform in (11.13).

than it gets input. In the case where the decimation of the scaling coefficients is dyadic, n_{j+1} fine scaling coefficients give rise to n_{j+1} detail coefficients and $n_j = \lceil n_{j+1}/2 \rceil$ coarse scaling coefficients. Starting from $n_J = n$ observations at the finest level, we have $n_j = \lceil n_J/2^{J-j} \rceil$ scaling coefficients at level j. The transform delivers n_L scaling coefficients at level L, plus $J-L$ levels of detail coefficients, which amounts to

$$\begin{aligned}\#\{s_{L,k}\} + \#\{d_{j,k}, j = L, \ldots, J-1\} &= n_L + \sum_{j=L}^{J-1} n_{j+1} = \sum_{j=L}^{J} \lceil n/2^{J-j} \rceil \\ &= \mathcal{O}(2n + log_2(n)).\end{aligned}$$

The transform as a whole is overcomplete, but less so than the nondecimated wavelet transform. Indeed, as analysed in Section 9.1.4, the nondecimated wavelet transform has $\mathcal{O}(n \log(n))$ coefficients.

The scaling functions of a multiscale local polynomial transform are just as numerous as their counterparts in a fast wavelet transform. On an equidistant vector of knots, they are all (rescaled) dilations and translations $c_j \varphi(2^j x - k)$

of a single father function. The redundancy of the multiscale local polynomial transform is situated in the detail functions. The detail functions follow from the fine scaling functions by an equation that can be established in a way similar to that of Section 2.2.4. More precisely, a substitution of the reconstruction (11.13) into the basis transform (11.9) leads to

$$\Phi_{j+1}(x)\left[(\mathbf{I}_{j+1}-\mathbf{Q}_{j+1})\widetilde{\mathbf{J}}_j(\boldsymbol{s}_j-\mathbf{U}_j\boldsymbol{d}_j)+\mathbf{Q}_{j+1}(\boldsymbol{d}_j+\mathbf{P}_j\boldsymbol{s}_j-\mathbf{P}_j\mathbf{U}_j\boldsymbol{d}_j)\right] = \Phi_j(x)\boldsymbol{s}_j+\Psi_j(x)\boldsymbol{d}_j,$$

from which a two-scale equation

$$\Phi_j(x)=\Phi_{j+1}(x)\left[(\mathbf{I}_{j+1}-\mathbf{Q}_{j+1})\widetilde{\mathbf{J}}_j+\mathbf{Q}_{j+1}\mathbf{P}_j\right] \tag{11.14}$$

and a detail equation

$$\Psi_j(x)=\Phi_{j+1}(x)\left[-(\mathbf{I}_{j+1}-\mathbf{Q}_{j+1})\widetilde{\mathbf{J}}_j\mathbf{U}_j+\mathbf{Q}_{j+1}(\mathbf{I}_{j+1}-\mathbf{P}_j\mathbf{U}_j)\right] \tag{11.15}$$

can be distillated, taking $\boldsymbol{s}_j=\mathbf{I}_j$, $\boldsymbol{d}_j=\mathbf{0}$ for the former and $\boldsymbol{s}_j=\mathbf{0}$, $\boldsymbol{d}_j=\mathbf{I}_{j+1}$ for the latter. The detail equation is of the form (2.65), $\Psi_j(x)=\Phi_{j+1}(x)\mathbf{G}_j$. According to (2.68), imposing one or more primal vanishing moments amounts to $\mathbf{G}_j^\top\boldsymbol{M}_{j+1}^{(q)}=\mathbf{0}$, with $\boldsymbol{M}_{j+1}^{(q)}$ the qth moments of $\Phi_{j+1}(x)$. As a result, vanishing moments imply that the left null space of $\mathbf{G}_j$ is nontrivial. In a wavelet transform, this observation brings nothing new, because $\mathbf{G}_j$ has more rows than columns anyway, so the rows cannot possibly constitute a linearly independent set. In a multiscale local polynomial transform, the matrix $\mathbf{G}_j$ is a square matrix. Being singular then implies that the functions in $\Psi_j(x)$ are not linearly independent, and hence not a basis.

In the presence of an update $\mathbf{U}_j$, the multiscale local polynomial detail functions can be conjectured to constitute a frame, although results about frame constants, especially in relation to the locations of the knots, are not available in the current stage of research.

In spite of the lesser redundancy, the performance of the multiscale local polynomial transform in tasks like image denoising seems to be quite close to that of comparable nondecimated wavelet transforms. On highly noisy images, experiments suggest that the multiscale local polynomial transform could even be the better alternative. This is illustrated in Figure 11.7, depicting a simulation with artificially added independent, homoscedastic, normal noise. The multiscale local polynomial transform is compared with the nondecimated version of the wavelet transform adopted in the jpeg image compression standard for lossy compression. This is the Cohen-Daubechies-Feauveau wavelet with less dissimilar length with four primal and four dual vanishing moments, discussed in Section 7.2.2. Both the nondecimated wavelet transform and the multiscale local polynomial transform are performed in a squared, tensor-product way, as explained in Section 10.2. Both methods apply thresholding at the three finest resolution levels, where the thresholds are chosen within each level and each component (horizontal, vertical, and diagonal) to minimise the sum of squared differences between the estimated and the true image coefficients. Of course, the true image remains unknown in real life applications, obliging the user to estimate the minimum squared loss thresholds. That sort of question is left for Chapter 12. For the multiscale local polynomial transform, we implement the local linear decomposition, which means that the number of dual vanishing moments equals two. We include an update for two primal vanishing moments, but also a prefilter implementing the same local linear smoothing as the prediction. The bandwidths are dyadic, i.e., $h_j = 2h_{j+1}$, with at the finest level of details $h_{J-1} = 2.5$ pixels.

Given all the tuning parameters, conclusions are premature. Nevertheless, in terms of the signal-to-noise ratio,

$$\mathrm{SNR}(\widehat{\boldsymbol{\mu}}) = 20 \log \left(\frac{\|\boldsymbol{\mu}\|}{\|\widehat{\boldsymbol{\mu}} - \boldsymbol{\mu}\|} \right), \tag{11.16}$$

we have $\mathrm{SNR}(\boldsymbol{Y}) = 3\mathrm{dB}$ for the observations $\boldsymbol{Y} = \boldsymbol{\mu} + \sigma \boldsymbol{Z}$. The nondecimated wavelet transform achieves $\mathrm{SNR}(\widehat{\boldsymbol{\mu}}) = 11.38\mathrm{dB}$, while the multiscale local polynomial approach achieves $\mathrm{SNR}(\widehat{\boldsymbol{\mu}}) = 11.17\mathrm{dB}$. The multiscale local polynomial method appears to be less noisy, at the price of being less precise in the reconstruction of small details and texture, such as the cobblestones in Figure 11.7. Preliminary experiments seem to confirm that the multiscale local polynomial method outperforms nondecimated wavelet transforms in cases of large images with heavy noise, but all this requires further research.

11.6 Variance propagation in a multiscale local polynomial transform

Controlling the variance propagation is of course a point of attention in the design of a MLPT. Just as in a wavelet transform, the update can be aimed at variance reduction, but also the choice of the bandwidth has an impact on the variance. The variance propagation can be analysed through the singular value decomposition of the matrix $\widetilde{\mathbf{P}}_{J-1} = \mathbf{H}_{J-1}\widetilde{\mathbf{H}}_{J-1}^{\top}$, defined in Section 3.1.3. Just as for the nondecimated wavelet transform, it is not a genuine projection matrix. The reconstruction matrix $\mathbf{H}_{J-1} = (\mathbf{I}_{j+1} - \mathbf{Q}_{j+1})\widetilde{\mathbf{J}}_j + \mathbf{Q}_{j+1}\mathbf{P}_j$ follows from the two-scale equation in (11.14), while the analysis matrix equals $\widetilde{\mathbf{H}}_{J-1}^{\top} = \widetilde{\mathbf{J}}_j^{\top} + \mathbf{U}_j - \mathbf{U}_j\mathbf{P}_j\widetilde{\mathbf{J}}_j^{\top}$ following a substitution of (11.11) into (11.12).

As illustrated in Figure 11.8, the singular values and hence the variation control depend on the choice of the bandwidths. Comparing the plots with those in Figure 9.3 for redudant spline wavelet transforms, the singular values of the MLPT are close to one, confirming the excellent variance control

Figure 11.6
Test image and image with artificial, additive, independent, homoscedastic, normal noise for use in nondecimated wavelet and multiscale local polynomial denoising.

Figure 11.7
Multiresolution noise reduction applied to the test image of Figure 11.6. Top: three levels of the squared nondecimated Cohen-Daubechies-Feauveau wavelets with less dissimilar lengths, four primal, and four dual vanishing moments. The decimated version of this wavelet transform is used in the jpeg loss compression standard. Bottom: three levels of the squared multiscale local linear transform (with two dual vanishing moments, that is) with two primal vanishing moment, and with local linear prefilter. The two methods achieve quite similar results.

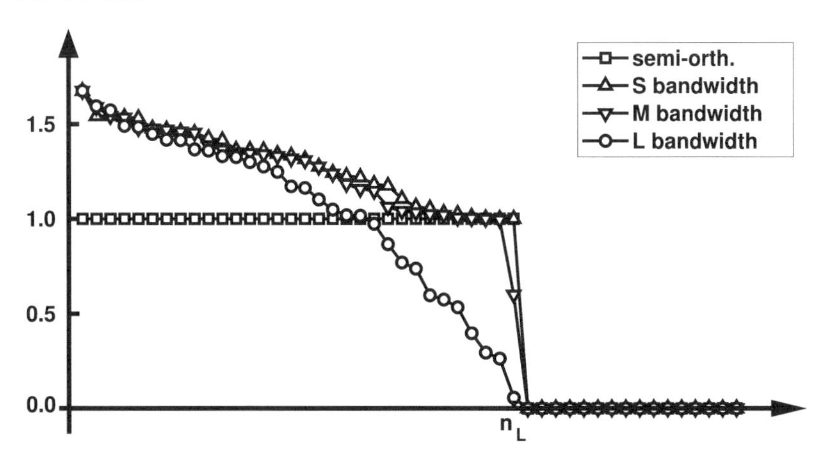

Figure 11.8
Singular values in descending order for projections $\widetilde{\mathbf{P}}_{J-1} = \mathbf{V}_{J-1}\widetilde{\mathbf{V}}_{J-1}$ in MLPT with two dual vanishing moments, using the same settings as in Figure 9.3. Three versions of the MLPT are used with small (S), medium (M), and large (L) bandwidths. Large bandwidths tend to oversmooth. Bandwidths should not be too small either for the sake of smoothness, as illustrated in Figure 11.3.

of the MLPT. On the other hand, oversmoothing takes place when the bandwidths are too large. Large bandwidths also lead to highly overlapping, widely supported scaling functions.

11.7 Scattered data

One of the benefits of an MLPT, compared to a wavelet transform, is the automatic definition of neighbourhood by the bandwidth h_j. Adjacent knots are those within distance h_j from the knot under consideration. As a consequence, a 2D-MLPT for scattered data can be constructed from bivariate local polynomial estimation without any need for a multiscale triangulation. A partitioning into even and odd knots can be based on a gradual peeling of the data cloud, referred to as split-by-peel, and illustrated in Figure 11.9.

Split-by-peel is rotation invariant. It first identifies the convex hull of the data cloud. The convex hull often has some long segments, situated at a much coarser scale than the typical distances between adjacent knots. Therefore, a line segment in the convex hull is taken out to be replaced by two segments linking at an inside point, whenever these two segments do not

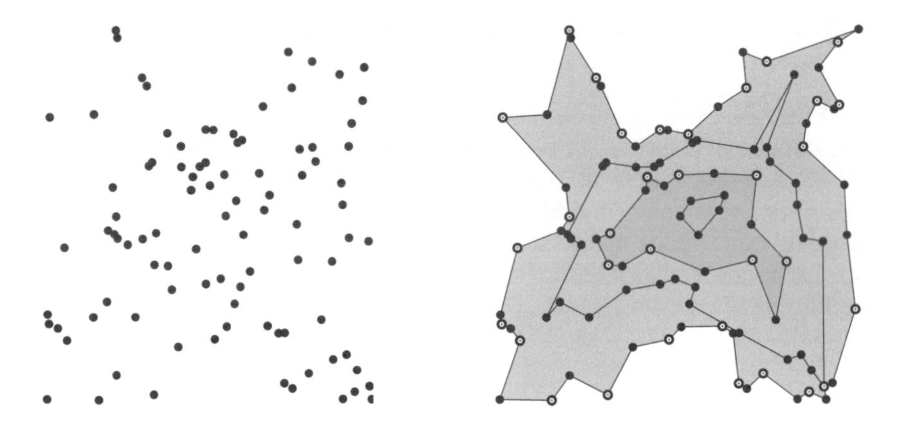

Figure 11.9
Split-by-peel: The right panel shows a partitioning of scattered data points, plotted in the left panel, into even and odds by a gradual peeling of the data cloud. The gradual peeling defines a sequence of boundaries. Even knots are displayed by black circles; odd knots are marked with a white circle and black center. In this example, the first and the third boundaries (counting from the outside in) are subsampled at rate one-half, while the second and fourth are not subsampled.

exclude another point and the angle in the inside point is obtuse and above a certain threshold, which is a first parameter of the procedure. The result of this step is a non-convex polyline marking the boundary of the data cloud. In the second step, that boundary is subsampled at a rate of, for instance, one-half. In a third step a new boundary is identified and subsampled at a possibly different rate, in order to obtain a global subsampling rate. As an example, it is possible to subsample the odd boundaries at a rate one half, while keeping all the knots on an even boundary.

12

Estimation in a wavelet basis

Sparse nonparametric estimation in a multiscale basis has been an important topic of research in the 1990's and early 2000's. It can be considered as the forerunner of the field of compressed sensing that emerged near the turn of the century. A related and also highly prolific domain, indebted to wavelet estimation, is high-dimensional, sparse modelling, including the work on ℓ_1 regularisation (lasso) and its variants.

Indeed, the reconstruction from a multiscale data transform $\boldsymbol{Y} = \mathbf{W}\boldsymbol{w}$ can be identified with a sparse, high-dimensional linear model

$$\boldsymbol{Y} = \boldsymbol{\mu} + \boldsymbol{\varepsilon} = \mathbf{X}\boldsymbol{\beta} + \boldsymbol{\varepsilon}, \tag{12.1}$$

where the design matrix corresponds to the reconstruction matrix, $\mathbf{X} = \mathbf{W}$, and the wavelet coefficients $\boldsymbol{w}$ are given by

$$\boldsymbol{w} = \widetilde{\mathbf{W}}\boldsymbol{Y} = \boldsymbol{\beta} + \widetilde{\mathbf{W}}\boldsymbol{\varepsilon}. \tag{12.2}$$

The noise in the wavelet domain will be denoted by $\boldsymbol{\zeta} = \widetilde{\mathbf{W}}\boldsymbol{\varepsilon}$. In the first instance, the wavelet coefficients are supposed to be homoscedastic, i.e., $\boldsymbol{\zeta} = \sigma\boldsymbol{Z}$, so that $\mathrm{var}(Z_{j,k}) = 1$, at all levels j and at all locations k. Wavelet coefficients are homoscedastic if the observations are homoscedastic, i.e., $\mathrm{var}(\varepsilon_i) = \sigma$, a constant for all $i = 1, 2, \ldots, n$, but also uncorrelated (referred to as white or uncoloured in some literature), and if the wavelet transform is orthogonal. This follows from filling in an orthogonal $\widetilde{\mathbf{W}}$ and $\boldsymbol{\Sigma}_{\boldsymbol{Y}} = \sigma^2\mathbf{I}$ into (10.4).

The model is **high-dimensional** if the size m of the parameter vector grows to infinity when the sample size n tends to infinity. This is in contrast to the classical low-dimensional problems, where the number of parameters is a constant or at least bounded by a constant. The model is **sparse** if the number of nonzero parameters is smaller than n. An oracle knowing the subset of nonzero parameters is capable of estimating these parameters from the n observations. The definition of sparsity is subject to wide variety in the literature, depending on the application.

In overcomplete multiscale transforms, the size of the coefficient vector m is larger than the sample size n. The nondecimated wavelet transform and wavelet packets have $m = \mathcal{O}(n\log(n))$, so for $n \to \infty$ we have $m \to \infty$,

In critically downsampled transforms, we have $m = n$, so this is still high-dimensional according to the definition, although we no longer have that

DOI: 10.1201/9781003265375-12

$n/m \to 0$. The multiscale local polynomial transform has the same order of dimensionality, as $m = \mathcal{O}(n)$.

There exist many strategies for the estimation of the parameters in a sparse high-dimensional model (12.1), depending on the exact definition of sparsity, further assumptions on parameters, design, and errors. One particular feature of sparse multiscale models is the availability of a fast forward transform $\widetilde{\mathbf{W}}$, leading to a simple signal-plus-noise model as in (12.2). Moreover, thanks to the Riesz constants of the wavelet basis, this signal-to-noise model is nearly equivalent to the original problem. Optimisation and processing can entirely take place in the wavelet domain, followed by a single final reconstruction.

12.1 A decision theoretic focus

The problem of estimation in a wavelet decomposition can be approached from a variety of viewpoints, each focusing on a specific aspect of it.

12.1.1 Oracular decision rules

In the first instance, we consider the effect of processing a single wavelet coefficient. This analysis does not assume any sparsity. Thanks to the Riesz constants of the wavelet basis, the accumulated effect on all wavelet coefficients approximates the effect on the reconstruction within well-defined bounds.

In a decision theoretic framework, several operations are proposed for the wavelet coefficient w_i. Every operation induces a **loss** with respect to the unobserved true value of that coefficient, β_i. The squared loss is defined by $L(\beta, w) = (\widehat{\beta} - \beta)^2$. A simple decision problem could be to choose to *keep or kill* the coefficient, i.e., to take $\widehat{\beta}_i \in \{0, w_i\}$. The squared loss is then $L(\beta_i, w_i) \in \{\beta_i^2, \zeta_i^2\}$, where $\zeta_i = w_i - \beta_i$ is the noise on the wavelet coefficient. The **risk** of the decision is the expected squared loss, $r(\beta) = EL(\beta_i, w_i) = E[(\widehat{\beta} - \beta)^2]$. In a keep or kill scenario, this becomes $r(\beta_i) = E(\zeta_i^2|\widehat{\beta}_i = w_i)P(\widehat{\beta}_i = w_i) + \beta_i^2 P(\widehat{\beta}_i = 0)$. At this point, we formalise the notion of an oracle as a benchmark for the performance of a decision rule.

Definition 12.1.1 *(oracle) An oracular estimator of a parameter is an estimator involving a variable selection or any other kind of decision, where the decision is based on knowledge of the true parameter value, while the estimation after decision uses observed values only.*

In an oracle based keep or kill approach, the decision to keep the value w_i is independent from the noise ζ_i. Hence, $r_{\text{or}}(\beta_i) = E(\zeta_i^2) = \sigma^2$, if the oracle decides to keep w_i, while $r_{\text{or}}(\beta_i) = \beta_i^2$ if the oracle decides to kill w_i

(i.e., replace it by zero). The oracle's risk is minimised if the decision to keep coincides with the case where $|\beta_i| \geq \sigma$.

12.1.2 Hard and soft thresholding

In practice, the decision to replace a coefficient by zero cannot be based on knowledge of its true value. Therefore, the magnitude of the true value in the decision rule is replaced by the magnitude of the observed coefficient, leading to a **threshold** scheme. Whereas the decision is based on a threshold, further details in the estimation need to be filled in. The most straightforward approach is again to keep or kill the coefficient, known as **hard thresholding**. The decision and subsequent estimation can be represented as a function of the wavelet coefficient w, depicted in Figure 12.1, left panel. The hard threshold function $\widehat{\beta} = \mathrm{HT}_\lambda(w)$ shows two discontinuities at the threshold's values λ and $-\lambda$, marking the transition between the zero (kill) and identity (keep) functions. A smoother, continuous transition is provided by the **soft threshold** alternative, where a coefficient with magnitude larger than threshold λ is shrunk by subtracting λ from its absolute value, leading to the function

$$\mathrm{ST}_\lambda(w) = \mathrm{HT}_\lambda(w) \cdot (|w| - \lambda) \cdot \mathrm{sign}(w),$$

plotted in the right panel of Figure 12.1. One of the benefits of a smooth

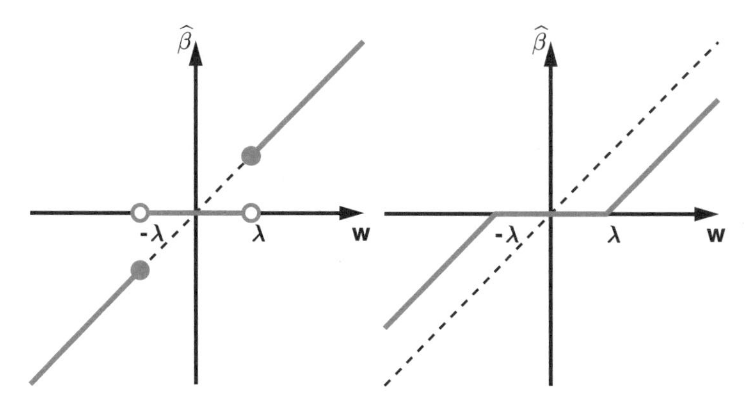

Figure 12.1
Left panel: hard threshold estimation. Right panel: soft threshold estimation.

transition at the threshold is the reduced impact of false positives. Indeed, if $|\beta_i| < \sigma$, the best choice would be to take $\widehat{\beta}_i = 0$, but if $|\zeta_i|$ is large, it may happen that $|w_i| > \lambda$. The probability of such a false positive depends also on the choice of λ of course, which is a topic of discussions further below. In any case, if the coefficient falsely survives the threshold, it is better to shrink its absolute value. Further discussions on the choice between soft and hard thresholding among other alternatives within a variable selection framework follow in Section 12.3.9.

12.1.3 Oracle mimicking by a universal threshold choice

The discussion in this section assumes that the noise η in the additive model of a wavelet coefficient $w = \beta + \eta$ is normally[1] distributed with zero mean and variance σ^2. The differential equation defining the zero mean normal density function,

$$x\phi_\sigma(x) = -\sigma^2\phi'_\sigma(x),$$

can be integrated into the Stein's Lemma

$$E[\eta(\mathrm{ST}_\lambda(w) - \beta)] = E[\eta\mathrm{ST}_\lambda(w)] = P(|w| > \lambda)\sigma^2, \tag{12.3}$$

which will be used throughout this chapter. In a vector of wavelet coefficients, we do not assume independence of the noise, nor do we assume any sparsity in the underlying vector $\boldsymbol{\beta}$ at this point. We do assume homoscedasticity though, with σ^2 denoting the variance of all wavelet coefficients. By $\phi(x)$ and $\Phi(x)$, we denote the standard normal density and standard normal cumulative distribution functions.

Let $r_{\mathrm{ST}}(\beta, \lambda) = E\left[(\mathrm{ST}_\lambda(w) - \beta)^2\right]$ be the soft threshold risk; then we have

$$\begin{aligned} r_{\mathrm{ST}}(\beta,\lambda) &= E\left[(\mathrm{ST}_\lambda(w) - w)^2\right] + E(w-\beta)^2 - 2E\left[(\mathrm{ST}_\lambda(w) - w)(w-\beta)\right] \\ &= E\left[(\mathrm{ST}_\lambda(w) - w)^2\right] + \sigma^2 - 2\sigma^2 P(|w| < \lambda), \end{aligned} \tag{12.4}$$

the last equality following from Stein's result and so holding for normal w only. The first term in (12.4) is bounded by λ^2, so we have

$$r_{\mathrm{ST}}(\beta, \lambda) \leq \lambda^2 + \sigma^2. \tag{12.5}$$

The first term in (12.4) is also bounded by $E(w^2)$, leading to the second upper bound

$$r_{\mathrm{ST}}(\beta,\lambda) \leq E(w^2) + \sigma^2 - 2\sigma^2 P(|w| < \lambda) = \beta^2 + 2\sigma^2 P(|w| > \lambda).$$

The second term on the right hand side is a function of β, as indeed

$$P(|w| > \lambda) = 2 - \Phi\left(\frac{\lambda+\beta}{\sigma}\right) - \Phi\left(\frac{\lambda-\beta}{\sigma}\right).$$

[1] Data with other than normally distributed noise may require some standardising preprocessing if that noise is heteroscedastic. This is the case with Poisson noise, which has a multiplicative character, meaning that the variance is proportional to the expected value. Poisson noise can be preprocessed with an Anscombe transform (Anscombe, 1948; Murtagh et al., 1995). Alternatively, a standardisation can be incorporated into the multiresolution analysis, leading to Haar-Fisz or wavelet-Fisz transforms (Fisz, 1955; Fryźlewicz and Nason, 2004; Jansen, 2006).

Denoting this function by $g(\beta)$, and noting that it has a minimum at $\beta = 0$, we find

$$g'(\beta) \leq g(0) + \sup_{-\beta<\xi<\beta} \frac{g''(\xi)}{2}\beta^2 = 2\left[1 - \Phi\left(\frac{\lambda}{\sigma}\right)\right] + \sup_{\zeta\in\mathbb{R}} \frac{\zeta\phi(\zeta)}{\sigma^2}\beta^2.$$

Since it can be proven that $1 - \Phi(t) \leq \phi(t)/t$ and $\zeta\phi(\zeta) \leq 1/\sqrt{2\pi e}$, we can conclude that

$$r_{\mathrm{ST}}(\beta, \lambda) \leq \beta^2 + 2\sigma^2\left[2\phi\left(\frac{\lambda}{\sigma}\right)\frac{\sigma}{\lambda} + \frac{1}{\sqrt{2\pi e}}\frac{\beta^2}{\sigma^2}\right]. \tag{12.6}$$

If a threshold is applied to all components of a vector $\boldsymbol{w}$, then the two upper bounds (12.5) and (12.6) can deal with the small and large coefficients respectively. In particular, choosing the **universal threshold**

$$\lambda_n = \sqrt{2\log(n)}\sigma, \tag{12.7}$$

the upper bound in (12.5) becomes $r_{\mathrm{ST}}(\beta, \lambda) \leq (2\log(n) + 1)\sigma^2 \leq (2\log(n) + 1)\sigma^2(n+1)/n$ while, for n sufficiently large, the upper bound in (12.6) can be further relaxed into

$$r_{\mathrm{ST}}(\beta, \lambda) \leq \left(1 + \frac{1}{\sqrt{2\pi e}}\right)\beta^2 + 2\sigma^2 2\phi\left(\frac{\lambda_n}{\sigma}\right)\frac{\sigma}{\lambda_n} \leq (2\log(n) + 1)\left(\frac{\sigma^2}{n} + \beta^2\right).$$

Now, adding the risks of all coefficients, we obtain a well known result (Donoho and Johnstone, 1994).

Theorem 12.1.2 *Let $\boldsymbol{w} = \boldsymbol{\beta} + \sigma\boldsymbol{Z}$ be an n-vector of homoscedastic normal random variables. Then the risk*

$$R_{\mathrm{ST}}(\boldsymbol{\beta}, \lambda) = \sum_{i=1}^{n} r_{\mathrm{ST}}(\beta_i, \lambda),$$

is bounded by $R_{\mathrm{ST}}(\boldsymbol{\beta}, \lambda) \leq (2\log(n) + 1)\left[\frac{\sigma^2}{n} + R_{\mathrm{or}}(\boldsymbol{\beta})\right]$, where

$$R_{\mathrm{or}}(\boldsymbol{\beta}) = \sum_{i=1}^{n} r_{\mathrm{or}}(\beta_i) = \sum_{i=1}^{n} \min(\beta_i^2, \sigma^2)$$

is the oracular risk.

Remark 12.1.3 *The threshold is* universal *in the sense that this result holds for any vector of length n. Sparsity is not required. Of course, sparsity may have an effect on the performance of the threshold estimator, but that effect is captured by the oracle. The universal threshold performs always within a short distance from the oracle.*

The result in Theorem 12.1.2 depends on the standard deviation σ. If σ is unknown, and if $\boldsymbol{\beta}$ has a large number of (near-)zeros, a robust estimator can be put in place

$$\widehat{\sigma} = \mathrm{MAD}(\boldsymbol{w})/\Phi^{-1}(3/4), \tag{12.8}$$

using the median absolute deviation $\mathrm{MAD}(\boldsymbol{X}) = \mathrm{median}(|\boldsymbol{X} - \overline{X}|)$ and the third quartile of the standard normal distribution $\Phi^{-1}(3/4) = 0.6745$.

If the vector $\boldsymbol{w}$ is heteroscedastic, then a coefficient dependent threshold can take $\lambda_{n;i} = \sqrt{2\log(n)}\sigma_i$, at least if σ_i can be estimated. This is the case if the covariance matrix of the observations $\boldsymbol{Y}$ is known up to a constant. Indeed, if $E(\boldsymbol{\varepsilon}\boldsymbol{\varepsilon}^\top) = \sigma^2\widetilde{\boldsymbol{\Sigma}}$, then the noise on the wavelet coefficients satisfies $E(\boldsymbol{\eta}\boldsymbol{\eta}^\top) = \sigma^2\widetilde{\mathbf{W}}\widetilde{\boldsymbol{\Sigma}}\widetilde{\mathbf{W}}^\top$. This is a rectangular 2D wavelet transform, introduced in Section 10.2. Pointwise division of $\boldsymbol{w}$ by the diagonal of $\widetilde{\mathbf{W}}\widetilde{\boldsymbol{\Sigma}}\widetilde{\mathbf{W}}^\top$ gives a homoscedastic vector with variance σ^2 which can be estimated.

An important special case is covered by the following result.

Theorem 12.1.4 *If the finest scaling coefficients in an equidistant wavelet transform on an infinite grid have stationary noise, meaning that the covariance* $\mathrm{cov}(s_{J,k}, s_{J,l})$ *depends on* $|k - l|$ *but not on* k *or* l *separately, then, by induction, all scaling coefficients and detail coefficients inherit this property. In particular, the coefficients are homoscedastic within each resolution level.*

Proof. The result follows from direct application of the expressions of the forward equidistant wavelet transform in (5.4) and (5.5). □

As a corollary of Theorem 12.1.4, the universal threshold for data with **correlated noise** (Johnstone and Silverman, 1997) or for data with uncorrelated noise, but analysed with a non-orthogonal wavelet transform, will depend on the resolution level, while it will be a constant within each resolution level.

12.1.4 Probabilistic upper bound

Although it operates on each coefficient individually, the value of the universal threshold depends on the number of coefficients n. This dependence on n reveals a sort of interaction between the components of the vector $\boldsymbol{w}$, even if these components are statistically independent. The interaction is described by a classical result from extreme value theory (Donoho, 1995b; Leadbetter et al., 1983).

Theorem 12.1.5 *Let* $\boldsymbol{X}$ *be an* n*-vector of i.i.d. random variables, with common distribution* $F_X(x)$ *and let* M_n *the maximum value; then for a real sequence* λ_n *it holds*

$$\lim_{n\to\infty} P(M_n \leq \lambda_n) = \exp(-a) \Leftrightarrow \lim_{n\to\infty} n\,(1 - F_X(\lambda_n)) = a. \tag{12.9}$$

Applied to i.i.d. normal vectors, i.e., $F_X(x) = \Phi(x/\sigma)$, we consider $\lambda_n = \alpha_n\sqrt{2\log(n)}\sigma$. Knowing that $\lim_{x\to\infty}[1-\Phi(x)]x/\phi(x) = 1$, we have

$$\begin{aligned} a &= \lim_{n\to\infty} n\left(1-F_X(\lambda_n)\right) = \lim_{n\to\infty} \frac{n\phi(\lambda_n/\sigma)}{\lambda_n/\sigma} = \lim_{n\to\infty} \frac{n\exp\left[-\alpha_n^2\log(n)\right]}{2\alpha_n\sqrt{\pi\log(n)}} \\ &= \lim_{n\to\infty} \frac{n^{1-\alpha_n^2}}{2\alpha_n\sqrt{\pi\log(n)}}. \end{aligned}$$

For $\alpha_n \to c < 1$, we find that $a = \infty$; hence $P(M_n \leq \lambda_n) \to 0$, but if $\alpha_n \to 1$, then $P(M_n \leq \lambda_n) \to 1$.

This means that the universal threshold, for which $\alpha_n = 1$, removes with probability tending to one all coefficients that carry only noise. The universal threshold is essentially the smallest threshold with this property. For this reason the universal threshold is called a *probabilistic upper bound.*

12.2 A multiple testing viewpoint

When the sample size n grows to infinity, the universal threshold takes away all coefficients with true value $\beta_i = 0$. For finite values of n, it is interesting to develop more data-adaptive thresholds. If the focus lies on avoiding coefficients with zero information surviving the threshold, the problem of finding a good threshold can be treated by a multiple hypothesis testing procedure, where the null hypotheses state that the coefficients carry no information and only noise, i.e., $H_0 : \beta_i = 0$. These hypotheses are tested against the classical two-sided alternative, $H_1 : \beta_i \neq 0$. A null hypothesis is rejected if the p-value of a test statistic is below the prespecified significance level α. Here the p-value is the probability, under the null hypothesis, of observing the observed value or another value which is more extreme in the direction of the alternative hypothesis. In the case of an observation from the signal-plus-noise model $w_i = \beta_i + \sigma Z_i$, the test statistic is simply w_i and the p-value is given by

$$\widehat{p}_i = P(\sigma|Z| > |w_i|),$$

as indeed, under the null hypothesis, w consists of σZ only. The null hypothesis is rejected if $\widehat{p}_i < \alpha$. If the density of Z is symmetric, this amounts to a threshold, $|w_i| > \lambda = F_Z^{-1}(1-\alpha/2)\sigma$. Unlike the universal threshold, this threshold is independent of the sample size.

False positives or **type I errors** occur when $\widehat{p}_i < \alpha$, while $\beta_i = 0$, i.e., given that H_0 holds exactly. If H_0 holds, then before observing w_i, the value of $\widehat{p}_i$ is a uniform random variable on $[0,1]$. Hence, the probability of a false positive is $P(\widehat{p}_i < \alpha|H_0) = \alpha$. At a significance level of α, independent hypothesis testing on n_0 coefficients with true value $\beta_i = 0$ would lead to an expected number of $n_0\alpha$ false positives. In most applications sparsity means that $n_0 \to \infty$ when $n \to \infty$; hence the fixed threshold has an unbounded number of false positives when n grows large. This is in contrast to the universal threshold, which, by Theorem 12.1.5 has asymptotically no false positives. An infinite number of false positives is problematic in a sparse model, as the false positives may easily outnumber the true positives.

Example 12.2.1 *Let n_1 denote the number of nonzero coefficients in a wavelet transform of a piecewise polynomial, including the n_L coarse scaling coefficients. The analysis in Section 8.2 found the number of nonzero detail coefficients to be given by (8.3), which is*

$$n_1 - n_L = \lceil \log_2(n_L)(2\widetilde{p} - 1)n_{11} \rceil = \left\lceil n_{11} \log_2 \left(n_L^{2\widetilde{p}-1} \right) \right\rceil,$$

on a total number of detail coefficients $n - n_L = 2^{J-L} = n_L^{2\widetilde{p}-1}$, meaning that $n_1 - n_L = \mathcal{O}(\log(n - n_L))$, much lower than the expected number of false positives from a constant threshold.

Even though the observations w_i may be independent, it is clear that controlling the number of the false positives requires a multiple testing approach. For this purpose, let $\widehat{\kappa}_F$ denote the number of false positives, $\widehat{\kappa}_T$ the number of true positives, and $\widehat{\kappa} = \widehat{\kappa}_T + \widehat{\kappa}_F$. Then the **family-wise error rate** is defined as $\mathrm{FWER} = P(\widehat{\kappa}_F > 0)$. The universal threshold has $\lim_{n\to\infty} \mathrm{FWER}(\lambda_n) = 0$. A classical procedure to control the FWER is the **Bonferroni correction**, where each individual hypothesis test is performed at a significance level of α/n. The correction is motivated by the pessimistic upper bound

$$\mathrm{FWER} = P\left(\bigcup_{i=1}^{n_0} \{\widehat{p}_i \leq \alpha/n\} \right) \leq \sum_{i=1}^{n_0} P(\widehat{p}_i \leq \alpha/n) \leq n\alpha/n = \alpha.$$

In the context of wavelet coefficients, the Bonferroni correction amounts to a threshold $\lambda_n = F_Z^{-1}(1-\alpha/2n)\sigma$, if the noise has a symmetric density function. For normal noise, this becomes $\lambda_n = \Phi^{-1}(1 - \alpha/2n)\sigma$, from which we find that $1 - \Phi(\lambda_n/\sigma) = \alpha/2n$. When $n \to \infty$, both sides of the equality converge to zero, leading to the conclusion that $\lambda_n \to \infty$. As a result, $1 - \Phi(\lambda_n/\sigma) \sim \phi(\lambda_n/\sigma)(\sigma/\lambda_n)$, so we find

$$\phi(\lambda_n/\sigma)(\sigma/\lambda_n) \sim \alpha/2n.$$

Substitution of λ_n by the universal threshold reveals that the Bonferroni threshold grows slightly slower than the universal one. Indeed, with $\lambda_n = \sqrt{2\log(n)}\sigma$, we have

$$\frac{\phi(\lambda_n/\sigma)(\sigma/\lambda_n)}{\alpha/2n} = \frac{2n}{\alpha n\sqrt{2\log(n)}} \to 0.$$

On the other hand, for finite n and for relatively small values of α, the Bonferroni threshold is a bit larger than the universal threshold.

A less conservative threshold is provided by the notion of False Discovery Rate (FDR), defined by (Benjamini and Hochberg, 1995; Abramovich and Benjamini, 1996),

$$\text{FDR} = E\left(\frac{\widehat{\kappa}_F}{\widehat{\kappa}}\right) = E\left(\frac{\widehat{\kappa}_F}{\widehat{\kappa}_F + \widehat{\kappa}_T}\right), \tag{12.10}$$

where $\widehat{\kappa}_T$ and $\widehat{\kappa}_F$ are the true and false positives, and $\widehat{\kappa} = \widehat{\kappa}_F + \widehat{\kappa}_T$, as before. The FDR can be controlled by the procedure adopted in the following theorem (Benjamini and Hochberg, 1995).

Theorem 12.2.2 *Let $\widehat{p}_{(l)}$ denote the lth order statistic of n independent p-values, i.e., the lth smallest p-value. Consider the procedure that rejects all null hypotheses with $\widehat{p}_{(i)} \leq \widehat{\kappa}a/n$, where $\widehat{\kappa}$ is taken to be*

$$\widehat{\kappa} = \max\{l = 1, 2, \ldots, n | \widehat{p}_{(l)} \leq la/n\}. \tag{12.11}$$

Then the FDR of this procedure is controlled at level a.

Note that if all null hypotheses are true, then the ordered p-values are expected to fluctuate around the identity line. More precisely, being the order statistics of an independent uniform sample on $[0, 1]$, the expected ordered p-values are given by $E(\widehat{p}_{(l)}) = l/(n+1)$. In order to control the FDR at level a, the procedure checks how far the sequence of ordered p-values remains below the lower line la/n. In the context of wavelet thresholding, assuming symmetrically distributed homoscedastic noise, the threshold is defined by $\lambda = F_Z^{-1}(\widehat{\kappa}a/n)\sigma$. Just as for the universal threshold, the construction of this threshold does not make any assumption on the sparsity of the coefficient vector. Unlike the universal threshold, however, the value of the FDR threshold depends on the coefficient vector and so on its sparsity. Indeed, the sparser the vector, the smaller $E(\widehat{\kappa})$, and so the larger the threshold is expected to be.

12.3 High-dimensional sparse variable selection

Hypothesis testing is an asymmetric procedure, where the presumption of innocence holds until proven significant. The asymmetric viewpoint is reinforced by the focus on controlling false discoveries. In many applications, for instance in biostatistics, false positives or type I errors are indeed a major concern. In multiscale models, however, avoiding type II errors or false negatives is often at least as important. This is because missing truly significant coefficients may lead to a blurred or less sharp reconstruction of a singularity in a piecewise smooth function.

Therefore, the problem of wavelet smoothing can be considered in the more symmetric framework of high-dimensional variable selection.

12.3.1 Variable selection in low-dimensional problems

In low dimensions, variable or model selection (Claeskens and Hjort, 2008) can be used as a preparatory step to statistical inference. One of the arguments for a variable selection is that the full model may be large compared to the sample size. Even if the sample size is sufficiently large for the estimation of the parameters, the estimators may be inaccurate, having high variances. Moreover, the parameter estimation will consume a lot of degrees of freedom, with few remaining degrees of freedom for the estimation of the variances. With variances high and hard to estimate, the subsequent hypothesis tests will have a low power, leaving really significant parameters undiscovered. A second argument for variable selection is that the true model is not always the best model for statistical inference. For example, the true model may contain a lot of minor parameters, whose estimation reduces less bias than it introduces variance. Third, keeping the model as simple as possible (but not simpler), reduces the chances of false positive test results. Fourth, variable selection may help to reduce collinearity and hence to control variance inflation. In particular, consider the full, linear model with homoscedastic, uncorrelated noise as in (12.1). Even in a low dimensional case, where the design has more rows than columns, the normal equation for the least squares solution $\mathbf{X}^\top \boldsymbol{Y} = (\mathbf{X}^\top\mathbf{X})\widehat{\boldsymbol{\beta}}$ may have an ill conditioned or even singular matrix $\mathbf{X}^\top\mathbf{X}$, pointing to a rank deficient or nearly rank deficient design matrix, or, in other words, to collinearity. The covariance of the least squares estimator, $\boldsymbol{\Sigma}_{\widehat{\boldsymbol{\beta}}} = \sigma^2(\mathbf{X}^\top\mathbf{X})^{-1}$, depends on the same matrix.

Singular or near singular matrices $\mathbf{X}$ typically appear in linear **inverse problems**, that is, problems where finding the responses $\boldsymbol{\mu}$ from the parameters $\boldsymbol{\beta}$ is relatively easy, while retrieving the parameters from the responses is difficult, even in the absence of noise. In particular, $\mathbf{X}$ may well be a sparse matrix, reducing the computational complexity of the calculation of $\boldsymbol{\mu}$. The inverse, however, is mostly a full matrix, making the retrieval of $\boldsymbol{\beta}$ much more complex. Moreover, the inverse problem is often ill-posed, meaning that two parameter vectors $\boldsymbol{\beta}_1$ and $\boldsymbol{\beta}_2$, relatively far from each other, may produce responses arbitrarily close to each other, so that from the responses alone it is impossible to identify the original parameter vector. Typical examples of inverse problems include deconvolution (for instance, deblurring images), tomographical reconstruction, seismic data, and so on. In this respect, multiscale decompositions are an important exception, having forward and inverse matrices that are both sparse, hence fast in computation, and also well conditioned thanks to the Riesz constants. An example of a deblurring problem occurs when $\mathbf{X}$ would be a simple moving average filter, i.e., $X_{i,j} = \frac{1}{l+r+1}$, for $j = i-l, i-l+1, \ldots, i+r$. The matrix $\mathbf{X}$ has a band structure, leading to linear computational complexity. For $l = 0, r = 1$, this computes the coarse scaling coefficients in a forward nondecimated Haar transform. Without the Haar detail functions, it is a computationally hard (quadratic) and ill conditioned task to recover the original fine scale data.

12.3.2 Regularisation by ridge regression

A regularised optimisation problem is designed to lead to solutions with smaller variances than the original least squares projection. A classical example is Tikhonov regularisation or ridge regression, defined by

$$\min_{\boldsymbol{\beta}} \|\boldsymbol{Y} - \mathbf{X}\boldsymbol{\beta}\|_2^2 + \lambda\|\boldsymbol{\beta}\|_2^2, \tag{12.12}$$

and solved by

$$\widehat{\boldsymbol{\beta}}_\lambda = \left(\mathbf{X}^\top\mathbf{X} + \lambda\mathbf{I}\right)^{-1}\mathbf{X}^\top\boldsymbol{Y}. \tag{12.13}$$

Ridge regression introduces some bias,

$$\begin{aligned}
\text{bias}(\widehat{\boldsymbol{\beta}}(\lambda)) &= E(\widehat{\boldsymbol{\beta}}(\lambda)) - \boldsymbol{\beta} = \left(\mathbf{X}^\top\mathbf{X} + \lambda\mathbf{I}\right)^{-1}\mathbf{X}^\top\mathbf{X}\boldsymbol{\beta} - \boldsymbol{\beta} \\
&= \left(\mathbf{X}^\top\mathbf{X} + \lambda\mathbf{I}\right)^{-1}\left[\mathbf{X}^\top\mathbf{X} - \left(\mathbf{X}^\top\mathbf{X} + \lambda\mathbf{I}\right)\right]\boldsymbol{\beta} \\
&= -\lambda\left(\mathbf{X}^\top\mathbf{X} + \lambda\mathbf{I}\right)^{-1}\boldsymbol{\beta},
\end{aligned}$$

while the covariance matrix becomes

$$\boldsymbol{\Sigma}_{\widehat{\boldsymbol{\beta}}(\lambda)} = \sigma^2(\mathbf{X}^\top\mathbf{X} + \lambda\mathbf{I})^{-1}(\mathbf{X}^\top\mathbf{X})(\mathbf{X}^\top\mathbf{X} + \lambda\mathbf{I})^{-1}.$$

Let $\mathbf{X}^\top\mathbf{X} = \mathbf{V}\mathbf{S}\mathbf{V}^\top$ the eigendecomposition; then $\mathbf{V}$ is an orthogonal matrix and $\mathbf{X}^\top\mathbf{X} + \lambda\mathbf{I} = \mathbf{V}(\mathbf{S} + \lambda\mathbf{I})\mathbf{V}^\top$. Hence

$$\boldsymbol{\Sigma}_{\widehat{\boldsymbol{\beta}}(\lambda)} = \sigma^2\mathbf{V}(\mathbf{S} + \lambda\mathbf{I})^{-1}\mathbf{V}^\top\mathbf{V}\mathbf{S}\mathbf{V}^\top\mathbf{V}(\mathbf{S} + \lambda\mathbf{I})^{-1}\mathbf{V}^\top = \sigma^2\mathbf{V}(\mathbf{S} + \lambda\mathbf{I})^{-2}\mathbf{S}\mathbf{V}^\top.$$

The risk of ridge regression becomes

$$\begin{aligned}
R_{\text{ridge}}(\boldsymbol{\beta}, \lambda) &= \frac{1}{n}\text{Tr}\left(\boldsymbol{\Sigma}_{\widehat{\boldsymbol{\beta}}(\lambda)}\right) + \frac{1}{n}\|E\widehat{\boldsymbol{\beta}}(\lambda) - \boldsymbol{\beta}\|_2^2 \\
&= \frac{\sigma^2}{n}\sum_{j=1}^m \frac{S_{jj}}{(S_{jj} + \lambda)^2} + \frac{\lambda^2}{n}\|\left(\mathbf{X}^\top\mathbf{X} + \lambda\mathbf{I}\right)^{-1}\boldsymbol{\beta}\|_2^2 \\
&\le \frac{\sigma^2}{n}\sum_{j=1}^m \frac{S_{jj}}{(S_{jj} + \lambda)^2} + \frac{\lambda^2}{n}\|\left(\mathbf{X}^\top\mathbf{X} + \lambda\mathbf{I}\right)^{-1}\|_F^2\|\boldsymbol{\beta}\|_2^2 \\
&= \frac{\sigma^2}{n}\sum_{j=1}^m \frac{S_{jj}}{(S_{jj} + \lambda)^2} + \frac{\lambda^2\|\boldsymbol{\beta}\|_2^2}{n}\sum_{j=1}^m \frac{1}{(S_{jj} + \lambda)^2} \\
&= \frac{1}{n}\sum_{j=1}^m \frac{S_{jj}\sigma^2 + \lambda^2\|\boldsymbol{\beta}\|_2^2}{(S_{jj} + \lambda)^2}.
\end{aligned}$$

All terms in this sum can be verified to have a negative derivative at the origin, meaning that there exists a positive regularisation parameter λ so that the risk of the regularised solution is at least somehow lower than the unbiased least squares solution without regularisation. Tikhonov regularisation is an example of Stein's phenomenon (Stein, 1956; Efron and Morris, 1977; Samworth, 2012):

> Biased estimators may have uniformly, for all true values, that is, lower risk than the best unbiased estimator.

This is to be expected for rank-deficient design matrices, but it is equally true for full rank design matrices, even for matrices with orthogonal columns.

12.3.3 Regularisation by variable selection

Ridge regression does not perform any variable selection, but other regularisations do. Assuming sparsity, i.e., the presence of many zeros or near-zeros in $\boldsymbol{\beta}$, the most straightforward example is to replace the ℓ_2 norm $\|\boldsymbol{\beta}\|_2^2$ in (12.12) by the number of nonzero components $\widehat{\kappa}$ in $\boldsymbol{\beta}$. The number of nonzeros can be considered to be the ℓ_0 norm. The regularised problem

$$\min_{\boldsymbol{\beta}} \|\boldsymbol{Y} - \mathbf{X}\boldsymbol{\beta}\|_2^2 + \lambda^2 \widehat{\kappa},$$

has, however, combinatorial complexity, meaning that for every value of $\widehat{\kappa}$, the minimisation procedure has to check every combination of $\widehat{\kappa}$ components out of m to find which choice has the least squared residual sum. If $\mathbf{X} = \mathbf{W}$ is an orthogonal wavelet reconstruction, then the problem becomes much simpler, as indeed,

$$\|\boldsymbol{Y} - \mathbf{X}\boldsymbol{\beta}\|_2^2 + \lambda^2\widehat{\kappa} = \|\mathbf{W}^\top \boldsymbol{Y} - \boldsymbol{\beta}\|_2^2 + \lambda^2\widehat{\kappa} = \sum_{i=1}^{n} \left[(w_i - \beta_i)^2 + \lambda^2 I(\beta_i \neq 0) \right].$$

The minimisation of this sum can proceed componentwise. If $|w_i| < \lambda$, then the best thing to do is to take $\beta_i = 0$. The contribution to the sum is then w_i^2, which is smaller than any alternative containing the term λ^2. On the other hand, if $|w_i| > \lambda$, then the best choice is $\beta_i = w_i$, thereby limiting the contribution to λ^2. The result is a hard threshold scheme. If $\mathbf{X} = \mathbf{W}$ is a biorthogonal inverse wavelet transform, then the hard threshold scheme is not exactly the solution to the ℓ_0 regularised minimisation problem, but thanks to the Riesz bounds, a scheme consisting of a forward wavelet transform, a hard thresholding, and a reconstruction makes a good approximation.

If $\mathbf{X}$ does not admit Riesz bounds and the availability of an analysis $\widetilde{\mathbf{W}}$, then the ℓ_0 regularised minimisation problem can be replaced by another regularisation problem

$$\min_{\boldsymbol{\beta}} \|\boldsymbol{Y} - \mathbf{X}\boldsymbol{\beta}\|_2^2 + 2\lambda^2 \|\boldsymbol{\beta}\|_1.$$

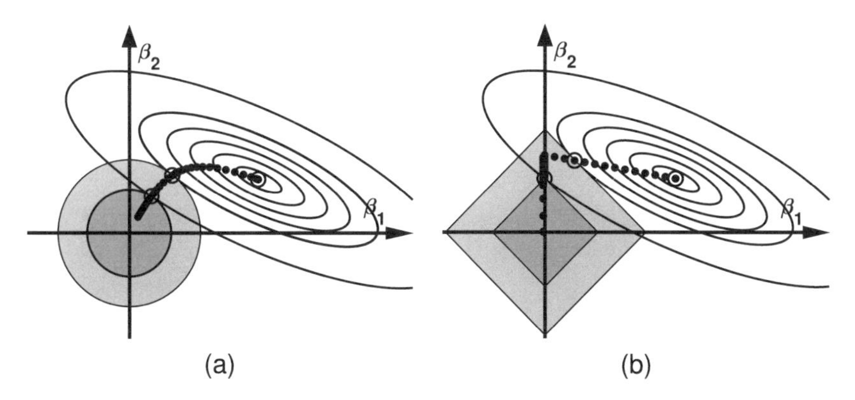

Figure 12.2
Regularised least squares estimation with a full rank design matrix. (a) ℓ_2 or Tikhonov regularisation, also known as ridge regression. (b) ℓ_1 regularisation, lasso.

This is a convex optimisation problem, well known as the **lasso** (least absolute shrinkage and selection operator), and mentioned already in Section 9.4. Lasso includes a variable selection, meaning that the outcome of the optimisation will contain lots of zeros. This can be visualised on a two-dimensional model ($m = 2$), for which Figures 12.2 and 12.4 provide two examples. The former has a regularisation on a problem which is already regular, while the latter has an example with a rank-deficient design matrix.

The regularised least squares problem can be thought of as the Lagrange multiplier version of the constrained optimisation problem, $\min_{\boldsymbol{\beta}} \|\boldsymbol{Y} - \mathbf{X}\boldsymbol{\beta}\|_2^2$ subject to $\|\boldsymbol{\beta}\|_1 = C$ or $\|\boldsymbol{\beta}\|_2^2 = C$, with C an appropriate constant. In Figure 12.2, regularised least squares estimation is applied to the model (12.1) with true values $\boldsymbol{\beta} = \begin{bmatrix} 2 \\ 1 \end{bmatrix}$ and with the full rank design matrix $\mathbf{X} = \begin{bmatrix} 3 & 1 \\ 1 & 4 \\ 3 & 5 \\ 1 & 5 \end{bmatrix}$.

The two panels depict, in black dots, the outcomes of the regularised minimisation problem for different values of the regularisation parameter λ. The ellipses are the contour lines of the residual sum of squares $\|\boldsymbol{Y} - \mathbf{X}\boldsymbol{\beta}\|_2^2$. The left panel shows two disks with boundary circles that are contour lines, defined by $\|\boldsymbol{\beta}\|_2^2 = C$ for two different values of C. The solution of the constrained minimisation problem can be found graphically as the tangent point of the circle and the innermost elliptic contour line that has a nonempty intersection with the disk. Every value C corresponds to a unique tangent point and to a unique value of λ. The higher λ, the smaller C, and the more stringent the constraint. The same construction applies to the right panel, where the Tikhonov regularisation is replaced by the ℓ_1 regularisation. As the con-

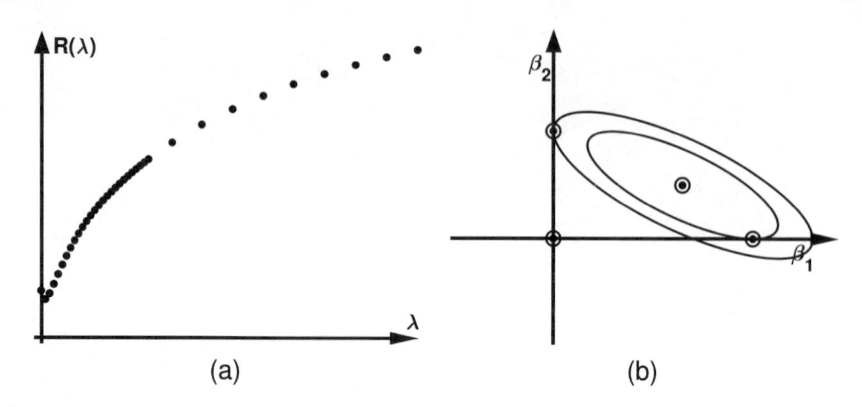

Figure 12.3
(a) Risk values of the ℓ_2-regularised least squares solutions in Figure 12.2(a).
(b) Best κ terms least squares solutions, i.e., ℓ_0 regularisation.

tour lines $\|\boldsymbol{\beta}\|_1 = C$ take the form of a diamond (rhombus), the tangent point may happen to be on one the diamond's vertices, especially when C is small and λ is large. In the edges, one of the two components of $\boldsymbol{\beta}$ is exactly zero. In high dimensions, the outcome of a lasso may have many exact zeros, especially for large values of λ. Given that the actual least squares problem is already regular, the regularisation adds little improvement, as can be seen in Figure 12.3(a), displaying the risk of a Tikhonov regularised solution as a function of the regularisation parameter. For values close to zero, the risk decreases a little before starting to grow.

Figure 12.3(b) illustrates the ℓ_0 regularisation, which is equivalent to the constrained optimisation $\min_{\boldsymbol{\beta}} \|\boldsymbol{Y} - \mathbf{X}\boldsymbol{\beta}\|_2^2$ subject to $\|\boldsymbol{\beta}\|_0 = \kappa$. Taking $\kappa = 2 = m$, the constraint puts no restrictions to the least squares solution. On the other hand, with $\kappa = 0$, the solution ends up at the origin $\beta_1 = 0 = \beta_2$. With $\kappa = 1$, we have to find all two tangent points of the axes with the innermost contour ellipses of the residual sum of squares. As the point on the horizontal axis lies on an innermore ellipse, we conclude that the best one term least squares solution has $\beta_2 = 0$. The regularisation of a singular problem is illustrated in Figure 12.4, which is the analogue of Figure 12.2, but now for a rank deficient design $\mathbf{X} = \begin{bmatrix} 3 & 2 \\ 3 & 2 \\ 3 & 2 \\ 3 & 2 \end{bmatrix}$.
The elliptic contour lines are now degenerated into parallel lines with one central line collecting all points with a minimum residual sum of squares.

In the example of Figures 12.2 and 12.3 the ℓ_0 regularisation leads to selecting β_1, whereas the lasso selects β_2. In general, however, there exist results stating that ℓ_1 regularisation leads to nearly the same degree of spar-

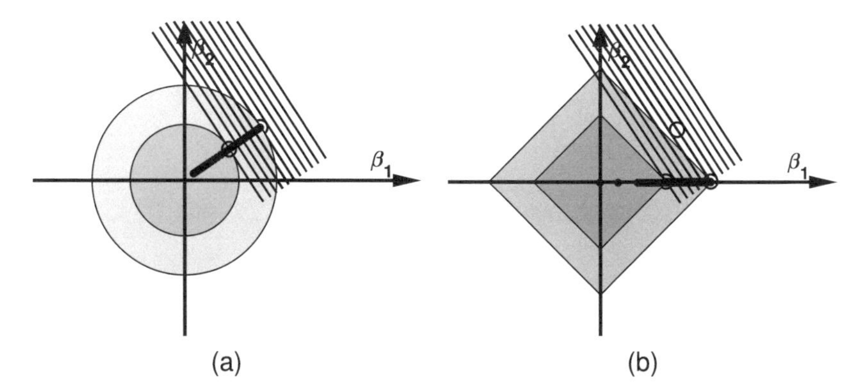

Figure 12.4
Regularised least squares estimation with a full rank design matrix. (a) ℓ_2 or Tikhonov regularisation, also known as ridge regression. (b) ℓ_1 regularisation, lasso.

sity as the ℓ_0 counterpart (Donoho, 2006). The case of orthogonal design matrices is illustrative. We know that the combinatorially complex ℓ_0 regularisation reduces to simple hard thresholding. The lasso, although far less computationally complex in its general version, reduces in exactly the same way to soft thresholding. Indeed, after rewriting

$$\|\boldsymbol{Y}-\mathbf{X}\boldsymbol{\beta}\|_2^2+2\lambda^2\|\boldsymbol{\beta}\|_1 = \|\mathbf{W}^\top\boldsymbol{Y}-\boldsymbol{\beta}\|_2^2+2\lambda^2\|\boldsymbol{\beta}\|_1 = \sum_{i=1}^{n}\left[(w_i-\beta_i)^2+2\lambda|\beta_i|\right],$$

each of the n terms can be minimised for β_i separately. The minimisation of each of these terms proceeds by minimising the function

$$f_i(\beta) = (w_i-\beta)^2+2\lambda|\beta| = w_i^2+|\beta|^2+[2\lambda-2w_i\text{sign}(\beta)]\,|\beta|,$$

which depends quadratically on $|\beta|$. The quadratic function reaches a global minimum at $w_i\text{sign}(\beta)-\lambda$. If this value is negative, then it cannot be used for the non-negative $|\beta|$. If $|w_i|>\lambda$, an admissible nonzero minimum is found by taking $\text{sign}(\beta)=\text{sign}(w_i)$, and $|\beta|=|w_i|-\lambda$, which amounts to soft thresholding. Obviously, hard and soft thresholding are two different estimators, but with the same threshold λ, at least the selection is the same in both cases.

This section has discussed regularisation as a tool to define a variable selection *procedure*. The procedure solves a constrained optimisation problem. The weight of the constraint, λ, is a tuning parameter controlling the size of the selected model. Choices of λ may focus on false positives as in Section 12.2. An alternative for fine-tuning a variable selection procedure is to work with an information criterion, which typically expresses a compromise between model complexity and capability of predicting the observed values. A few examples are developed in Section 12.3.5.

12.3.4 The wavelet-vaguelette decomposition

With $\mathbf{X} = \mathbf{U}\mathbf{S}\mathbf{V}^\top$ the singular value decomposition of $\mathbf{X}$, we have $\mathbf{X}^\top\mathbf{X} = \mathbf{V}\mathbf{S}^2\mathbf{V}^\top$. If the system of normal equations is non-singular, then the least squares solution of the system $\mathbf{X}\boldsymbol{\beta} = \boldsymbol{Y}$ can be written as

$$\begin{aligned}\widehat{\boldsymbol{\beta}} &= \mathbf{V}\mathbf{S}^{-2}(\mathbf{X}\mathbf{V})^\top\boldsymbol{Y}\\ &= \sum_{j=1}^{m}\frac{1}{S_{jj}^2}[(\mathbf{X}\boldsymbol{V}_j)^\top\boldsymbol{Y}]\boldsymbol{V}_j\\ &= \sum_{j=1}^{m}\frac{1}{S_{jj}}[\widetilde{\boldsymbol{v}}_j^\top\boldsymbol{Y}]\boldsymbol{V}_j,\end{aligned}$$

where m denotes the length of the parameter vector $\boldsymbol{\beta}$, $\boldsymbol{V}_j$ is the jth column of $\mathbf{V}$, $\widetilde{\boldsymbol{v}}_j = \mathbf{X}\boldsymbol{V}_j/\|\mathbf{X}\boldsymbol{V}_j\|$. The last step in the argument above follows from the observation that

$$\|\mathbf{X}\boldsymbol{V}_j\| = \sqrt{(\mathbf{X}\boldsymbol{V}_j)^\top(\mathbf{X}\boldsymbol{V}_j)} = \sqrt{\boldsymbol{V}_j^\top\mathbf{V}\mathbf{S}^2\mathbf{V}^\top\boldsymbol{V}_j} = \sqrt{S_{jj}^2}.$$

If the system of normal equations is singular or near-singular, it can be regularised by taking an integer $\kappa \leq m$ so that all S_{jj} for $j = 1, 2, \ldots, \kappa$ are sufficiently large. The smaller the κ, the smaller the variance of the estimator

$$\widehat{\boldsymbol{\beta}} = \sum_{j=1}^{\kappa}\frac{1}{S_{jj}}[\widetilde{\boldsymbol{v}}_j^\top\boldsymbol{Y}]\boldsymbol{V}_j.$$

Just like in the ridge regression case, this estimator assumes no sparsity in $\boldsymbol{\beta}$. The design $\mathbf{X}$ need not be a wavelet reconstruction. When $\boldsymbol{\beta}$ is not sparse, then a wavelet or other multiscale transform $\boldsymbol{\theta} = \widetilde{\mathbf{W}}\boldsymbol{\beta}$ with reconstruction $\boldsymbol{\beta} = \mathbf{W}\boldsymbol{\theta}$ may lead to a sparse representation in $\boldsymbol{\theta}$. The model for the observations is then $\boldsymbol{Y} = \mathbf{X}\mathbf{W}\boldsymbol{\theta}+\sigma\boldsymbol{X}$. The sparse representation can be retrieved from the noise-free observations $\boldsymbol{\mu} = \mathbf{X}\boldsymbol{\beta}$ by the *wavelet-vaguelette decomposition* (Donoho, 1995a) $\boldsymbol{\theta} = \widetilde{\mathbf{W}}\boldsymbol{\beta}\mathbf{V}\mathbf{S}^{-2}(\mathbf{X}\mathbf{V})^\top(\mathbf{X}\boldsymbol{\beta})$. An estimator of $\boldsymbol{\theta}$ is then found from thresholding the empirical value

$$\widehat{\boldsymbol{\theta}} = \mathrm{ST}_\lambda(\widetilde{\mathbf{W}}\boldsymbol{\beta}\mathbf{V}\mathbf{S}^{-2}(\mathbf{X}\mathbf{V})^\top\boldsymbol{Y}),$$

from which $\widehat{\boldsymbol{\beta}} = \mathbf{W}\widehat{\boldsymbol{\theta}}$ can be reconstructed. The roles of wavelets $\mathbf{W}$ in the reconstruction and vagulettes, $\mathbf{X}\mathbf{W}$, in the analysis (learning), can be interchanged (Abramovich and Silverman, 1998).

12.3.5 Expected log-likelihood in variable selection

Many variable selection procedures adopt the notion of likelihood in their fine-tuning, but they do so in a different way than statistical inference. In statistical inference, a likelihood can be used in several stages. First, for the estimation of parameters *within a given model*, maximum likelihood provides a well established and well founded tool. Second, in hypothesis testing, likelihood ratios can be used as test statistics. Third, after testing, validation of the model with the significant parameters may be based on likelihood or related measures such as coefficients of determination.

In variable selection, taking place before statistical inference, the likelihood can be used to *compare different models*. The objective is to select the model in which the maximum likelihood estimator maximises the **expected (log-)likelihood** with respect to the true, underlying **data generating process (DGP)**, denoted by $g_{\boldsymbol{Y}}$. The expected maximum likelihood appears as one of two terms in the definition of the **Kullback-Leibler** divergence

$$\mathrm{KL}\left(g_{\boldsymbol{Y}}, f_{\boldsymbol{Y}}(\cdot;\widetilde{\boldsymbol{\beta}}_p,\widetilde{\sigma}_p^2)\right) = \frac{1}{n}\sum_{i=1}^{n}\left[E\log g_{Y_i}(Y_i) - E\log f_{Y_i}(Y_i;\widetilde{\boldsymbol{\beta}}_p,\widetilde{\sigma}_p^2)\right], \tag{12.14}$$

measuring the distance between two models. In this definition, the marginal densities $f_{Y_i}(y_i;\widetilde{\boldsymbol{\beta}}_p,\widetilde{\sigma}_p^2)$ represent independent observations in the model under consideration, indexed by $p \in \mathcal{M} = \{1,2,\ldots,\widetilde{m}\}$, where $\widetilde{m}$ is the number of candidate models. All models are supposed to plug in a value $\widetilde{\boldsymbol{\beta}}_p$ of the parameter vector $\boldsymbol{\beta}$ in the full model (12.1), namely $\boldsymbol{Y} = \mathbf{X}\boldsymbol{\beta} + \boldsymbol{\varepsilon}$. The value $\widetilde{\boldsymbol{\beta}}_p$ satisfies the constraints imposed by the model under consideration, typically meaning that a large subset of the components of $\widetilde{\boldsymbol{\beta}}_p$ must be zero. The values in $\widetilde{\boldsymbol{\beta}}_p$ may follow from a constrained optimisation of the log-likelihood within that model. The density $g_{Y_i}(y)$, on the other hand, models the true, unknown data generating process. The expected values in (12.14) are computed with respect to $g_{Y_i}(y)$. The expression in (12.14) can also be written as

$$\mathrm{KL}\left(g_{\boldsymbol{Y}}, f_{\boldsymbol{Y}}(\cdot;\widetilde{\boldsymbol{\beta}}_p,\widetilde{\sigma}_p^2)\right) = \frac{1}{n}E\left[\frac{g_{\boldsymbol{Y}}(\boldsymbol{Y})}{f_{\boldsymbol{Y}}(\boldsymbol{Y};\widetilde{\boldsymbol{\beta}}_p,\widetilde{\sigma}_p^2)}\right].$$

The form in (12.14) has the advantage that the first term, $E\log g_{Y_i}(Y_i)$, is common to all models. The second term is exactly the expected log-likelihood: the higher the expected log-likelihood, the smaller the Kullback-Leibler divergence from the true model.

12.3.6 Balance between bias and variance

In sparse multiscale models, it is fairly easy to estimate the nuisance parameter σ^2 independently from the variable selection process, for example, using the median absolute deviation as in (12.8). In that case, the expected log-likelihood for a model with normal errors becomes

$$\begin{aligned}
E\left[-\log L(\widetilde{\boldsymbol{\beta}}_p;\boldsymbol{Y})\right] &= \frac{1}{2\sigma^2}E\left[(\boldsymbol{Y}-\widetilde{\boldsymbol{\mu}}_p)^\top(\boldsymbol{Y}-\widetilde{\boldsymbol{\mu}}_p)\right]+\frac{n}{2}\log(2\pi\sigma^2)\\
&= \frac{1}{2\sigma^2}E\|\boldsymbol{Y}-\widetilde{\boldsymbol{\mu}}_p\|_2^2+\frac{n}{2}\log(2\pi\sigma^2)\\
&= \frac{1}{2\sigma^2}\|\boldsymbol{\mu}-\widetilde{\boldsymbol{\mu}}_p\|_2^2+\frac{1}{2\sigma^2}E\|\boldsymbol{Y}-\boldsymbol{\mu}\|_2^2+\frac{n}{2}\log(2\pi\sigma^2)\\
&= \frac{1}{2\sigma^2}\|\boldsymbol{\mu}-\widetilde{\boldsymbol{\mu}}_p\|_2^2+\frac{1}{2\sigma^2}n\sigma^2+\frac{n}{2}\log(2\pi\sigma^2)\\
&= n\mathrm{PE}(\boldsymbol{\beta};p)+\frac{1}{2\sigma^2}n\sigma^2+\frac{n}{2}\log(2\pi\sigma^2),
\end{aligned}$$

where the **prediction error (PE)** is defined as

$$\mathrm{PE}(\boldsymbol{\beta};p)=\frac{1}{n}E\left(\|\widehat{\boldsymbol{\mu}}_p-\boldsymbol{\mu}\|^2\right)=\frac{1}{n}E\left(\|\mathbf{X}\widehat{\boldsymbol{\beta}}_p-\mathbf{X}\boldsymbol{\beta}\|_p^2\right). \tag{12.15}$$

Note that the definition of the prediction error allows $\widehat{\beta}_p$ to be an estimator (hence, a sample dependent variable), whereas the above equivalence between the prediction error and the expected log-likelihood holds for a fixed choice of $\widetilde{\beta}_p$ only.

The prediction error measures the error on the response side, whereas the risk operates on the parameter side of the design. In a wavelet decomposition, the risk is expressed in terms of wavelet coefficients, while the prediction error uses the reconstructed values. Thanks to the Riesz constants, both measures are nearly equivalent. Full equivalence is attained in an orthogonal wavelet decomposition.

Both the prediction error and the risk can be decomposed into bias and variance components as indeed,

$$\begin{aligned}
\mathrm{PE}(\boldsymbol{\beta};p) &= \frac{1}{n}E\left(\|\widehat{\boldsymbol{\mu}}_p-E(\widehat{\boldsymbol{\mu}}_p)+E(\widehat{\boldsymbol{\mu}}_p)-\boldsymbol{\mu}\|^2\right)\\
&= \frac{1}{n}\left(\sum_{i=1}^n\mathrm{var}(\widehat{\mu}_i)+E\left[\widehat{\mu}_i-\mu_i\right]^2\right).
\end{aligned} \tag{12.16}$$

For the risk this becomes

$$R(\boldsymbol{\beta};p)=\frac{1}{n}E\left(\|\widehat{\boldsymbol{\beta}}_p-\boldsymbol{\beta}\|_p^2\right)=\frac{1}{n}\left(\sum_{j=1}^m\mathrm{var}(\widehat{\beta}_j)+E\left[\widehat{\beta}_j-\beta_j\right]^2\right).$$

The variance is related to false positives, while false negatives induce bias. Indeed, coefficients with magnitudes below the threshold are set to zero, and therefore do not contribute to the variance. On the other hand, large coefficients survive the threshold, thus bringing noise back into the reconstruction. In a soft threshold scheme, the impact of false positives is tempered, at the price of introducing a shrinkage bias. This form of bias comes on top of the false negatives.

12.3.7 The closeness-complexity balance

The prediction error can also be decomposed in a way different from that in (12.16). Introducing the vector of **residuals** $\boldsymbol{e}_p = \boldsymbol{Y} - \widehat{\boldsymbol{\mu}}_p$, and its squared norm, $\mathrm{SS_E}(\widehat{\boldsymbol{\beta}}_p) = \|\boldsymbol{e}_p\|_2^2$, termed the **sum of squared residuals** (or residual sum of squares), we find

$$\begin{aligned}
\mathrm{PE}(\boldsymbol{\beta}; p) &= \frac{1}{n} E\|\widehat{\boldsymbol{\mu}}_p - \boldsymbol{\mu}\|_2^2 = \frac{1}{n} E\|\widehat{\boldsymbol{\mu}}_p - \boldsymbol{Y} + \boldsymbol{Y} - \boldsymbol{\mu}\|_2^2 = \frac{1}{n} E\| - \boldsymbol{e}_p + \boldsymbol{\varepsilon}\|_2^2 \\
&= \frac{1}{n} \left[E\|\boldsymbol{e}_p\|_2^2 + 2E(-\boldsymbol{e}_p^\top \boldsymbol{\varepsilon}) + E\|\boldsymbol{\varepsilon}\|_2^2 \right] \\
&= \frac{1}{n} E\left[\mathrm{SS_E}(\widehat{\boldsymbol{\beta}}_p)\right] + \frac{2}{n} E\left[(\boldsymbol{\varepsilon} - \boldsymbol{e}_p)^\top \boldsymbol{\varepsilon}\right] - \frac{1}{n} E\|\boldsymbol{\varepsilon}\|_2^2 \\
&= \frac{1}{n} E\left[\mathrm{SS_E}(\widehat{\boldsymbol{\beta}}_p)\right] + \frac{2\nu_p}{n}\sigma^2 - \sigma^2, \qquad (12.17)
\end{aligned}$$

where ν_p stands for the number of (generalised) degrees of freedom, defined as (Ye, 1998)

$$\nu_p = \frac{1}{\sigma^2} E\Big[\boldsymbol{\varepsilon}^\top(\boldsymbol{\varepsilon} - \boldsymbol{e}_p)\Big]. \qquad (12.18)$$

The result in (12.17) can be read as a compromise between the closeness of fit, expressed by the expected sum of squared residuals, and the complexity of the adopted model, expressed by the degrees of freedom.

To understand that the degrees of freedom are a measure of model complexity, we first develop an alternative expression

$$\nu_p = \frac{1}{\sigma^2} E(\boldsymbol{\varepsilon}^\top \widehat{\boldsymbol{\mu}}_p), \qquad (12.19)$$

which follows from a straightforward calculation

$$\nu_p = \frac{1}{\sigma^2} E\Big[\boldsymbol{\varepsilon}^\top(\boldsymbol{\varepsilon} - \boldsymbol{e}_p)\Big] = \frac{1}{\sigma^2} E\Big[\boldsymbol{\varepsilon}^\top(\boldsymbol{Y} - \boldsymbol{\mu} - \boldsymbol{Y} + \widehat{\boldsymbol{\mu}}_p)\Big] = \frac{1}{\sigma^2} E\Big[\boldsymbol{\varepsilon}^\top(\widehat{\boldsymbol{\mu}}_p - \boldsymbol{\mu})\Big].$$

Let $\widehat{\boldsymbol{\mu}}_p = \mathbf{P}_p \boldsymbol{Y}$ denote a linear estimator within a fixed, admissible model, indexed by $p \in \mathcal{M}$ as in (12.14). Depending on the context, the matrix $\mathbf{P}_p$ is termed a **projection matrix** or an **influence matrix**. Then it holds that

$$\nu_p = \mathrm{Tr}(\mathbf{P}_p), \qquad (12.20)$$

as indeed,

$$\begin{aligned}
\nu_p &= E(\boldsymbol{\varepsilon}^\top \widehat{\boldsymbol{\mu}}_p)/\sigma^2 = E(\boldsymbol{\varepsilon}^\top \mathbf{P}_p \boldsymbol{Y})/\sigma^2 = E\left[\boldsymbol{\varepsilon}^\top \mathbf{P}_p(\boldsymbol{\mu} + \boldsymbol{\varepsilon})\right]/\sigma^2 \\
&= 0 + E\left[\mathrm{Tr}(\mathbf{P}_p \boldsymbol{\varepsilon}\boldsymbol{\varepsilon}^\top)\right]/\sigma^2 = \mathrm{Tr}(\mathbf{P}_p \sigma^2 \mathbf{I})/\sigma^2 = \mathrm{Tr}(\mathbf{P}_p).
\end{aligned}$$

As a special case, we consider a submodel $p \in \mathcal{M}$ of the full regression model (12.1), defined by $\beta_i = 0$ for $i \notin \mathrm{S}_p$. The set S_p is the pth subset in an enumeration of all subsets of $\{1, 2, \ldots, m\}$, where m is the length of vector $\boldsymbol{\beta}$ in (12.1), i.e., the full model size. Let κ denote the cardinality of S_p. If $\mathbf{P}_p$ represents the orthogonal projection onto S_p, then it is a symmetric, idempotent matrix, as explained in Section 3.1.4, for which $\mathrm{Tr}(\mathbf{P}_p) = \kappa$.

Soft thresholding in an orthogonal wavelet decomposition is a prototype of a nonlinear scheme, including the selection of the model. The model is fully determined by the threshold value, which is reflected in the notation by replacing index p by λ. By direct application of Stein's lemma in (12.3) the expression (12.19) for the degrees of freedom becomes

$$\nu_\lambda = \frac{1}{\sigma^2} E(\boldsymbol{\varepsilon}^\top \widehat{\boldsymbol{\mu}}_\lambda) = \frac{1}{\sigma^2} E(\boldsymbol{\eta}^\top \mathrm{ST}_\lambda(\boldsymbol{w})) = \sum_{i=1}^{n} P(|w_i| > \lambda) = E(\widehat{\kappa}),$$

where $\widehat{\kappa}$ is the number of nonzero coefficients (the union of false and true positives, as in (12.10) that is). This result can be extended towards general ℓ_1 regularised linear regression, lasso that is (Zou et al., 2007; Tibshirani and Taylor, 2012).

As a conclusion, the orthogonal projection onto a fixed subspace and the shrinkage coefficient selection using soft thresholding have similar expressions for the number of degrees of freedom. The number equals the model size or the expected model size, making it indeed a measure of the model complexity.

12.3.8 Estimating the closeness-complexity balance

Define

$$\Lambda_\lambda = \frac{1}{n} \mathrm{SS_E}(\widehat{\boldsymbol{\beta}}_\lambda) + \frac{2\nu_\lambda}{n} \sigma^2 - \sigma^2; \tag{12.21}$$

then, in expected value, the closeness-complexity balance is equivalent to the original bias-variance balance,

$$E(\Lambda_\lambda) = \mathrm{PE}(\boldsymbol{\beta}; \lambda). \tag{12.22}$$

If ν_λ and σ can be easily estimated, then Λ_λ can be used to estimate the optimal bias-variance and closeness-complexity balances. In the assessment of a fixed regression model, where the ν_λ equals the model size κ, the expression of Λ_λ can be divided by the variance or a model independent estimator of it, leading to a studentised version of (12.21), known as Mallows' C_p (Mallows, 1973). In the context of soft thresholding, the degrees of freedom can be estimated unbiasedly by $\widehat{\nu}_\lambda = \widehat{\kappa}$, leading to

$$\mathrm{SURE}(\lambda) = \frac{1}{n} \mathrm{SS_E}(\widehat{\boldsymbol{\beta}}_\lambda) + \frac{2\widehat{\kappa}}{n} \sigma^2 - \sigma^2, \tag{12.23}$$

known as Stein's unbiased risk estimator (Stein, 1981; Donoho and Johnstone, 1995; Luisier et al., 2007).

An alternative with implicit variance estimator is provided by **generalised cross validation** (GCV), which comes from a formalisation, followed by a simplification of the leave-out-one cross validation (CV) procedure (Wahba, 1990).

$$\mathrm{GCV}(\lambda) = \frac{\frac{1}{n}\mathrm{SS_E}(\widehat{\boldsymbol{\beta}}_\lambda)}{\left(1 - \frac{\nu_\lambda}{n}\right)^2}. \tag{12.24}$$

GCV combines the benefits from CV (Nason, 1996) and C_p: like C_p, it is much faster than ordinary cross validation. Like CV, it is does not rely on an explicit variance estimator. The implicit variance estimator is reported to be quite robust (Jansen, 2015).

The working of GCV is based on its relation with C_p

$$\mathrm{GCV}(\lambda) - \sigma^2 = \frac{\Lambda_\lambda - \left(\frac{\nu_\lambda}{n}\right)^2 \sigma^2}{\left(1 - \frac{\nu_\lambda}{n}\right)^2},$$

which leads to two results supporting its use.

Theorem 12.3.1 *(uniform efficiency) Under certain conditions of* ***sparsity*** *(Jansen, 2015), GCV is* ***uniformly efficient****. More precisely, there exists a sequence $\mathbb{L}_n$ of (large) subsets of $\mathbb{R}$, so that*

$$\sup_{\lambda \in \mathbb{L}_n} \frac{|\mathrm{GCV}(\lambda) - \sigma^2 - \Lambda_\lambda|}{\Lambda_\lambda + V_n} \xrightarrow{\mathrm{P}} 0 \text{ as } n \to \infty, \tag{12.25}$$

where

$$V_n = \max\left(0, \sup_{\lambda \in \mathbb{L}_n} (\mathrm{PE}(\boldsymbol{\beta}; \lambda) - \Lambda_\lambda)\right).$$

This result states that on a large range of regularisation parameter values in a sparse selection procedure within a linear regression model, GCV is "equivalent" to Mallows' C_p, meaning that, roughly speaking, $\mathrm{GCV}(\lambda) \approx \Lambda_\lambda + \sigma^2$. The approximation holds in a relative sense with respect to the value of the prediction error, so that the minimiser of $\mathrm{GCV}(\lambda)$ is asymptotically efficient.

Corollary 12.3.2 *Let $\widehat{\lambda}_n^* = \arg\min_{\lambda \in \mathbb{L}_n} \Lambda_\lambda$ and $\widehat{\widehat{\lambda}}_n = \arg\min_{\lambda \in \mathbb{L}_n} \mathrm{GCV}(\lambda)$, with $\mathbb{L}_n$ defined in Theorem 12.3.1; then it holds that*

$$\frac{\Lambda_{\widehat{\widehat{\lambda}}_n, p} + V_n}{\Lambda_{\widehat{\lambda}_n^*, p} + V_n} \xrightarrow{\mathrm{P}} 1, \tag{12.26}$$

with V_n as in Theorem 12.3.1.

In contrast to the link between C_p and the prediction error, the equivalence between GCV and C_p holds asymptotically; it does not require the use of expectations, and it relies on sparsity assumptions.

The second result (Jansen, 2015) is about the behaviour of GCV for very small thresholds in a lasso or soft threshold scheme.

Theorem 12.3.3 *In a sparse variable selection scheme defined by an ℓ_1 regularised least squares problem,*

$$\lim_{\lambda \to 0} E\left[\mathrm{GCV}(\lambda)\right] = c_n > 0. \tag{12.27}$$

Under mild assumptions on the error density $f_Z(z)$ and the sparsity of $\boldsymbol{\beta}$, it holds, in a signal-plus-noise model $\boldsymbol{w} = \boldsymbol{\beta} + \sigma \boldsymbol{Z}$ for wavelet coefficients in an orthogonal wavelet basis, and using soft thresholding, that

$$\lim_{n \to \infty} c_n = \frac{\sigma^2}{4 f_Z(0)}, \tag{12.28}$$

which is, for $f_Z(z)$ the standard normal density, $c_n \to \sigma^2 \pi/2$.

This theorem guarantees that the behaviour of $\mathrm{GCV}(\lambda)$ for λ near zero does not disturb the minimisation. Near zero, $E\mathrm{GCV}(\lambda)$ deviates from being a vertical translation of the prediction error, but the deviation is not so strong that it creates a local optimum.

The foregoing results are illustrated in Figure 12.5, plotting the values of the $C_p(\lambda)$, $\mathrm{GCV}(\lambda)$ and $\mathrm{PE}(\boldsymbol{\beta}; \lambda)$ as a function of λ in a simulation study with a sparse vector of size 2000 of which 5% are nonzero. The nonzeros are sampled from a double exponential (i.e., Laplace) distribution. All plots have approximately the same minimum.

12.3.9 The shape of the balance curve

For the same simulation study as in Figure 12.5, Figure 12.6 compares the soft and hard thresholding prediction errors. Hard and soft thresholding are two representatives of a wider classes of ℓ_0 and ℓ_1 regularised least squares approaches. Although the two regularisations lead to similar selections, the estimator for a given selection behaves quite differently. As a result, the best threshold for soft thresholding is not the same as for hard thresholding.

In particular, for small threshold values, soft thresholding performs better. This is in line with Stein's phenomenon (see Section 12.3.2) stating that the introduction of some shrinkage bias may outperform an unbiased orthogonal projection. On the other hand, since all coefficients above the threshold are shrunk by the threshold value, the shrinkage bias increases when the threshold value grows larger. As a result, the optimal soft threshold is smaller than the optimal hard threshold. In this example the optimal hard threshold slightly outperforms the optimal soft threshold in terms of prediction error.

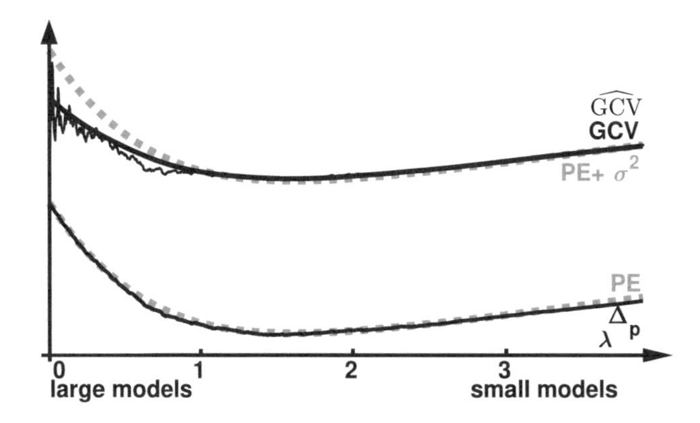

Figure 12.5
Plots of C_p, GCV, and the prediction error as a function of the threshold value. The GCV is a vertical shift of the prediction error, except near the origin. The deviation of the GCV curve near the origin does not hinder the optimisation.

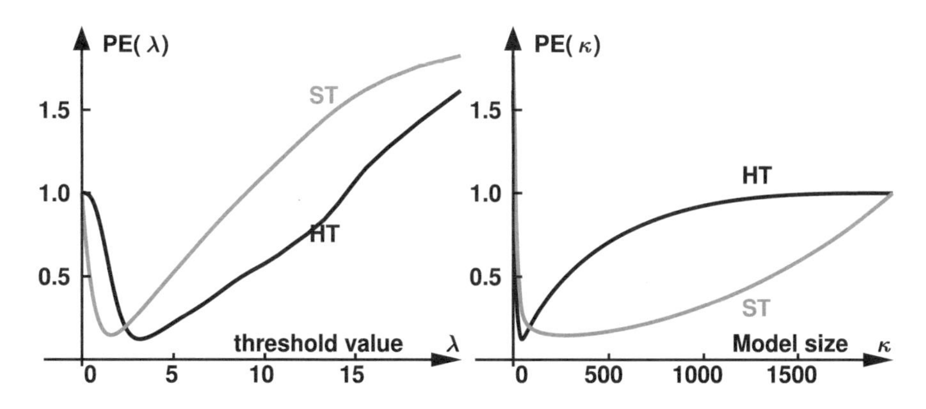

Figure 12.6
Plots of the soft and hard threshold prediction errors for the same simulation study as in Figure 12.5. Left panel: prediction errors as a function of the threshold value. Right panel: as a function of the model size, i.e., the number of nonzero coefficients.

The right panel of Figure 12.6 depicts the prediction errors as a function of the number of selected components. Now the curves of hard and soft thresholding show an even more different behaviour. From the hard threshold curve, it becomes clear that the choice of the model size is a delicate problem. Choosing the size a bit too large or too small results in a projection with rapidly increasing prediction errors. The soft threshold problem is far more relaxed, but the optimal size is much larger, a few hundreds versus a few dozens in the hard threshold case. Given the fact that the true model has 5% nonzeros out of 2000 components, the optimal hard threshold is much closer to the true model. The explanation for the gap between soft and hard threshold behaviour lies in the false positives. As the effect of a false positive selection is tempered by the shrinkage, soft thresholding is tolerant towards false positives, leading to models that are way too large. This comes on top of the shrinkage bias in large components: the smaller the model, the larger the threshold, and so the larger the bias on the selected components.

In order to reduce the shrinkage bias, the ℓ_1 regularised least squares method can be relaxed to a method of the form

$$\min_{\boldsymbol{\beta}} \|\boldsymbol{Y} - \mathbf{X}\boldsymbol{\beta}\|_2^2 + \lambda \sum_{j=1}^{m} q(|\beta_j|),$$

where the function $q(x) \sim x$ for $x \to 0$ while $q(x)$ is bounded when $x \to \infty$. In other words, for small values of β_j, the regularisation behaves as an ℓ_1 regularisation, while for large values of β_j, the regularisation does not depend on the actual value, just like in the ℓ_0 regularised problem. Popular examples include the smoothly clipped absolute deviation (SCAD) (Fan and Li, 2001) and a minimax concave penalty (MCP) (Zhang, 2010). Other shrinkage rules that combine benefits from hard and soft thresholding include the non-negative garrote (Gao, 1998).

Alternatively, the degrees of freedom in a hard threshold scheme, for use in (12.21), can be approximated by (Jansen, 2014)

$$\widehat{\nu}_\lambda = \widehat{\kappa} + n \int_{-\lambda/\sigma}^{\lambda/\sigma} (1 - z^2) f_Z(z) dz, \tag{12.29}$$

where $f_Z(z)$ is the standardised density of the noise, as in Theorem 12.3.3. This becomes

$$\widehat{\nu}_\lambda = \widehat{\kappa} + 2(n\lambda/\sigma)\phi(\lambda/\sigma)$$

when $Z \sim N(0, 1)$.

12.4 Estimating in sparse function classes

The methods discussed in Section 12.3 consider the selection of wavelet coefficients as a high-dimensional variable selection problem. In contrast to the decision theoretic approach of Section 12.1 and the multiple testing viewpoint in Section 12.2, the high-dimensional variable selection hinges on the assumption of sparsity. Without sparsity, the problem would not be regularised.

This section further explores sparsity from another angle, phrasing results in function classes, such as Besov balls, that express sparsity.

12.4.1 Minimax thresholds

The performance of a threshold rule in a function class $\mathcal{F}$ can be measured by comparison with the minimax risk. This benchmark is defined as follows.

Definition 12.4.1 *Let $V_J \subset L_2([0,1])$ be the space spanned by scaling functions $\Phi_J(x)$ and let $\mathcal{F} \subset V_J$. Denote by $\boldsymbol{\beta}$ the wavelet coefficients of $f \in \mathcal{F}$, and by $\widehat{\boldsymbol{\beta}} = \widehat{\mathrm{B}}(\boldsymbol{w})$ an estimation procedure. Then the minimax risk of the estimation problem in that function class is given by*

$$\mathcal{R}(\mathcal{F}) = \inf_{\widehat{\mathrm{B}}} \sup_{f \in \mathcal{F}} R_{\widehat{\mathrm{B}}}(\boldsymbol{\beta}), \tag{12.30}$$

where $n = 2^J$ and $R_{\widehat{\mathrm{B}}}(\boldsymbol{\beta}) = \sum_{j=1}^{m} E(\widehat{\beta}_j - \beta_j)^2$ is the risk of the estimation procedure.

The minimax risk thus measures the complexity of the estimation problem. The notations do not reflect the fact that the complexity depends on the variances or the covariance matrix of the observations.

A first result (Donoho and Johnstone, 1998, Theorem 7) in this context states that wavelet thresholding is a near-optimal strategy in estimating functions with bounded Besov norm.

Theorem 12.4.2 *(near-minimax thresholding) Let $\mathcal{F}_n(p,q,\alpha,R) = V_n \cap B^{\alpha}_{p,q}(R)$ where $B^{\alpha}_{p,q}(R)$ is a Besov ball as defined in Section 8.3.4. Let $\widehat{\beta}_j = \mathrm{T}_{\lambda_i}(w_i)$ be a threshold procedure, where $\mathrm{T}_{\lambda_i} \in \{\mathrm{ST}_{\lambda_i}, \mathrm{ST}_{\lambda_i}\}$; then there exists a constant $K_{p,q}$, dependent on p and q, so that for n sufficiently large,*

$$\inf_{\boldsymbol{\lambda}} \sup_{f \in \mathcal{F}_n(p,q,\alpha,R)} R_{\widehat{\mathrm{T}}}(\boldsymbol{\beta}, \boldsymbol{\lambda}) \leq K_{p,q} \cdot \mathcal{R}(\mathcal{F}_n(p,q,\alpha,R)). \tag{12.31}$$

In other words, a soft or a hard threshold scheme with well chosen coefficient dependent thresholds $\boldsymbol{\lambda} = (\lambda_1, \lambda_2, \ldots, \lambda_n)$ performs nearly as well as the overall minimax estimator.

The next result (Donoho and Johnstone, 1995) states that the near-minimax thresholds in their turn are well mimicked by the level-dependent SURE thresholds.

Theorem 12.4.3 *(SureShrink) Consider a resolution level dependent implementation of a minimum Stein unbiased risk estimator, as defined in (12.23), leading to a level dependent threshold $\boldsymbol{\lambda}_{\mathrm{SURE}}$. At levels with very sparse coefficients, defined in (Donoho and Johnstone, 1995), the SURE threshold is replaced by the universal threshold. Then, for any $p, q \geq 1$, $R > 0$ and under mild conditions for α,*

$$\sup_{f \in \mathcal{F}_n(p,q,\alpha,R)} R_{\widehat{\mathrm{ST}}}(\boldsymbol{\beta}, \boldsymbol{\lambda}_{\mathrm{SURE}}) \leq K'_{p,q} \cdot \inf_{\boldsymbol{\lambda}} \sup_{f \in \mathcal{F}_n(p,q,\alpha,R)} R_{\widehat{\mathrm{T}}}(\boldsymbol{\beta}, \boldsymbol{\lambda}). \tag{12.32}$$

The level dependent minimum risk (or its Stein unbiased estimator) thresholds automatically performs near-minimax in any Besov ball, even without knowing the parameters p, q, α, and R of that ball. For this we need resolution level dependent thresholds, even if the variance is the same at all levels. This level dependency thus serves a different goal than that in the problem of non-orthogonal transforms or correlated noise, although all adaptivity to level dependent variances and level dependent sparsity are of course dealt with at the same time by the level dependent minimum risk thresholds.

12.4.2 Tree structured selection

Threshold assessment aiming at minimum risk, such as SURE (Theorem 12.4.3) and GCV (Theorem 12.3.1) suffer from visual artifacts due to false positives. These false positives are inherent to the method, precisely because minimisation of the risk amounts to finding a good compromise between bias and variance, or false negatives and false positives. Methods that focus on false positives, such as the universal threshold, have far less — if any at all — of these spurious spikes or other artifacts. Within the framework of Besov smoothness, the near absence of false positives leads to the following result (Donoho, 1995b; Donoho et al., 1995).

Theorem 12.4.4 *Let $f \in L_2([0,1])$ and let $\Psi(x)$ be a wavelet basis whose elements are Lipschitz-γ continuous and have $\widetilde{p}$ dual vanishing moments. Then there exists a constant C, not depending on f or n, and a sequence p_n tending to one for sample size n growing to infinity, so that the reconstruction $\widehat{f}_n$ from using a universal threshold in a soft thresholding scheme satisfies*

$$P\left(\|\widehat{f}_n\|_{B^{\alpha}_{p,q}} \leq C\|f\|_{B^{\alpha}_{p,q}}, \forall (p,q,\alpha) \in A\right) \geq p_n, \tag{12.33}$$

where A is defined by $\alpha > 1/p$, $\widetilde{p} > \alpha$ and $\gamma > \alpha$.

In other words, if the wavelet reconstruction has sufficiently smooth basis functions and if the wavelet analysis has enough vanishing moments providing an unconditional basis for Besov spaces (as in Section 8.3.5), then with a probability tending to one for growing sample sizes, the estimator is at least as smooth as the underlying, true function, the property holding simultaneously in any Besov smoothness norm.

The smoothness of the universal threshold comes at the price of blur near the singularities due to false negatives. The false positives in minimum risk based selections can be reduced by imposing a tree structured selection. The imposed structure fits with the binary tree that can be associated to a decimated wavelet transform. The rooted binary tree of a complete wavelet decomposition is defined by taking the single detail at coarsest level, $d_{0,0}$ as root and letting $d_{j+1,2k}$ and $d_{j+1,2k+1}$ be the children of $d_{j,k}$. The rationale behind this construction is that the coefficients in the father-child relationship describe features at different scales but at the same location. More precisely, it can be verified that the union of the supports of the fine scaling wavelet functions $\psi_{j+1,2k}$ and $\psi_{j+1,2k+1}$ is a subset of the support of the basis function $\psi_{j,k}$ at coarse scale.

For the sake of simplification and generalisation, we introduce a notation with a single index, $X_i = d_{j,k}$, where $i = 2^j + k \in \{1, 2, \ldots, 2^J - 1\}$. The children of X_i are then X_{2i} and X_{2i+1}. Within the tree structure, admissible selections of wavelet coefficients are restricted to sets where each element also has its father element in the set. This means that the admissible selections S belong to $\mathcal{T}_1$, the set of binary subtrees rooted at the first coefficient, defined by taking $i = 1$ in

$$\mathcal{T}_i = \left\{\mathrm{S} \subset \{i, i+1, \ldots, n\}; i \in \mathrm{S} \text{ and } \{2l, 2l+1\} \cap \mathrm{S} \neq \emptyset \Rightarrow l \in \mathrm{S}\right\}.$$

The idea is that important features in a function, signal, or image show contributions at several scales. Imposing a binary subtree selection thus removes most of the isolated false positives.

Searching for the tree structure selection with minimum C_p value proceeds in two steps. First, the best κ-subtrees are found for all values of κ, using a **backtracking** routine, related to Best Orthogonal Basis selection (Coifman and Wickerhauser, 1992). The backtracking routine fulfills the role of a selection procedure, just like a thresholding or lasso selection routine. Assuming that selected coefficients are not shrunk in one way or another, the routine runs as follows. The expression of C_p, as given in (12.21), is minimised by maximising

$$T(\mathrm{S}; \boldsymbol{X}) = \sum_{l \in \mathrm{S}} |X_l|^2.$$

Now, define $T^*_{i,\kappa} = \max_{\mathrm{S} \in \mathcal{T}_{i,\kappa}} T(\mathrm{S}; \boldsymbol{X})$, where $\{\mathcal{T}_{i,\kappa}, \kappa = 0, 1, \ldots, n\}$ is a partitioning of $\mathcal{T}_i$ according to the size κ of the subtrees. We have the property that

$$T^*_{i,\kappa+1} = |X_i|^2 + \max_{l=0,1,\ldots,\kappa} \left(T^*_{2i,l} + T^*_{2i+1,\kappa-l}\right). \tag{12.34}$$

As a result, with

$$\mathrm{S}^*_{i,\kappa} = \arg \max_{\mathrm{S} \in \mathcal{T}_{i,\kappa}} T(\mathrm{S}; \boldsymbol{X}),$$

and $\ell = \arg\max_{l=0,1,\ldots,\kappa} \left(T^*_{2i,l} + T^*_{2i+1,\kappa-l}\right)$, we have

$$\mathrm{S}^*_{i,\kappa+1} = \{i\} \cup \mathrm{S}^*_{2i,\ell} \cup \mathrm{S}^*_{2i+1,\kappa-\ell}. \tag{12.35}$$

A simultaneous branch and bound algorithm calculating all best κ-subtrees in knots i thus proceeds by applying (12.35) backwards, from fine to coarse scales that is, ending up with all values of best κ subtrees rooted at the first point.

In the second step of the minimisation of C_p, the best κ-subtrees are compared among each other, by taking the complexity dependent penalty of (12.21) into account. Since the tree structured selection involves no shrinkage, the degrees of freedom need to be estimated in a way similar to that of (12.29).

Remark 12.4.5 *The binary tree structure of a wavelet transform and the clustering of large coefficients across scales within that tree are not captured by smoothness spaces or norms, such as the Besov sequence norm of (8.9). Function classes defined by Besov sequence norms therefore contain objects composed of non-structured wavelet vectors.*

12.4.3 Block thresholding

Another example of group structured coefficient selection is given by block thresholding (Cai, 1999). As in tree structured selection, block thresholding reduces the number of isolated false positives. Let the vector of detail coefficients, $\boldsymbol{w} = \boldsymbol{\beta} + \sigma \boldsymbol{Z}$, be partitioned into subvectors of length n_B. These blocks may contain adjacent coefficients at a single scale, or they may group coefficients at different scales and locations that are believed to share a common behaviour. A tree based cluster of fixed size n_B would fit within this framework. Let B_l denote the partition, so that $\bigcup_l \mathsf{B}_l = \{1, 2, \ldots, n\}$; then a block soft threshold scheme estimates

$$\widehat{\boldsymbol{\beta}}_{\mathsf{B}_l} = \max\left(0, 1 - \lambda \frac{n_B \sigma^2}{\sum_{i \in \mathsf{B}_l} |d_i|^2}\right) \boldsymbol{d}_{\mathsf{B}_l}. \tag{12.36}$$

An interesting oracle inequality can be established for this group selection (Cai, 1999, Theorem 1).

Theorem 12.4.6 *Consider the additive, homoscedastic model of n-vectors $\boldsymbol{w} = \boldsymbol{\beta} + \sigma \boldsymbol{Z}$ with Z_i $N(0, 1)$. Let $R_{\mathrm{BT}}(\boldsymbol{\beta}; \lambda, n_B)$ denote the risk of the block thresholding scheme defined in (12.36). On the other hand, let $R_{\mathrm{Bor}}(\boldsymbol{\beta}; n_B)$ be the risk of an oracle block selection scheme, keeping all blocks where $\sum_{i \in \mathsf{B}_l} |\beta_i|^2 > n_B \sigma^2$ and annihilating all others. Then*

$$R_{\mathrm{BT}}(\boldsymbol{\beta}; \lambda, n_B) \leq \lambda R_{\mathrm{Bor}}(\boldsymbol{\beta}; n_B) + 4\sigma^2 P(\chi^2_{n_B} > n_B \lambda). \tag{12.37}$$

In particular, for $n_B = \log(n)$ and $\lambda_0 = 4.50524$ defined by $\lambda_0 - \log(\lambda_0) = 3$, this becomes

$$R_{\mathrm{BT}}(\boldsymbol{\beta}; \lambda_0, n_B) \leq \lambda_0 R_{\mathrm{Bor}}(\boldsymbol{\beta}; n_B) + \frac{2\sigma^2}{n}.$$

In contrast to the oracle inequality for ordinary thresholding in Theorem 12.1.2, the threshold can be taken to be a constant and the inequality for block thresholding does not involve a logarithmic factor. From the practical viewpoint, a logarithmic factor is not dramatic anyway. At a more theoretical level, this remarkable effect is due to the presence of two tuning parameters in the block thresholding approach: the threshold and the block size. By letting the block size grow logarithmically with the sample size, the threshold can be kept at a constant level.

Obviously, the oracle inequality says nothing about the quality of the oracular block selection itself, in particular when the block size grows. That quality depends on the data at hand, but also on the definition of the blocks.

12.5 Bayesian shrinkage of wavelet coefficients

12.5.1 Bayesian models of sparsity

Another ample class of wavelet estimators is based on a prior model for wavelet coefficients without noise. Let $W_{j,k}$ and $V_{j,k}$ be the random variables modelling the wavelet coefficients with and without noise at resolution level j and location k, and assume an additive model $W_{j,k} = V_{j,k} + \sigma Z_{j,k}$, to which a Bayesian analysis is applied. The coefficient without noise is thought to come from a population of coefficients modelled by a prior density $f_{V_{j,k}}(v)$. This density describes the typical features of the coefficients, including the sparsity. A vast subset of the literature in the field adopts a **a mixture distribution** as a framework for modelling sparsity. More precisely the coefficient without noise is written as

$$V_{j,k} = S_{j,k}\widehat{V}_{j,k} + (1 - S_{j,k})\,\overset{\circ}{V}_{j,k},$$

where the latent binary variable $S_{j,k}$ indicates whether $V_{j,k}$ is drawn from the density of the large coefficients $\widehat{V}_{j,k}$, or whether it comes from the class of small coefficients $\overset{\circ}{V}_{j,k}$. For both classes the literature has proposed a variety of models. The small coefficients may be supposed to be exactly zero, i.e., $\overset{\circ}{V}_{j,k}= 0$, meaning that $f_{V_{j,k}}(v)$ is a **zero-inflated density**, i.e., a density with a point mass at the origin (Clyde et al., 1998; Abramovich et al., 1998; Johnstone and Silverman, 2004, 2005; Jansen et al., 2009). Alternatively, small nonzero coefficients may be modelled as normal zero mean random variables with a small variance (Chipman et al., 1997). The large coefficients, carrying the bulk of the information, may then be modelled as normal random variables with zero mean and large variance (Chipman et al., 1997; Clyde et al., 1998; Abramovich et al., 1998). More often in the literature, the large coefficients are supposed to fit into a model with slightly heavier tails, such as a double exponential (or Laplace) distribution or even a tail as heavy as that of the Cauchy distribution (Johnstone and Silverman, 2004, 2005; Jansen et al., 2009). The prior probability of $p_{j,k} = P(S_{j,k} = 1)$ controls the sparsity of the coefficient vector. It is reasonable to take $p_{j,k} = p_j$ constant at each resolution level. The latent vector $\boldsymbol{S}$ of binary values $S_{j,k}$ can be used to model tree structured dependencies, leading to a hidden Markov model (Crouse et al., 1998; Romberg et al., 2001). As another example of a latent dependence structure an Ising model may impose dependence between spatially adjacent coefficients, especially in two-dimensional wavelet decompositions, for use in, for instance, image processing (Jansen and Bultheel, 2001). The Ising or other Markov random field models are an alternative to the anisotropic basis functions mentioned in Section 10.4 when dealing with line singularities in two-dimensional data. The hyperparameters p_j describing the degree of sparsity and the hyperparameters appearing in the models for small and large coefficients in some models can be associated to the underlying function $f(x)$ being a member of a Besov or other smoothness space (Abramovich et al., 1998).

12.5.2 Bayesian decision rules

Application of Bayes' rule to the mixture distribution proceeds in two steps. First, the posterior probability of the latent labels is found by

$$p^*_{j,k} = P(S_{j,k} = 1|W_{j,k} = w) = \frac{P(S_{j,k} = 1)f_{\widehat{W}_{j,k}}(w)}{P(S_{j,k} = 1)f_{\widehat{W}_{j,k}}(w) + P(S_{j,k} = 0)f_{\overset{\circ}{W}_{j,k}}(w)}, \tag{12.38}$$

where $\widehat{W}_{j,k} = \widehat{V}_{j,k} + \sigma Z_{j,k}$ and $\overset{\circ}{W}_{j,k} = \overset{\circ}{V}_{j,k} + \sigma Z_{j,k}$. If the latent variables $S_{j,k}$ are modelled by a Markov or Gibbs random field, then Bayes' rule is applied to a joint probability mass function to find first $P(\boldsymbol{S} = \boldsymbol{s}|\boldsymbol{W} = \boldsymbol{w})$, before computing the marginal probabilities by summing up. The expression of Bayes' rule in this case involves the multivariate conditional densities $f_{\boldsymbol{W}|\boldsymbol{S}}(\boldsymbol{w}|\boldsymbol{s})$ which are assumed to factor into a product of marginal conditional densities

$$f_{\boldsymbol{W}|\boldsymbol{S}}(\boldsymbol{w}|\boldsymbol{s}) = \prod_{j,k} f_{W_{j,k}|S_{j,k}}(w_{j,k}|s_{j,k}),$$

where $f_{W_{j,k}|S_{j,k}}(w_{j,k}|s_{j,k})$ follows from the convolution of $f_{V_{j,k}|S_{j,k}}(v_{j,k}|s_{j,k})$ with the density of the noise Z.

The posterior density of $V_{j,k}$ is still a mixture, given by

$$\begin{aligned} f_{V_{j,k}|\boldsymbol{W}}(v|\boldsymbol{w}) &= f_{\widehat{V}_{j,k}|\boldsymbol{W}}(v|\boldsymbol{w})P(S_{j,k} = 1|\boldsymbol{W} = \boldsymbol{w}) \\ &\quad + f_{\overset{\circ}{V}_{j,k}|\boldsymbol{W}}(v|\boldsymbol{w})P(S_{j,k} = 0|\boldsymbol{W} = \boldsymbol{w}), \end{aligned}$$

which can be further simplified by imposing conditional independence in the sense

$$f_{\widehat{V}_{j,k}|\boldsymbol{W}}(v|\boldsymbol{w}) = f_{\widehat{V}_{j,k}|\widehat{W}_{j,k}}(v|w_{j,k}) \text{ and } f_{\overset{\circ}{V}_{j,k}|\boldsymbol{W}}(v|\boldsymbol{w}) = f_{\overset{\circ}{V}_{j,k}|\overset{\circ}{W}_{j,k}}(v|w_{j,k}).$$

A second application of Bayes' rule is required to find the posterior density of $V_{j,k}$,

$$f_{\widehat{V}_{j,k}|\widehat{W}_{j,k}}(v|w) = \frac{f_{\widehat{V}_{j,k}}(v)f_{\widehat{W}_{j,k}|\widehat{V}_{j,k}}(w|v)}{f_{\widehat{W}_{j,k}}(w)}.$$

As an example, if $V_{j,k}$ is a mixture of Gaussians (Chipman et al., 1997), with $\widehat{V}_{j,k} \sim N(0,(c\tau)^2)$ and $\overset{\circ}{V}_{j,k} \sim N(0,\tau^2)$, then $\widehat{W}_{j,k} \sim N(0,(c\tau)^2+\sigma^2)$ and $\overset{\circ}{W}_{j,k} \sim N(0,\tau^2+\sigma^2)$. The posterior density is then

$$\begin{aligned}(V_{j,k}|W_{j,k}=w) \quad &\sim \quad p^*_{j,k} N\left(\frac{(c\tau)^2}{(c\tau)^2+\sigma^2}w, \frac{(c\tau)^2\sigma^2}{(c\tau)^2+\sigma^2}\right) \\ &\quad +(1-p^*_{j,k})N\left(\frac{\tau^2}{\tau^2+\sigma^2}w, \frac{\tau^2\sigma^2}{\tau^2+\sigma^2}\right),\end{aligned}$$

where $p^*_{j,k}$ is the posterior sparsity as found in (12.38).

A straightforward estimator of the noise-free coefficient using this posterior density would be the **posterior mean**

$$E(V_{j,k}|W_{j,k}=w) = \left[p^*_{j,k}\cdot\frac{(c\tau)^2}{(c\tau)^2+\sigma^2} + (1-p^*_{j,k})\cdot\frac{\tau^2}{\tau^2+\sigma^2}\right]w,$$

which is clearly a shrinkage estimator.

By taking a zero-inflated prior for $f_{V_{j,k}}(v)$, the **posterior median** leads to **threshold rule**, because for small values of $w_{j,k}$, the point mass at zero causes the posterior distribution to jump from below to above 50% at the origin (Abramovich and Benjamini, 1996).

12.5.3 Filling in the hyperparameters

In a fully Bayesian approach, the hyperparameters of the prior and conditional models are again supposed to be random variables with prior distributions. Bayes' rule takes care of the full estimation process. Empirical Bayes methods (Johnstone and Silverman, 2005) use frequentist methods, i.c. maximum likelihood for finding good values of the hyperparameters. More precisely, the method maximises the marginal log-likelihood,

$$\log L(\boldsymbol{p},\boldsymbol{\theta}) = \sum_{j=L}^{J-1}\sum_{k=0}^{n_j-1}\log\left[p_j f_{\widehat{W}_{j,k}}(w_{j,k};\boldsymbol{\theta}) + (1-p_j) f_{\overset{\circ}{W}_{j,k}}(w_{j,k};\boldsymbol{\theta})\right].$$

12.6 Confidence bands

12.6.1 Pointwise and global confidence bands

After estimating the response vector $\boldsymbol{\mu}$ in the high-dimensional, sparse linear regression model of (12.1) the natural question arises regarding what statements can be made about the confidence of each of these values μ_i. Let $[\widehat{a}_{n,i},\widehat{b}_{n,i}]$, for $i=1,2,\ldots,n$ be a set of random intervals; then the simplest approach would be to establish intervals such that the coverage probabilities satisfy

$$\liminf_{n\to\infty} \min_{i=1,\ldots,n} P(\mu_i \in [\widehat{a}_{n,i}, \widehat{b}_{n,i}]) \geq 1-\alpha. \quad (12.39)$$

These are pointwise confidence intervals, constructed for each component μ_i separately. A global kind of confidence bands follows by imposing that the vector $\boldsymbol{\mu}$ *as a whole* stays within the random band with probability at least $1-\alpha$, i.e.,

$$\liminf_{n\to\infty} P(\mu_i \in [\widehat{a}_{n,i}, \widehat{b}_{n,i}]; \forall i = 1,\ldots,n) \geq 1-\alpha. \quad (12.40)$$

A slight variation replaces the fine scaling coefficients in $\boldsymbol{\mu}$ by the function $f(x) = \Phi(x)\boldsymbol{\mu}$. Assuming to work on $[0,1]$ we look for functions $\widehat{a}_n(x)$ and $\widehat{b}_n(x)$ so that

$$\liminf_{n\to\infty} P\big(f(x) \in [\widehat{a}_n(x), \widehat{b}_n(x)]; \forall x \in [0,1]\big) \geq 1-\alpha. \quad (12.41)$$

Rather than working with confidence bands, the wavelet coefficients may lend themselves to the definition of other forms of confidence sets. As an example, one may define a random set from the estimator $\widehat{\boldsymbol{\mu}}$ within a function or sequence class $\mathcal{F}_n$ by $S_n = \{\widetilde{\boldsymbol{\mu}} \in \mathcal{F}_n : \|\widetilde{\boldsymbol{\mu}} - \widehat{\boldsymbol{\mu}}\| \leq \delta_n\}$. The squared norm of the difference is relatively easy to find from an orthogonal wavelet decomposition or to approximate using a biorthogonal decomposition.

Confidence intervals are one form of statistical inference, and as such, they are closely related to the testing of hypotheses. In the light of this connection, it should be pointed out that pointwise confidence bands cannot possibly be of any help in deciding on any hypothesis about the curve as a whole. Confidence bands or regions for the curve as a whole are needed, for instance, to test the smoothness class to which it belongs, or even to test whether the function is a constant or a slope with an intercept against some more complex alternative.

12.6.2 The construction of pointwise confidence bands

In theory, confidence bands for $\boldsymbol{\mu}$ or for $f(x)$ can be constructed from the estimators $\widehat{\boldsymbol{\mu}}$ or $\widehat{f}(x)$, if it is known how the distributions of the estimators parametrically depend on the values in $\boldsymbol{\mu}$ or $f(x)$. Since the estimation includes forward and inverse wavelet transforms, the construction in practice relies on approximations using variances and, if possible, higher order moments. In the Bayesian framework of Chipman et al. (1997), where credibility bands take the place of the frequentist confidence bands, the bands are defined by three times the posterior standard deviation. The posterior standard deviation follows rather straightforwardly from the posterior covariance matrix in the wavelet domain. Moreover, given the setting of a Gaussian mixture prior and posterior, the covariance matrices carry important information about the full posterior density. The posterior variance is less satisfactory in combination with other, non-Gaussian models for wavelet coefficients in Bayesian processing (Barber et al., 2002). Recent developments in high-dimensional post-selection inference (Leeb and Pötscher, 2006; Berk et al., 2013; van de Geer et al., 2014; Zhang and Zhang, 2014; Lee et al., 2016; Charkhi and Claeskens, 2018) may be applied to wavelet coefficient estimation in order to obtain better coverage probabilities.

12.6.3 Adaptive confidence bands

For both pointwise confidence intervals (Picard and Tribouley, 2000) and more general confidence sets (Genovese and Wasserman, 2005), adaptivity results w.r.t. unobserved smoothness classes are available, meaning that

$$\liminf_{n\to\infty} \lim_{f\in\mathcal{F}} P\big(f \in S_n\big) \geq 1-\alpha. \tag{12.42}$$

In the literature, this form of coverage probability is called uniformly convergent, where uniform refers to all functions in the function class, not to all points in the domain of a single function. It is a much harder task to find confidence sets in the form of nontrivial confidence bands so that

$$\liminf_{n\to\infty} \lim_{f\in\mathcal{F}} P\big(f(x_i) \in [\widehat{a}_n(x_i), \widehat{b}_n(x_i)]; i = 1, 2, \ldots, n\big) \geq 1-\alpha. \tag{12.43}$$

A trivial confidence band could be proposed by

$$[\widehat{a}_n(x), \widehat{b}_n(x)] = [\widehat{f}_n(x) \pm c_n\sigma],$$

where c_n follows from a Bonferroni correction for multiple testing in n observation points. The width of that band, $2c_n\sigma$, grows to infinity when $n \to \infty$, making it useless for large sample sizes. Moreover, the band does not depend on the smoothness of $f(x)$. While it can be expected that for functions in smooth classes, the confidence bands can be narrower than for less smooth functions, the construction of adaptive confidence bands is far from trivial (Genovese and Wasserman, 2008; Hoffmann and Nickl, 2011; Cai et al., 2014).

Outlook

The main objective of this book has been to demonstrate that wavelets or other multiscale analyses can be constructed on a wide variety of data. Spline methods and local polynomial smoothing approaches can be lifted higher to a multiscale version. This book has discussed how to construct a wavelet analyis of irregular point sets from scratch. Many types of data and data structures have been not been mentioned in this book, and as a matter of fact, many of these data seem to be largely unexplored from the multiscale point of view. Possible domains of further multiscale research include graphical and network data, neural networks. In functional data, each observation is a different function, possibly showing discontinuities at different locations, for which adaptive non-equispaced multiscale analyses could be developed.

Of course, the first question to ask oneself should always be if a multiscale analysis would add valuable insights to what can be learned from a uniscale analysis. Some problems are naturally multiscale, for instance if a multiscale analysis induces a sparse representation, but this is obviously not always the case.

The origin of the images

Images are a classical, nice, and visual illustration in a text on wavelets. As a personal touch to this book, I have decided not to use the all too well known examples, used in the literature, but take my own pictures instead. As an additional advantage, the sizes of the images could be set to arbitrary values by cropping the original before starting the analysis.

The three pictures in Figures 1.1, 3.2, and 11.6 have been taken in the *Grand Béguinage*, a Unesco world heritage ensemble of seventeenth century houses, arranged along a dozen narrow streets and two canals, situated in the city of Leuven, Belgium. More information on this site and its remarkable history can be found, for instance on, `https://en.wikipedia.org/wiki/Groot_Begijnhof,_Leuven`. The map in Figure 7 shows the exact location of the pictures.

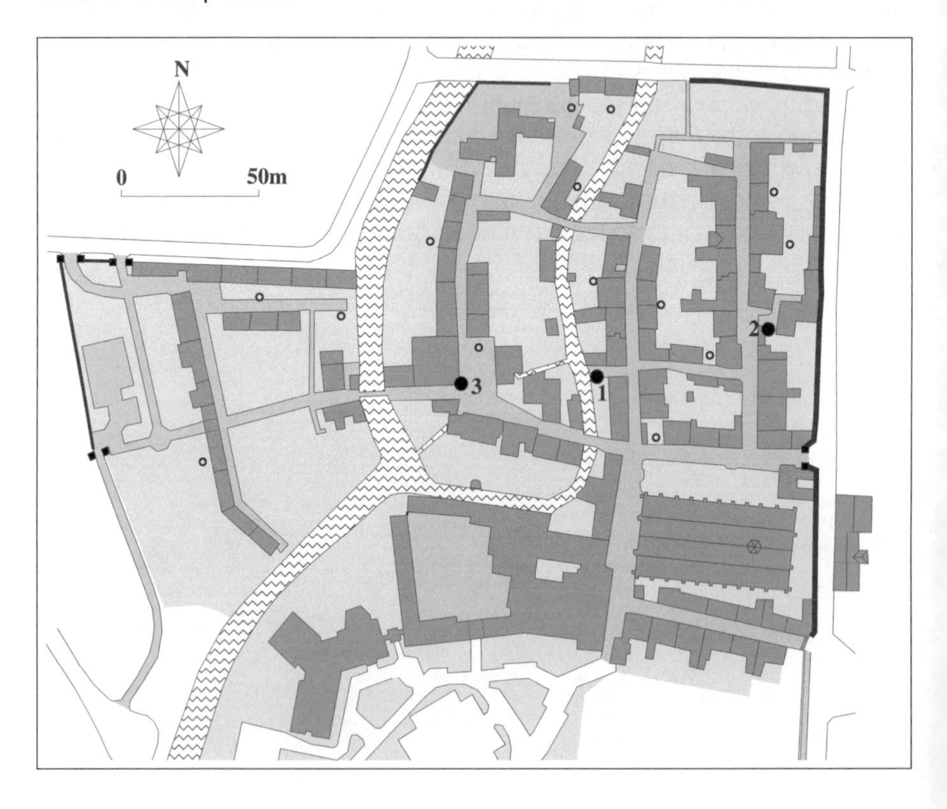

Figure 7
Locations of the pictures. **1**: picture in Figure 1.1; **2**: picture in Figure 3.2; **3**: picture in Figure 11.6.

References

F. Abramovich and Y. Benjamini. Adaptive thresholding of wavelet coefficients. *Computational Statistics and Data Analysis*, 22:351–361, 1996.

F. Abramovich and B. W. Silverman. Wavelet decomposition approaches to statistical inverse problems. *Biometrika*, 85:115–129, 1998.

F. Abramovich, F. Sapatinas, and B. W. Silverman. Wavelet thresholding via a Bayesian approach. *Journal of the Royal Statistical Society, Series B*, 60:725–749, 1998.

C. Angelini, D. De Canditiis, and F. Lablanc. Wavelet regression estimation in nonparametric mixed effect models. *J. Multivariate Analysis*, 85:267–291, 2003.

F. Anscombe. The transformation of Poisson, binomial and negative binomial data. *Biometrika*, 35:246–254, 1948.

A. Arneodo, E. Bacry, S. Jaffard, and J. F. Muzy. Singularity spectrum of multifractal functions involving oscillating singularities. *J. of Fourier Analysis and Applications*, 4(2):159–174, 1998.

S. Barber, G. P. Nason, and B. W. Silverman. Posterior probability intervals for wavelet thresholding. *Journal of the Royal Statistical Society, Series B*, 64:189–205, 2002.

G. Battle. A block spin construction of ondelettes. Part I: Lemarié functions. *Communications In Mathematical Physics*, 110(4):601–615, 1987.

Y. Benjamini and Y. Hochberg. Controlling the false discovery rate: A practical and powerful approach to multiple testing. *Journal of the Royal Statistical Society, Series B*, 57:289–300, 1995.

R. Berk, L. Brown, K. Zhang, and L. Zhao. Valid post-selection inference. *The Annals of Statistics*, 41(2):802–837, 2013.

K. Berkner and R. O. Wells. Smoothness estimates for soft-threshold denoising via translation-invariant wavelet transforms. *Applied and Computational Harmonic Analysis*, 12(1):1–24, 2002.

P. J. Burt and E. H. Adelson. Laplacian pyramid as a compact image code. *IEEE Trans. Commun.*, 31(4):532–540, 1983.

T. T. Cai. Adaptive wavelet estimation: a block thresholding and oracle inequality approach. *The Annals of Statistics*, 27:898–924, 1999.

T. T. Cai, M. Low, and Z. Ma. Adaptive confidence bands for nonparametric regression functions. *J. American Statistical Association*, 109(507):1054–1070, 2014.

R. Calderbank, I. Daubechies, W. Sweldens, and B.-L. Yeo. Wavelet transforms that map integers to integers. *Applied and Computational Harmonic Analysis*, 5(3):332–369, 1998.

E. J. Candès and D. L. Donoho. Recovering edges in ill-posed inverse problems: Optimality of curvelet frames. *The Annals of Statistics*, 30:784–842, 2000.

R. A. Carmona. Extrema reconstructions and spline smoothing: Variations on an algorithm of Mallat & Zhong. In A. Antoniadis and G. Oppenheim, editors, *Wavelets and Statistics*, volume 103 of *Lecture Notes in Statistics*, pages 83–94. Springer, 1995.

C. Chang, Y. Chen, and R. T. Ogden. Functional data classification: a wavelet approach. *Computational Statistics*, 29:1497–1513, 2014.

X. W. Chang and L. Qu. Wavelet estimation of partially linear models. *Computational Statistics and Data Analysis*, 47(1):31–48, 2004.

A. Charkhi and G. Claeskens. Asymptotic postselection inference for the Akaike information criterion. *Biometrika*, 105(3):645–664, 2018.

S. Chen, D. L. Donoho, and M. A. Saunders. Atomic decomposition by basis pursuit. *SIAM J. on Scientific Computing*, 20(1):33–61, 1998.

H. Chipman, E. Kolaczyk, and R. McCulloch. Adaptive Bayesian wavelet shrinkage. *J. American Statistical Association*, 92:1413–1421, 1997.

H. Choi and R. G. Baraniuk. Wavelet statistical models and Besov spaces. In M. A. Unser, A. Aldroubi, and A. F. Laine, editors, *Wavelet Applications in Signal and Image Processing VII*, volume 3813 of *SPIE Proceedings*, pages 489–501, July 1999.

G. Claeskens and N. L. Hjort. *Model Selection and Model Averaging*. Cambridge University Press, first edition, 2008.

G. Claeskens, H. Ding, and M. Jansen. Lack-of-fit tests in linear mixed models with application to wavelet tests. *Journal of Nonparametric Statistics*, 23(4):853–865, 2011.

R. Claypoole, G. M. Davis, W. Sweldens, and R. Baraniuk. Nonlinear wavelet transforms for image coding via lifting. *IEEE Transactions on Image Processing*, 12(12):1449–1459, 2003.

R. L. Claypoole, R. G. Baraniuk, and R. D. Nowak. Lifting construction of non-linear wavelet transforms. In *Proceedings of the IEEE-SP Int. Symp. on Time-Frequency and Time-Scale Analysis*, 1998.

M. Clyde, G. Parmigiani, and B. Vidakovic. Multiple shrinkage and subset selection in wavelets. *Biometrika*, 85:391–401, 1998.

A. Cohen, I. Daubechies, and J. Feauveau. Bi-orthogonal bases of compactly supported wavelets. *Comm. on Pure and Applied Mathematics*, 45:485–560, 1992.

I. Cohen, I. Raz, and D. Malah. Orthonormal shift-invariant adaptive local trigonometric decomposition. *Signal Processing*, 57(1):43–64, 1997a.

I. Cohen, I. Raz, and D. Malah. Orthonormal shift-invariant wavelet packet decomposition and representation. *Signal Processing*, 57(3):251–270, 1997b.

R. R. Coifman and M. V. Wickerhauser. Entropy based algorithms for best basis selection. *IEEE Transactions on Information Theory*, 38(2):713–718, 1992.

R. R. Coifman, Y. Meyer, and M. V. Wickerhauser. Wavelet analysis and signal processing. In M. B. Ruskai, G. Beylkin, R. R. Coifman, I. Daubechies, S. Mallat, Y. Meyer, and L. Raphael, editors, *Wavelets and their Applications*, pages 153–178. Jones and Bartlett, Boston, 1992.

R. R. Coifman Coifman and D. L. Donoho. Translation-invariant de-noising. In A. Antoniadis and G. Oppenheim, editors, *Wavelets and Statistics*, Lecture Notes in Statistics, pages 125–150. Springer, 1995.

M. S. Crouse, R. D. Nowak, and R. G. Baraniuk. Wavelet-based signal processing using hidden Markov models. *IEEE Transactions on Signal Processing*, 46, Special Issue on Wavelets and Filterbanks:886–902, 1998.

W. Dahmen. Stability of multiscale transformations. *J. Fourier Analysis and Applications*, 2(4):341–361, 1996.

I. Daubechies. *Ten Lectures on Wavelets*. CBMS-NSF Regional Conf. Series in Appl. Math., Vol. 61. Society for Industrial and Applied Mathematics, Philadelphia, PA, 1992.

I. Daubechies. Orthonormal bases of compactly supported wavelets II: Variations on a theme. *SIAM J. on Mathematical Analysis*, 24(2):499–519, 1993.

I. Daubechies and W. Sweldens. Factoring wavelet transforms into lifting steps. *J. of Fourier Analysis and Applications*, 4(3):245–267, 1998.

I. Daubechies, I. Guskov, and W. Sweldens. Regularity of irregular subdivision. *Constructive Approximation*, 15(3):381–426, 1999.

I. Daubechies, I. Guskov, and W. Sweldens. Commutation for irregular subdivision. *Constructive Approximation*, 17(4):479–514, 2001.

C. de Boor. *A Practical Guide to Splines*. Springer, New York, 2001. Revised Edition.

S. Dekel and D. Leviatan. Adaptive multivariate approximation using binary space partitions and geometric wavelets. *SIAM Journal on Numerical Analysis*, 43(2):707–732, 2005.

G. Deslauriers and S. Dubuc. Interpolation dyadique. In *Fractales, Dimensions Non-entières et Applications*, pages 44–55. Masson, Paris, 1987.

E. Di Nezza, G. Palatucci, and E. Valdinoci. Hitchhiker's guide to the fractional Sobolev spaces. *Bulletin des Sciences Mathématiques*, 136(5):521–573, 2012.

M. N. Do and M. Vetterli. Contourlets. In G. Welland, editor, *Beyond Wavelets*, pages 83–107. Academic Press, 2003.

D. L. Donoho. Nonlinear solution of linear inverse problems by wavelet-vaguelette decomposition. *Applied and Computational Harmonic Analysis*, 2(2):101–126, 1995a.

D. L. Donoho. De-noising by soft-thresholding. *IEEE Transactions on Information Theory*, 41(3):613–627, 5 1995b.

D. L. Donoho. Wedgelets: Nearly minimax estimation of edges. *The Annals of Statistics*, 27(3):859–897, 1999.

D. L. Donoho. For most large underdetermined systems of linear equations the minimal ℓ_1-norm solution is also the sparsest solution. *Comm. on Pure and Applied Mathematics*, 59:797–829, 2006.

D. L. Donoho and X. Huo. Beamlets and multiscale image processing. Technical report, Department of Statistics, Stanford University, 2001.

D. L. Donoho and I. M. Johnstone. Ideal spatial adaptation via wavelet shrinkage. *Biometrika*, 81(3):425–455, 1994.

D. L. Donoho and I. M. Johnstone. Adapting to unknown smoothness via wavelet shrinkage. *J. American Statistical Association*, 90(432):1200–1224, 1995.

D. L. Donoho and I. M. Johnstone. Minimax estimation via wavelet shrinkage. *The Annals of Statistics*, 26:879–921, 1998.

D. L. Donoho and T. P. Y. Yu. Deslauriers-Dubuc: Ten years after. In S. Dubuc and G. Deslauriers, editors, *Spline Functions and the Theory of Wavelets*, CRM Proceedings and Lecture Notes. American Mathematical Society, 1999.

D. L. Donoho, I. M. Johnstone, G. Kerkyacharian, and D. Picard. Wavelet shrinkage: Asymptopia? *Journal of the Royal Statistical Society, Series B*, 57(2):301–369, 1995.

B. Efron and C. Morris. Stein's paradox in statistics. *Scientific American*, 5: 119–127, 1977.

P. H. C. Eilers and B. D. Marx. Flexible smoothing with B-splines and penalties. *Statistical Science*, 7(2):89–121, 1996.

J. Fan and I. Gijbels. *Local Polynomial Modelling and its Applications*. Chapman and Hall, London, 1996.

J. Fan and R. Li. Variable selection via nonconcave penalized likelihood and its oracle properties. *J. American Statistical Association*, 96(456):1348–1360, 12 2001.

R. M. Figueras i Ventura, P. Vandergheynst, and P. Frossard. Low rate and flexible image coding with redundant representations. *IEEE Transactions on Image Processing*, 15(3):726–739, 2006.

M. Fisz. The limiting distribution of a function of two independent variables and its statistical application. *Colloquium Mathematicum*, 3:138–146, 1955.

P. Fryźlewicz and G. Nason. A wavelet-Fisz algorithm for Poisson intensity estimation. *Journal of Computational and Graphical Statistics*, 13(3):621–638, 2004.

H.-Y. Gao. Wavelet shrinkage denoising using the non-negative garrote. *Journal of Computational and Graphical Statistics*, 7(4):469–488, December 1998.

Ch. R. Genovese and L. Wasserman. Confidence sets for nonparametric wavelet regression. *The Annals of Statistics*, 33(2):698–729, 2005.

Ch. R. Genovese and L. Wasserman. Adaptive confidence bands. *The Annals of Statistics*, 36:875–905, 2008.

P. Goupillaud, A. Grossmann, and J. Morlet. Cycle-octave and related transforms in seismic signal analysis. *Geoexploration*, 23(1):85–102, 10 1984.

J. Goutsias and H. J. A. M. Heijmans. Multiresolution signal decomposition schemes. Part 1: Linear and morphological pyramids. *IEEE Transactions on Image Processing*, 9:979–995, 2000a.

J. Goutsias and H. J. A. M. Heijmans. Multiresolution signal decomposition schemes. Part 2: Morphological wavelets. *IEEE Transactions on Image Processing*, 9:1897–1913, 2000b.

A. Grossmann and J. Morlet. Decomposition of hardy functions into square integrable wavelets of constant shape. *SIAM J. on Mathematical Analysis*, 15(4):723–736, 1984.

W. Härdle, G. Kerkyacharian, D. Picard, and A. Tsybakov. *Wavelets, Approximation and Statistical Applications*. Lecture Notes in Statistics, Number 129. Springer, 1998.

H. J. A. M. Heijmans, G. Piella, and B. Pesquet-Popescu. Adaptive wavelets for image compression using update lifting: Quantisation and error analysis. *International Journal of Wavelets, Multiresolution and Information Processing*, 4(1):41–65, 2006.

Ch. Heil. *A Basis Theory Primer – Expanded Edition*. Birkhäuser, Boston, 2011.

M. Hoffmann and R. Nickl. On adaptive inference and confidence bands. *The Annals of Statistics*, 39(5):2383–2409, 2011.

S. Jaffard. Pointwise smoothness, two-microlocalisation and wavelet coefficients. *Publicacions Matemàtiques*, 35:155–168, 1991.

M. Jansen. Multiscale poisson data smoothing. *Journal of the Royal Statistical Society, Series B*, 68(1):27–48, 2006.

M. Jansen. Information criteria for variable selection under sparsity. *Biometrika*, 101(1):37–55, 2014.

M. Jansen. Generalized cross validation in variable selection with and without shrinkage. *Journal of Statistical Planning and Inference*, 159:90–104, 2015.

M. Jansen. Non-equispaced B-spline wavelets. *International Journal of Wavelets, Multiresolution and Information Processing*, 14(6), 2016. doi: 10.1142/S0219691316500569.

M. Jansen and M. Amghar. Multiscale local polynomial decompositions using bandwidths as scales. *Statistics and Computing*, 27(5):1383–1399, 2017. doi: 10.1007/s11222-016-9692-8.

M. Jansen and A. Bultheel. Empirical Bayes approach to improve wavelet thresholding for image noise reduction. *J. American Statistical Association*, pages 629–639, June 2001.

M. Jansen and P. Ooninckx. *Second Generation Wavelets and Applications*. Springer, 2005.

M. Jansen, R. Baraniuk, and S. Lavu. Multiscale approximation of piecewise smooth two-dimensional functions using normal triangulated meshes. *Applied and Computational Harmonic Analysis*, 19(1):92–130, 2005.

M. Jansen, G. Nason, and B. Silverman. Multiscale methods for data on graphs and irregular multidimensional situations. *Journal of the Royal Statistical Society, Series B*, 71(1):97–125, 2009.

I. M. Johnstone and B. W. Silverman. Wavelet threshold estimators for data with correlated noise. *Journal of the Royal Statistical Society, Series B*, 59: 319–351, 1997.

I. M. Johnstone and B. W. Silverman. Needles and straw in haystacks: Empirical Bayes estimates of possibly sparse sequences. *The Annals of Statistics*, 32(4):1594–1649, 2004.

I. M. Johnstone and B. W. Silverman. Empirical Bayes selection of wavelet thresholds. *The Annals of Statistics*, 33(4):1700–1752, 2005.

M. Knight and G. P. Nason. Improving prediction of hydrophobic segments along a transmembrane protein sequence using adaptive multiscale lifting. *SIAM Journal on Multiscale Modeling and Simulation*, 5:115–129, 2006.

T. H. Koornwinder. *Wavelets: An elementary treatment of theory and applications.* World Scientific, Singapore, 1993.

D. La Gall and A. Tabatabai. Sub-band coding of digital images using symmetric short kernel filters and arithmetic coding techniques. In *Proceedings of ICASSP 1988, International Conference on Acoustics, Speech, and Signal Processing*, pages 761–764. IEEE, 1988.

E. Le Pennec and S. Mallat. Sparse geometrical image representations with bandelets. *IEEE Transactions on Image Processing*, 14(4):423–438, 2005.

M. R. Leadbetter, G. Lindgren, and H. Rootzén. *Extremes and Related Properties of Random Sequences and Processes.* Springer Series in Statistics. Springer, 175 Fifth Avenue, New York 10010, USA, 1983.

E. T. Y. Lee. Marsden's identity. *Computer Aided Geometric Design*, 13(4): 287–305, June 1996.

J. D. Lee, D. L. Sun, and J. E. Taylor. Exact post-selection inference, with application to the lasso. *The Annals of Statistics*, 44(3):907–927, 2016.

H. Leeb and B. M. Pötscher. Can one estimate the conditional distribution of post-model-selection estimators? *The Annals of Statistics*, 34(5):2554–2591, 2006.

P.-G. Lemarié. Ondelettes à localisation exponentielles. *Journal des Mathématiques Pures et Appliquées*, 67(3):227–236, 1988.

R. Lorentz and P. Oswald. Criteria for hierarchical bases in Sobolev spaces. *Applied and Computational Harmonic Analysis*, 8(1):32–85, 2000.

F. Luisier, T. Blu, and M. Unser. A new SURE approach to image denoising: Interscale orthonormal wavelet thresholding. *IEEE Transactions on Image Processing*, 16(3):593–606, March 2007.

S. Mallat. *A Wavelet Tour of Signal Processing.* Academic Press, 2nd edition, 2001.

S. Mallat and Z. Zhang. Matching pursuits with time-frequency dictionaries. *IEEE Transactions on Signal Processing*, 41(12):3397–3415, 1993.

S. Mallat and S. Zhong. Characterization of signals from multiscale edges. *IEEE Transactions on Pattern Analysis and Machine Intelligence*, 14:710–732, 1992.

S. G. Mallat. A theory for multiresolution signal decomposition: The wavelet representation. *IEEE Transactions on Pattern Analysis and Machine Intelligence*, 11(7):674–693, 1989.

C. L. Mallows. Some comments on C_p. *Technometrics*, 15:661–675, 1973.

F. G. Meyer. Wavelet-based estimation of a semiparametric generalized linear model of fMRI time-series. *IEEE Transactions on Image Processing*, 22(3):315–322, 2003.

Y. Meyer. *Wavelets and Operators*, volume 37 of *Cambridge Studies in Advanced Mathematics.* Cambridge University Press, 1992.

J. S. Morris and R. J. Carroll. Wavelet-based functional mixed models. *Journal of the Royal Statistical Society, Series B*, 68(2):179–199, 2006.

F. Murtagh, J. L. Starck, and A. Bijaoui. Image restoration with noise suppression using a multiresolution support. *Astronomy and Astrophysics, Suppl. Ser.*, 112:179–189, 1995.

G. P. Nason. Wavelet shrinkage using cross validation. *Journal of the Royal Statistical Society, Series B*, 58:463–479, 1996.

G. P. Nason. *Wavelet Methods in Statistics with R.* Springer, first edition, 2008.

G. P. Nason and B. W. Silverman. The stationary wavelet transform and some statistical applications. In A. Antoniadis and G. Oppenheim, editors, *Wavelets and Statistics*, Lecture Notes in Statistics, pages 281–299. Springer, 1995.

M. Nunes, M. Knight, and G. P. Nason. Adaptive lifting for nonparametric regression. *Statistics and Computing*, 16(2):143–159, 2006.

R. T. Ogden. *Essential Wavelets for Statistical Applications and Data Analysis*. Birkhauser, Boston, 1997.

P. J. Ooninckx and P. M. de Zeeuw. Adaptive lifting for shape-based image retrieval. *Pattern Recognition*, 36(11):2663–2672, November 2003.

C. G. Park, M. Vannucci, and J. D. Hart. Bayesian methods for wavelet series in single-index models. *Journal of Computational and Graphical Statistics*, 14(4):770–794, 2005.

J. Peetre. On spaces of Triebel-Lizorkin type. *Arkiv för Matematik*, 13(1-2): 123–130, 1975.

D. B. Percival and A. T. Walden. *Wavelet Methods for Time Series Analysis*. Cambridge University Press, first edition, 2008.

D. Picard and K. Tribouley. Adaptive confidence interval for pointwise curve estimation. *The Annals of Statistics*, 28(1):298–335, 2000.

G. Piella and H. J. A. M. Heijmans. Adaptive lifting schemes with perfect reconstruction. *IEEE Transactions on Signal Processing*, 50(7):1620–1630, July 2002.

R. Qu and J. Gregory. A subdivision algorithm for non-uniform B-splines. In S. P. Singh, editor, *Approximation Theory, Wavelets and Applications*, volume 356 of *NATO ASI Series C*, pages 423–436, 1992.

J. Romberg, M. Wakin, and R. Baraniuk. Approximation and compression of piecewise smooth images using a wavelet/wedgelet geometric model. In *Proc. IEEE Int. Conf. on Image Proc. — ICIP '03*, 2003.

J. K. Romberg, H. Choi, and R. G. Baraniuk. Bayesian tree structured image modeling using wavelet-domain hidden Markov models. *IEEE Transactions on Image Processing*, 10(7):1056–1068, July 2001.

D. Ruppert, M. P. Wand, and R. Carroll. *Semiparametric Regression*. Cambridge University Press, Cambridge, UK, 2003.

R. Samworth. Stein's pardox. *Eureka*, 62:38–41, 2012.

M. J. Shensa. The discrete wavelet transform: Wedding the à trous and Mallat algorithms. *IEEE Transactions on Information Theory*, 40:2464–2482, 1992.

R. Shukla, P. L. Dragotti, M. Do, and M. Vetterli. Rate distortion optimized tree structured compression algorithms. *IEEE Transactions on Image Processing*, 14(3):343–359, 2005.

R. Sibson. A vector identity for Dirichlet tessellation. *Mathematical Proceedings of the Cambridge Philosophical Society*, 87:151–155, 1980.

N. Simon, J. Friedman, T. J. Hastie, and R. J. Tibshirani. A sparse-group LASSO. *Journal of Computational and Graphical Statistics*, 22(2):231–245, 2013.

C. Stein. Inadmissibility of the usual estimator for the mean of a multivariate distribution. In *Proc. Third Berkeley Symp. Math. Statist. Prob.*, pages 197–206. University of California Press, 1956.

C. Stein. Estimation of the mean of a multivariate normal distribution. *The Annals of Statistics*, 9(6):1135–1151, 1981.

G. Strang. Wavelet transforms versus Fourier transforms. *Bulletin Amer. Math. Soc.*, 28:288–305, 1993.

G. Strang and T. Nguyen. *Wavelets and Filter Banks.* Wellesley-Cambridge Press, Box 812060, Wellesley MA 02181, fax 617-253-4358, 1996.

W. Sweldens and R. Piessens. Quadrature formulae and asymptotic error expansions for wavelet approximations of smooth functions. *SIAM J. on Numerical Analysis*, 31(4):1240–1264, 1994.

W. Sweldens and P. Schröder. Building your own wavelets at home. In *Wavelets in Computer Graphics*, ACM SIGGRAPH Course Notes, pages 15–87. ACM, 1996.

R. J. Tibshirani. Regression shrinkage and selection via the lasso. *Journal of the Royal Statistical Society, Series B*, 58(1):267–288, 1996.

R. J. Tibshirani and J. E. Taylor. Degrees of freedom in lasso problems. *The Annals of Statistics*, 40(2):1198–1232, 2012.

A. A. Tsiatis. *Semiparametric Theory and Missing Data.* Springer Series in Statistics. Springer-Verlag, New-York, 2006. ISBN 0-387-32448-8 (hard cover), 978-0387-32448-7 (paperback).

S. van de Geer, P. Bühlmann, Y. Ritov, and R. Dezeure. On asymptotically optimal confidence regions and tests for high-dimensional models. *The Annals of Statistics*, 42(3):1166–1202, 2014.

M. Vetterli and C. Herley. Wavelets and filter banks: Theory and design. *IEEE Transactions on Signal Processing*, 40(9):2207–2232, 1992.

B. Vidakovic. *Statistical Modeling by Wavelets.* Wiley Series in Probability and Mathematical Statistics - Applied Probability and Statistics Section. John Wiley & Sons, 605 Third Avenue, New York, NY 10158-0012, USA, 1999.

G. Wahba. *Spline Models for Observational Data*, chapter 4, pages 45–65. CBMS-NSF Regional Conf. Series in Appl. Math. Society for Industrial and Applied Mathematics, Philadelphia, PA, 1990.

M. P. Wand and J. T. Prmerod. Penalized wavelets: Embedding wavelets into semiparametric regression structure. *Electronic Journal of Statistics*, 5:1654–1717, 2011.

R. M. Willett and R. D. Nowak. Platelets: A multiscale approach fot recovering edges and surfaces in photon-limited medical imaging. *IEEE Transactions on Medical Imaging*, 22(3):332–350, March 2003.

J. Ye. On measuring and correcting the effects of data mining and model selection. *J. Amer. Statist. Assoc.*, 93:120–131, 1998.

H. Yserentant. On the multi-level splitting of finite element spaces. *Numerische Mathematik*, 49:379–412, 1986.

Ming Yuan and Yi Lin. Model selection and estimation in regression with grouped variables. *Journal of the Royal Statistical Society, Series B*, 68(1): 49–67, 2007.

C. H. Zhang. Nearly unbiased variable selection under the minimax concave penalty. *The Annals of Statistics*, 38(2):894–942, 2010.

C.-H. Zhang and S. S. Zhang. Confidence intervals for low dimensional parameters in high dimensional linear models. *Journal of the Royal Statistical Society, Series B*, 76:217–242, 2014.

H. Zou, T. J. Hastie, and R. J. Tibshirani. On the “degrees of freedom” of the lasso. *The Annals of Statistics*, 35(5):2173–2192, 2007.

Subject index

Contact information

Maarten Jansen, Université libre de Bruxelles
Departments of Mathematics and Computer Science
Campus Plaine, B-1050 Brussels, Belgium.